TRANSPORTATION LAND USE, PLANNING, AND AIR QUALITY

PROCEEDINGS OF THE 2007 TRANSPORTATION LAND USE, PLANNING, AND AIR QUALITY CONFERENCE

July 9–11, 2007
Orlando, Florida

SPONSORED BY

Federal Highway Administration (FHWA)

Transportation and Air Quality Committee
of the Transportation Research Board (TRB)

Center for Transportation Research and Education
at Iowa State University

Department of Civil & Environmental Engineering
at The University of North Carolina at Charlotte

Transportation and Development Institute (T&DI)
of the American Society of Civil Engineers

Edited by
Srinivas S. Pulugurtha, Ph.D., P.E.
Robert O'Loughlin
Shauna Hallmark, Ph.D.

ASCE

T&DI
TRANSPORTATION AND DEVELOPMENT INSTITUTE

Published by the American Society of Civil Engineers

Library of Congress Cataloging-in-Publication Data

Transportation Land Use, Planning, and Air Quality Conference (2007 : Orlando, Fla.)
Transportation Land Use, Planning, and Air Quality Conference 2007 : proceedings of the 2007 Transportation Land Use, Planning, and Air Quality Conference : July 9-11, 2007, Orlando, Florida / sponsored by Federal Highway Administration (FHWA) … [et al.] ; edited by Srinivas S. Pulugurtha, Robert O'Loughlin, Shauna Hallmark.
p. cm.
Includes bibliographical references and index.
ISBN-13: 978-0-7844-0960-2
ISBN-10: 0-7844-0960-9
1. Transportation--Environmental aspects--Congresses. 2. Land use--Environmental aspects--Congresses. 3. Air quality management--Congresses. 4. Air--Pollution--Congresses. 5. Transportation--Planning--Congresses. I. Pulugurtha, Srinivas S. II. O'Loughlin, Robert. III. Hallmark, Shauna L. IV. United States. Federal Highway Administration. V. Title. VI. Title: Proceedings of the 2007 Transportation Land Use, Planning, and Air Quality Conference.

TD195.T7T747 2008
363.739'2--dc22 2008015229

American Society of Civil Engineers
1801 Alexander Bell Drive
Reston, Virginia, 20191-4400

www.pubs.asce.org

ISBN 978-0-7844-0960-2 Manufactured in the United States of America.

Cover photographs courtesy of (clockwise from top right) Bert Glibbery/morguefile, Kevin Connors/morguefile, LaRae/morguefile, and U.S. Army Corps of Engineers.

Preface

The 2007 Transportation Land-use, Planning, and Air Quality Conference held on July 9, 10, and 11 in Orlando, FL provided a forum to address pertinent issues and to keep practicing engineers, planners and researchers abreast of the latest developments and innovative practices in the field of transportation planning, land-use, and air quality.

The conference brought together more than 110 planners, designers, engineers, managers, and researchers. It provided ample opportunities to learn from experts worldwide as well as exchange of ideas with peers in the field. The three day conference included concurrent technical sessions on the first two days and workshops on the third day of the conference. The concurrent sessions include (but are not limited to) community visioning and design, impacts due to development, SAFETEA-LU planning requirements, transit innovations, climate change, emissions, and air quality models, congestion and mitigation strategies, land-use models, and advancements in computing technology and modeling,. The workshops topics include US EPA MOVES Model, Land-use Models/Tools, next generation transportation models, FHWA Mobile Source Air Toxics Methodology, and TRANSCAD.

The conference participants benefited considerably from the numerous presenters who shared their ideas, experiences, and insights. All the presenters are thanked for their contribution. Julie Kieffer and Jennifer Tabke of University Conference Services at Iowa State University are thanked for their assistance with conference services.

The proceedings of the 2007 Transportation/Land-use Planning Air Quality Conference includes selected papers that were presented at the conference. The papers included in the proceedings went through a two step peer review process. Shauna Hallmark, Robert O'Loughlin and Srinivas Pulugurtha coordinated the peer review process. The authors are thanked for their hard work in preparation of manuscripts and for cooperation and patience during the peer review process and the publication of proceedings. All the reviewers are thanked for their input and suggestions to the authors. Arpan Desai and Kuvleshay Patel, graduate students in the Department of Civil & Environmental Engineering at the University of North Carolina at Charlotte, are thanked for their help in formatting the papers to the required format.

A special thanks to Federal Highway Administration who provided a significant financial contribution which made the conference possible. The Center for Transportation Research and Education (CTRE) at Iowa State University is thanked for its financial contribution. The Department of Civil and Environmental Engineering at the University of North Carolina at Charlotte is thanked for its computing resources and office supplies for conference proceedings.

Srinivas S. Pulugurtha, Ph.D., P.E., M.ASCE
Robert O'Loughlin
Shauna Hallmark, Ph.D., M. ASCE

Acknowledgments

Conference Steering Committee

Robert O' Loughlin, Federal Highway Administration, Conference Chair
Shauna Hallmark, Iowa State University, Technical Program Committee Chair
Srinivas S. Pulugurtha, University of North Carolina at Charlotte, Conference Proceedings Editor

Conference Planning Committee

James B. Martin, North Carolina State University
John Koupal, US Environmental Protection Agency
John Suhrbier, Cambridge Systematics, Inc.
Randall L. Guensler, Georgia Institute of Technology
Reginald Souleyrette, Iowa State University
Vicky McLane, North Front Range MPO

Sponsors

Federal Highway Administration (FHWA)
TRB's Transportation Air Quality Committee (ADC20)
T&DI's Planning and Economics Committee
Center for Transportation Research and Education (CTRE) at Iowa State University
Department of Civil and Environmental Engineering at the University of North Carolina at Charlotte
Center for Transportation and the Environment (CTE) at North Carolina State University
Air and Waste Management Association (AWMA)
AWMA's On and Off-Road Mobile Source Committee
Georgia Institute of Technology
T&DI's Land-use Committee
Transportation and Development Institute of ASCE (T&DI)
Transportation Research Board (TRB)
TRB Subcommittee on Air Quality in Small and Medium Communities
TRB's Access Management Committee (ADA70)
US Environmental Protection Agency (USEPA)

Contents

Transit Innovations

Transit and Air Quality

Integrating Transportation, Land Use and Air Quality

Environmental Impacts of Freight

Climate Change

Indexes

Neighborhood Design as a Strategy for Improving Air Quality: Evidence from Northern California

X. Cao,[1] S. Handy,[2] and P. Mokhtarian[3]

[1]Associate Research Fellow, Upper Great Plains Transportation Institute, North Dakote State, University Fargo, ND, 58105, email: xinyucao@gmail.com
[2]Professor, Department of Environmental Science and Policy, University Of California Davis, CA-95616; PH (530) 752-5878: FAX (530) 752-3350; email: slhandy@ucdavis.edu
[3]Professor, Department of Environmental Science and Policy, University Of California Davis, CA-95616; email: plmokhtarian@ucdavis.edu

Abstract

The sprawling patterns of land development common to metropolitan areas of the US have been blamed for high levels of automobile travel, and thus for air quality problems. In response, smart growth programs – designed to counter sprawl – have gained popularity in the US. Studies show that residents of neighborhoods with higher levels of density, land-use mix, transit accessibility, and pedestrian friendliness drive less than residents of neighborhoods with lower levels of these characteristics. However, these studies have shed little light on the underlying direction of causality – whether neighborhood design influences travel behavior or whether travel preferences influence the choice of neighborhood. The available evidence thus leaves a key question largely unanswered: if cities use land use policies to bring residents closer to destinations and provide viable alternatives to driving, will people change their behavior in ways that reduce emissions? This study examines evidence from a study of residents of eight neighborhoods in Northern California on the link between neighborhood design and two behaviors that affect emissions: driving and choice of vehicle type. The study used multivariate modeling techniques to control for socio-demographic characteristics as well as attitudes and preferences. The results support the premise that land use policies have at least some potential to reduce driving and ownership of light duty trucks, thereby reducing emissions.

Key words: built environment, self-selection, smart growth, travel behavior, vehicle type choice

Introduction

The sprawling patterns of land development common to metropolitan areas of the United States have been blamed for high levels of automobile travel, and thus for air quality problems. The defining characteristics of "sprawl" include: low-density development, unlimited outward expansion, and "leapfrog" development (Burchell et al. 2002, p. 39). This low-density pattern of growth has two important effects on travel: longer trip distances and greater reliance on the car. In response to these travel effects and other impacts, "smart growth" programs designed to counter sprawl have gained popularity in the United States. The hope is that these strategies will bring residents closer to destinations and provide viable alternatives to driving and thus help to reduce automobile use.

Recognizing this potential, the U.S. Environmental Protection Agency (EPA) now accepts land-use policies as an effective tool for improving air quality and allows state and local communities to account for the air quality benefits of smart growth strategies in state air quality plans (EPA 2001). However, the estimation of the emissions effects of land use policies is based on limited empirical evidence, and little is known about the sensitivity of air quality to changes in land use. Studies in the U.S. show that, all else equal, residents of neighborhoods with higher levels of urban density, land-use mix, transit accessibility, and pedestrian friendliness (among other characteristics) drive less than residents of neighborhoods with lower levels of these characteristics (e.g., Cervero and Radisch 1996). These studies have not established the underlying direction of causality – whether neighborhood design influences travel behavior or whether preferences towards travel and land use patterns influence the choice of neighborhood. The available evidence thus leaves a key question largely unanswered: If cities use land use policies to bring residents closer to destinations and provide viable alternatives to driving, will at least some people drive less, thereby reducing emissions?

Further, the increasing share of light-duty trucks (LDTs, including minivans and pickup trucks as well as sport utility vehicles (SUVs)) in the passenger vehicle fleet contributes to the problems of air quality and oil dependence, due to the differential fuel efficiency and emissions standards between passenger cars and LDTs. According to the 2004 Fuel Economy Guide (www.fueleconomy.gov), for example, on average a 2WD Ford F150 (a pickup truck) consumes 35% more gasoline per mile than a Ford Taurus (a passenger car), and produces 30% more greenhouse gases and 200% more air pollutants. A few recent studies have found that suburban development is associated with the unbalanced choice of LDTs. For example, an analysis of the 1995 Nationwide Personal Transportation Survey (NPTS) showed that suburban residents own a disproportionate share of LDTs (Niemeier et al. 1999). After examining data from the San Francisco Bay Area, Bhat and Sen (2006) found that households living in denser areas are less inclined to drive SUVs and pickup trucks. However, these studies seldom reveal which specific characteristics of the built environment matter to vehicle type choice. Further, they have not shed much light on the underlying direction of causality: Whether neighborhood design itself, as opposed to preferences for neighborhood characteristics or attitudes towards travel, more strongly influences individuals' decisions regarding vehicle type. Similarly, the available evidence leaves unanswered questions: If policies encourage more compact development, will more people choose to drive passenger cars over LDTs, with a corresponding benefit to air quality?

The purpose of this study is to investigate the role of neighborhood design in influencing driving behavior and vehicle type choice using a sample from Northern California. This paper addresses the following questions: (1) Do changes in the built environment have an influence on changes in driving? (2) What aspects of neighborhood design influence vehicle type choice after controlling for socio-demographics and attitudes? Answering these questions helps us answer the ultimate question of interest: Can land use policies contribute to air quality improvement by influencing driving and vehicle type choice?

Data and Variables

This paper uses data collected from residents of four "traditional" neighborhoods and four "suburban" neighborhoods in Northern California (see Handy et al. 2004 for details). These designations were made based on neighborhood age, the street network, age and design of houses, and the location and type of commercial centers. The neighborhoods chosen as "traditional" included Mountain View (Downtown), Sacramento (Midtown), Santa Rosa (Junior College area), and Modesto (Central). The neighborhoods chosen as "suburban" were Sunnyvale (I-280 area), Sacramento (Natomas area), Santa Rosa (Rincon Valley area), and Modesto (suburban area). For each neighborhood, we purchased two databases of residents from New Neighbors Contact Service (www.nncs.com): a database of "movers" and a database of "nonmovers." The "movers" included all current residents of the neighborhood who had moved within the previous year. From this database, we drew a random sample of 500 residents for each of the eight neighborhoods. The database of "nonmovers" consisted of a random sample of 500 residents not included in the "movers" list for each neighborhood.

The survey was administered in October and November 2003, using a mail-out, mail-back approach. This approach resulted in 1682 responses, a 24.9% response rate based on the valid addresses only. It is important to note that median household income for survey respondents was higher than the census median for all but Rincon Valley area in Santa Rosa, a typical result for voluntary self-administered surveys. Further, survey respondents tend to be older on average than residents of their neighborhood as a whole, a reasonable result for the survey including only adult respondents. Although these univariate distributions may not be representative, we expect conditional relationships (e.g. travel behaviors given age and income) to be reasonably well estimated (Babbie 1998). The variables used here were classified into five groups: travel behavior, neighborhood characteristics, neighborhood preferences, travel attitudes, and socio-demographics.

Travel behavior

One dependent variable presented in Section 3.2 is the possible category for the vehicles that respondents drove most frequently. These vehicles were classified into four categories: passenger car, minivan, SUV, and pickup truck, based on reported make, model, and year combinations. The survey also asked respondents to report their weekly vehicle miles driven (VMD). Change in driving from either just before the move (for the movers) or from one year ago (for the nonmovers) was measured using 5-point scales from "a lot less now" to "a lot more now."

Neighborhood characteristics and neighborhood preferences

Respondents were asked to indicate how true 34 characteristics are for their current and previous neighborhoods, on a four-point scale from 1 ("not at all true") to 4 ("entirely true"). The characteristics of these neighborhoods as perceived by survey respondents reflect fundamental differences in neighborhood design. Further, the importance of these items to respondents when/if they were looking for a new place to live were measured on a four-point scale from 1 ("not at all important") to 4 ("extremely important"). A factor analysis on perceptions and preferences of neighborhood characteristics reduced these items (after dropping some) to six factors: accessibility, physical activity options, safety, socializing, attractiveness, and outdoor spaciousness (Table 1). For the movers, changes in the built environment were measured by taking the difference in perceived characteristics between the current and previous neighborhoods, while the built environment was assumed unchanged for the sample of non-movers. Commute distances were also measured in the survey.

Following the survey, objective measures of accessibility were estimated for each respondent, based on distance along the street network from home to a variety of destinations classified as institutional (bank, church, library, and post office), maintenance (grocery store and pharmacy), eating-out (bakery, pizza, ice cream, and take-out), and leisure (health club, bookstore, bar, theater, and video rental). Accessibility measures included the number of different types of businesses within specified distances, the distance to the nearest establishment of each type, and the number of establishments of each business type within specified distances.

Travel attitudes

To measure attitudes regarding travel, the survey asked respondents whether they agreed or disagreed with a series of 32 statements on a 5-point scale from 1 ("strongly disagree") to 5 ("strongly agree"). Factor analysis was then used to extract the relatively uncorrelated fundamental dimensions underlying these 32 items. Six were identified: pro-bike/walk, pro-transit, pro-travel, travel minimizing, car dependent, and safety of car (Table 1).

Socio-demographics

Finally, the survey contained a list of socio-demographic variables. These variables include gender, age, employment status, educational background, household income, household size, the number of children in the household, mobility constraints, residential tenure, and so on. Some changeable socio-demographics such as household structure and income were measured before residential relocation and currently.

Table 1. Key Variables Loading on the Neighborhood Characteristic and Travel Attitude Factors that are Significant in the Final Model

Factor	Statement
Perceived and Preferred Neighborhood Characteristics	
Accessibility	Easy access to a regional shopping mall (0.854); easy access to downtown (0.830); other amenities such as a pool or a community center available nearby (0.667); shopping areas within walking distance (0.652); easy access to the freeway (0.528); good public transit service (bus or rail) (0.437)
Safety	Quiet neighborhood (0.780); low crime rate within neighborhood (0.759); low level of car traffic on neighborhood streets (0.752); safe neighborhood for walking (0.741); safe neighborhood for kids to play outdoors (0.634); good street lighting (0.751)
Outdoor spaciousness	Large back yards (0.876); large front yards (0.858); lots of off-street parking (garages or driveways) (0.562); big street trees (0.404)
Travel Attitudes	
Pro-bike/walk	I like riding a bike (0.880); I prefer to bike rather than drive whenever possible (0.865); biking can sometimes be easier for me than driving (0.818); I prefer to walk rather than drive whenever possible (0.461); I like walking (0.400); walking can sometimes be easier for me than driving (0.339)
Pro-transit	I like taking transit (0.778); I prefer to take transit rather than drive whenever possible (0.771); public transit can sometimes be easier for me than driving (0.757); I like walking (0.363); walking can sometimes be easier for me than driving (0.344); traveling by car is safer overall than riding a bicycle (0.338)
Safety of car	Traveling by car is safer overall than riding a bicycle (0.489); traveling by car is safer overall than walking (0.753); traveling by car is safer overall than taking transit (0.633); the region needs to build more highways to reduce traffic congestion (0.444); the price of gasoline affects the choices I make about my daily travel (0.357)
Car dependent	I need a car to do many of the things I like to do (0.612); getting to work without a car is a hassle (0.524); we could manage pretty well with one fewer car than we have (or with no car) (-0.418);traveling by car is safer overall than riding a bicycle(0.402); I like driving (0.356)

Note: The numbers in parentheses are the pattern matrix loadings for the obliquely rotated factors. Source: Handy et al. (2004).

Results

In this section, models for change in driving and for vehicle type choice are presented to illustrate the potential of neighborhood design as a strategy for reducing air pollution.

Change in driving

A stronger test of a causal relationship between the built environment and travel behavior than the typical cross-sectional design involves an examination of the association between a change in the built environment and a change in driving. Such an approach addresses the time-order criterion for establishing causal validity: if the change in the built environment precedes the change in driving, then a causal relationship is more certain (Singleton and Straits 2005). In the quasi-longitudinal approach used here, changes are measured for residents who have recently moved from before to after their move, and for non-movers from one year earlier to the present time. The quasi-longitudinal model estimated from these data tests the hypothesis that moves to environments where residents are closer to destinations and have viable alternatives to driving are associated with a decrease in driving after accounting for neighborhood preferences and travel attitudes; decreases in driving might result from a decrease in driving distances and/or a decrease in driving trips.

The relationship between changes in the built environment and changes in driving while controlling for attitudes (and changes in socio-demographics) was estimated using an ordered probit model. This technique is appropriate for an ordinal dependent variable, and its model structure is parsimonious. In developing this model, the following sets of variables were tested: current socio-demographic characteristics, changes in socio-demographic characteristics, travel attitudes (assumed constant over this period), preferences for neighborhood characteristics (also assumed constant), objective accessibility measures for the current neighborhood, perceived neighborhood characteristics for the current neighborhood, and changes in perceived neighborhood characteristics. Non-movers were also included in the model, with changes in driving and socio-demographic characteristics measured from one year ago and changes in perceived neighborhood characteristics assumed to be zero. The resulting equation can be interpreted as representing the propensity of an individual to have a numerically larger change –either less of a decrease or more of an increase – in driving following the move. A statistically significant association between a change in the built environment and change in travel behavior provides evidence of a causal relationship.

Change in the accessibility factor was the most important factor in explaining changes in driving, as indicated by the standardized coefficients, with an increase in accessibility associated with either a smaller increase or a larger decrease in driving (Table 2). Change in the safety factor was also significant, with an increase in safety associated with either a smaller increase or a larger decrease in driving. Three accessibility measures were also significant: number of grocery stores and number of pharmacies within 1600m and number of theaters within 400m. Note that objective accessibility was measured for the current neighborhood only, rather than as the change in accessibility; however, a high current level of accessibility is more likely to be associated with an increase in accessibility than a decrease as a result of a move. For all these measures of accessibility, an increase in accessibility is associated with a higher propensity to drive less. Two travel attitudes were also significant: car dependent, with a positive effect on the propensity to drive more, and pro-bike/walk, with a negative effect on the propensity to drive more. These results support the hypothesis that changes in the built environment are associated with changes in driving and point to increases in accessibility as the factor having the greatest negative effect on driving.

Table 2. Ordered Probit Model for Change in Driving

Variables	Coefficient	Standardized Coefficient [a]	p-value
Constant	1.508	1.147	0.000
Socio-demographics			
Current age	-0.006	-0.084	0.014
Currently working	0.155	0.059	0.065
Current # of kids (<18)	0.070	0.057	0.051
Limitations on driving	-0.678	-0.074	0.000
Change in income (k$)	0.008	0.155	0.000
Neighborhood characteristics			
# of groceries within 1600 m	-0.014	-0.066	0.048
# of pharmacies within 1600 m	-0.028	-0.069	0.041
# of theaters within 400 m	-0.703	-0.057	0.055
Change in accessibility	-0.269	-0.226	0.000
Change in safety	-0.088	-0.086	0.000
Travel attitudes			
Car dependent	0.115	0.111	0.000
Pro-bike/walk	-0.070	-0.070	0.020
Threshold parameter 1	0.543	0.543	0.000
Threshold parameter 2	2.142	2.142	0.000
Threshold parameter 3	2.589	2.589	0.000
N	1490		
Log-likelihood at 0	-2378.038		
Log-likelihood at constant	-1954.785		
Log-likelihood at convergence	-1869.302		
Pseudo R-square	0.214		
Adjusted pseudo R-square	0.209		

a. All independent variables except constant term were standardized and model was reestimated; dependent variable was not standardized.

Vehicle type choice

The survey was originally developed to explore the relationship between neighborhood design and travel behavior, so not all neighborhood characteristics have meaningful connections with vehicle type choice. In addition, the survey only asked about current vehicle type. As a result, only cross-sectional analysis is possible and the time-order criterion for causality cannot be tested. However, if neighborhood characteristics are significantly associated with vehicle type choice after accounting for attitudes and preferences, the criterion of non-spuriousness has addressed and the results provide evidence of a causal effect.

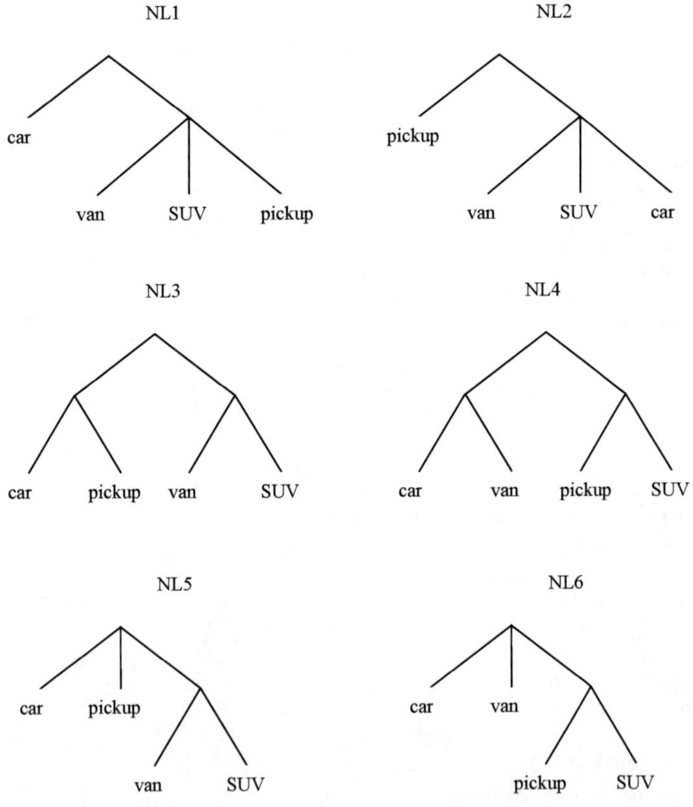

Figure 1. Nested Logit Model Structures Tested

Since vehicle type consists of four nominal categories and some categories share common characteristics, we attempted to estimate various nested logit (NL) models (Figure 1). The inclusive value (IV) parameters for most of these models were either outside the permitted range (i.e., greater than 1) or not significantly different from 1 (i.e., not different from the multinomial logit model). Initial structure NL4 performed the best, but its IV parameter for the pickup-SUV nest was not significantly different from 1 and hence that nest collapsed. The parameter for the car-van nest, however, was estimated at 0.299 and significantly different from 1, so it is that final structure whose model we present in Table 3. Since the fundamental function of both passenger cars and minivans is more to carry passengers than to carry goods, it is reasonable that these two vehicle types share the same nest.

Table 3. Nested Logit Model for Vehicle Type Choice (base alternative: passenger car)

Variables	**Coefficients**		
	Minivan	SUV	Pickup
Constant	-1.383 [0.000]	-2.884 [0.000]	-0.664 [0.081]
Socio-demographics			
Home owner	0.202 [0.077]		
Number of children (<18)	0.296 [0.000]	0.296 [0.000]	
Age	0.009 [0.016]		
Household income (k$)		0.012 [0.000]	
Female			-1.313 [0.000]
Education			-0.303 [0.000]
Number of vehicles			0.233 [0.038]
Neighborhood preferences			
Accessibility	-0.106 [0.013]		-0.106 [0.013]
Outdoor spaciousness	0.176 [0.001]	0.176 [0.001]	0.176 [0.001]
Travel attitudes			
Pro-bike/walk		0.287 [0.000]	0.287 [0.000]
Pro-transit			-0.423 [0.001]
Safety of car		0.331 [0.000]	
Neighborhood characteristics			
Outdoor spaciousness			0.199 [0.060]
Commute distance (miles)		0.008 [0.018]	
IV parameter		0.299 [0.000]	
Number of observations		1387	
Log-likelihood at 0: LL(0)		-2238.4	
Log-likelihood at constants: LL(C)		-1331.5	
Log-likelihood at convergence: LL		-1198.3	
Model improvement χ^2 = -2[LL-LL(C)]		266.4	
ρ^2 = 1-LL/LL(0)		0.472	
Adjusted ρ^2 = 1-(LL-18)/LL(0)		0.457	

Note: The number in brackets indicates the p-value for that coefficient.

Several socio-demographic characteristics significantly affect vehicle type choice. For example, those who are more affluent and have more children under 18 years old in the household tend to drive SUVs. Men and people with less education are more likely to choose pickup trucks. Those owning more vehicles are also more likely to drive pickup trucks, suggesting that a pickup is often the second, third, or later vehicle acquired in order to diversify the household's transportation options. Individuals who are home owners and have more children under 18 years old are more likely to drive minivans. Perhaps surprisingly, age is positively associated with the choice of minivans; the mean age of minivan drivers (48.6) is highest among the four vehicle types.

Attitudinal factors also play an important role in influencing vehicle type choice. Individuals who prefer living in less accessible areas are more likely to drive minivans and pickup trucks, and those preferring large yards and off-street parking have an inclination for all three types of LDTs. Interestingly, a preference for walking and biking is positively associated with the choice of SUVs and pickup trucks. It is possible that this association results from a preference for outdoor activities of various kinds, which is linked to both a preference for SUVs and pickups and a preference for walking and biking. Further, those vehicle types may be consciously chosen for their capacities to carry cycling, hiking, and camping gear. By contrast, those who have positive attitudes toward public transit are less likely to choose pickup trucks. It is possible that this association is also spurious and results from a concern for the environment that is positively linked to a preference for transit and negatively linked to driving pickups, which get relatively poor gas mileage. Underlying differences in lifestyle between transit users and pickup drivers might also help to explain this association. In addition, people who think the car is a safer mode are more likely to drive SUVs.

After accounting for the influences of socio-demographics and attitudes, two neighborhood characteristics appear significant in the model. Individuals who live in areas with more space are more likely to drive pickup trucks (significant at the 0.1 level). Note that a *preference* for more space has already been accounted for, suggesting that availability of parking space *itself* exerts some influence toward choosing a pickup truck. Further, workers living farther from their employment locations are more likely to drive SUVs. This result may be attributable to a greater concern for safety for respondents who spend more time on the road commuting. However, the effects of these characteristics are limited in comparison to the extensive influence of socio-demographics and attitudinal factors. While the results support a causal relationship between the built environment and vehicle type choice, they also suggest that the primary role of neighborhood design is to accommodate vehicle type preferences.

Conclusions

This study explored the potential of land use policies as a strategy for improving air quality. In particular, we investigated the influences of neighborhood design on driving behavior and vehicle type choice. Overall, the results support the premise that land use policies have at least some potential to reduce driving and ownership of light duty trucks, thereby reducing emissions.

Specifically, we found that changes in the built environment are significantly associated with changes in driving, controlling for current attitudes and changes in socio-demographics. This result provides some encouragement that land-use policies designed to put residents closer to destinations and provide them with viable alternatives to driving will actually lead to less driving. In particular, it appears that an increase in accessibility may lead to a decrease in driving, all else equal. With respect to vehicle type choice, we found that after controlling for attitudes and socio-demographics, outdoor spaciousness (a factor based on yard sizes and off-street parking availability) and commute distances were significant in the model. Thus, the built environment appears to play a separate, though modest, role in vehicle type choice: suburban development itself has an incremental impact on encouraging the acquisition of LDTs and hence contributes to the deterioration of air quality. Given the fact that LDT owners drive more miles, on average, than do passenger car drivers (as shown in our data as well as by Kockelman and Zhao 2000), this contribution is compounded. These finding thus support the idea that land use policies have at least some potential to reduce both driving and the choice of LDTs, thereby reducing emissions.

In general, two approaches might be taken: creating more neighborhoods that offer traditional characteristics associated with less driving and lower LDT use, or modifying the characteristics of suburban neighborhoods that are associated with more driving and higher LDT use. Policies that could lead to the creation of more neighborhoods offering traditional characteristics include mixed-use zoning that allows for retail and other commercial within close proximity to residential areas and street connectivity ordinances that ensure more direct routes between residential and commercial areas. The success of this approach depends on the ability of such areas to attract residents who would otherwise live in neighborhoods with suburban characteristics, choose to own LDTs, and drive a lot. Recent studies suggest that traditional neighborhoods are undersupplied relative to the demand (Levine and Inam 2004). In contrast, the modification of suburban neighborhoods could include restrictions on the provision of off-street parking in new suburban developments and efforts to bring more matched jobs to suburban areas to reduce commute distances, thereby reducing LDT ownership. Policies that could increase accessibility in existing suburban neighborhoods in order to reduce driving include incentives for infill development and redevelopment of underutilized shopping centers, and the like. Although this study does not definitively prove that land use policies can reduce emissions, it provides new evidence that supports the adoption of such policies.

Future studies that adopt research designs that more closely resemble a true experimental design will provide more definitive evidence yet. Two types of studies are possible: true panel studies of residents who move from one type of neighborhood to another (with measurements of attitudes as well as driving and vehicle type before and after the move, and further exploration of the reasons behind the move), and natural experiments that examine the impact on driving and vehicle type in response to a change in the built environment, such as the implementation of a traffic calming program. Changes in driving are likely in the short run, with changes in vehicle type choice more likely over the longer term. Only with such evidence can we be sure that by increasing opportunities for driving less through land use policies, cities will help to reduce driving and thus emissions.

Acknowledgements

This research was funded by the UC Davis-Caltrans Air Quality Project and was supported by grants from the Robert Wood Johnson Foundation and the University of California Transportation Center. Thanks to Ted Buehler, Gustavo Collantes, and Sam Shelton for their work on the implementation of the survey. We thank WenChao Huang for his help on vehicle type classifications.

Note

This paper is substantially based on the following two previously-published articles:
Cao, X., Mokhtarian, P. L., and Handy, S. L. (2006). "Neighborhood design and vehicle type choice: Evidence from Northern California." *Transportation Research D*, 11 (2), 133-145.
Handy, S. L., Cao, X., and Mokhtarian, P. L. (2005). "Correlation or causality between the built environment and travel behavior? Evidence from Northern California." *Transportation Research D*, 10 (6), 427-444.

References

Babbie, E. (1998). "The practice of social research, 8th Ed." *Wadsworth Publishing Company*, Belmont, CA.

Bhat, C. R., and Sen, S. (2006). "Household vehicle type holdings and usage: An application of the multiple discrete-continuous extreme value (MDCEV) model." *Transportation Research B*, 40 (1), 35-53.

Burchell, R.W., Lowenstein, G., Dolphin, W.R., Galley, C.C., Downs, A., Seskin, S., Gray Still, K., and Moore, T. (2002). "Costs of sprawl – 2000. TCRP Report 74, Transit Cooperative Research Program." *Transportation Research Board*, Washington, DC.

Cervero, R., and Radisch, C. (1996). "Travel choices in pedestrian versus automobile oriented neighborhoods." *Transport Policy*, 3 (3), 127-141.

Environmental Protection Agency (EPA) (2001). "EPA Guidance: Improving Air Quality through Land Use Activities." *EPA420-R-01-001, Transportation and*

Regional Programs Division, Office of Transportation and Air Quality, January.

Handy, S., Mokhtarian, P., Buehler, T., and Cao, X. (2004). "Residential Location Choice and Travel Behavior: Implications for Air Quality." *Research Report, UC Davis – Caltrans Air Quality Project*, December. <http://aqp.engr.ucdavis.edu/Documents/Final_report_editted_updated1.pdf>.

Kockelman, K. M., and Zhao, Y. (2000). "Behavioral distinctions: The use of light-duty trucks and passenger cars." *Journal of Transportation and Statistics*, 3 (3), 47-60.

Levine, J., and Inam, A. (2004). "The market for transportation-land use integration: Do developers want smarter growth than regulations allow?" *Transportation*, 31 (4), 409-427.

Niemeier, D. A., Hendren, P., Foresman, E., Morey, J. E., Redmond, L., Lin, J., Hicks, J., and Zheng, Y. (1999). "Redefining conventional wisdom: An exploration of auto ownership and travel behavior in the United States." *Transportation Research Circular E-C026*, <http://gulliver.trb.org/publications/circulars/ec026/07_niemeier.pdf> (April 10, 2007).

Singleton, R. A. Jr., and Straits, B. C. (2005). "Approaches to social research, 4th Ed." *Oxford University Press*, New York.

Designing Inside the Box- Strategies to Successfully Marry Smart Growth and Context-Sensitive Transportation Initiatives

G. Wade Walker, P.E.[1]
[1]Kubilins Transportation Group, Inc., 800 West Hill Street, Suite 202, Charlotte, NC 28208; PH (980) 321-0108; email: Ewwalker@kubilins.com

Introduction

With the increasing realization that transportation facilities can no longer be designed as "reactionary" measures to land use decisions, many communities are turning to a more integrated land use and transportation planning process adhering to Smart Growth principles. Unfortunately, many local and state design standards often require roadway design parameters that are completely out of context with the surrounding environment; implementation of these "standards" within a Smart Growth framework often results in roadways that encourage higher vehicular speeds at the expense of other modes, creating a pedestrian-hostile environment in an area that should be better balanced for all modes and sensitive to the context in which it exists. This paper examines the issues facing designers wishing to create context-sensitive roadways and the flexibility afforded by nationally-accepted design practices and guidelines. It also presents some examples where this flexibility has allowed for facilities to be constructed that blend into and enhance the surrounding environment. Finally, it answers the question of "what's next?" in terms of mainstreaming the practice of context-sensitive design and solutions.

The Issue: You Can't Build That Here!

In the years following World War II, America saw a dramatic increase in mobility and the resultant growth of the suburban lifestyle. As Americans became more mobile due to the increasing availability to attain auto ownership, many citizens realized that they no longer had to live within a walk or train ride of their workplace. Houses were built further and further out from City centers, and retail development followed the rooftops. Local and state jurisdictions helped enable this mass exodus from the traditional urban centers by extending infrastructure to these far-reaching locales; in the realm of transportation, this new era dictated that automobile delays were to be kept to a minimum, oftentimes as if nothing else mattered. Similar to the way in which building codes were modified to apply to this type of new suburban development, state and local jurisdictions established land development codes and ordinances that usually included guidance about roadway design for suburban areas. These roadway design "standards" as they were called often precluded the design and construction of roadways except of a type that was heavily geared toward moving traffic as quickly as possible, sometimes at the expense of the surrounding landscape, environment, or communities. Even though the design guidelines from which these local "standards" were derived addressed lower speed urban facilities in which provisions were made for pedestrians and accessibility, the lack of those urban context types within the new suburbia often resulted in the inability to provide more context-sensitive facilities in these areas. The local "standards" often only addressed facilities with desired vehicular speeds in excess of 35 or even 45 miles per hour; since lower speed urban facilities were not present in these areas, they were usually not addressed within the "standards." Characteristics of these type "standards" usually included design elements such as 3.6 meter (12 foot) travel lanes, large corner radii, prohibition of on-street parking, and lane encroachment prohibitions.

Beginning in the mid 1980's, America began to rediscover their great urban places, with people moving back into the cities and even a more urban development type beginning to occur in the suburbs. This movement has coincided with the rise of Traditional Neighborhood Development, which emulates the pre-War development type by creating mixed-use communities with a variety of housing types located within proximity to a commercial core or town center. Residents of these communities are often drawn to them by the true sense of community that exists, and the ability to live, work, shop, and have recreational and social activities within the same area, without the constant dependence of the automobile. "Smart Growth" was also an offshoot of this movement, with the increasing realization that environmental and community impacts were far less severe by clustering development types rather than allowing them to "strip out" swathes of countryside along highways that moved people further and further from the traditional city centers. Both of these development types rely heavily on a street network that does things other than just move cars quickly; streets need to be walkable, friendly to retail uses, allow for amenities such as on-street parking, and encourage slower vehicular speeds. Unfortunately, many local street design "standards simply do not allow for narrower, slower streets; designers struggle with being able to provide facilities with these more livable characteristics in a regulatory framework where they are simply "illegal" to construct.

Flexibility within Accepted Design Guidelines

One avenue that designers can utilize when confronted with the challenge of local standards not allowing a slower, urban-type street is to go back to the accepted design guidelines form which these standards were likely derived. Many local and state standards were derived from the "bible" of transportation engineering: ***A Policy on Geometric Design of Highways and Streets***, published by the American Association of State and Highway Transportation Officials (AASHTO). Known within the industry as the "Green Book" this publication has been republished at several intervals during its lifespan, with the most recent version published in 2004. The document is very clear that it is intended to be used as a set of guidelines for roadway design, and that engineering judgment should be taken into account for designing in context-rich environments to insure that the proposed roadway design has acceptable impacts on the landscape, environment, or surrounding community. The Green Book allows for use of varying parameters such as lower design speeds to produce designs that have less of an impact on the surrounding context; the tradeoff is that vehicular operating speeds will decrease somewhat, with a corresponding increase in through vehicular delay. The Green Book directs the designer to weigh the impacts of these type decisions in considering alternative designs. Although most state and local standards were in fact derived from the Green Book, this direction toward flexibility was often lost due to the fact that many communities at the time were constructing development that was geared more toward moving automobiles quickly, with little or no regard to non-motorized forms of transportation. In 1997, the Federal Highway Administration published ***Flexibility in Highway Design***, intended to serve as a companion to the Green Book and to highlight the ability of designers to consider alternate designs that may have an impact on operations but not safety when the surrounding context dictated that a roadway should blend more harmoniously into the environment. Since that time, the practice of "context sensitive design" (CSD) and "context sensitive solutions" (CSS) have become more mainstream in the practice of transportation engineering, as designers strive to provide facilities that enhance, rather than overwhelm, their surroundings.

Collaboration: The Key to Context-Sensitive Design

A result of the CSD movement has been a rethink of how facilities are designed. Historically, designers faced with a transportation issue have designed a facility away from the public eye, and then come back to the community to present the solution. Predictably, many of those designs are met with criticism due to the fact that the designer did not consider elements that were important to the host community; oftentimes, these elements are outside the right-of-way, but directly influence the roadway and its impact on the community. One way that CSD avoids these potential pitfalls is to integrate the public into the design process. Throughout the design process, community "stakeholders" are consulted and brought along in the design decisions. The result is a project that has consensus built along the way, without the surprises that happen when a conventionally-designed project is presented to the community for the first time.

An increasing number of transportation projects are being designed using the charrette process in which citizens and community stakeholders are an integral part of the design team. The principles of designing within the context of a charrette are described as follows:

- *Collaborative and Interdisciplinary* – Every successful project has many talented hands that have touched it at some point. In a charrette environment, this involves assembling a team of professionals who understand land use and development patterns, historical implications, ecological considerations, marketing and economics, and landscape and aesthetics, as well as being grounded in the art and science of traffic and transportation engineering. Secondly, assembling community stakeholders from the business, residential, and jurisdictional realms and having them intimately involved in the design process allows the team to quickly flesh out issues and opportunities, develop a slate of alternatives, analyze those alternatives, and come up with a consensus-driven solution to a given transportation issue. By using this process, projects that typically take months to build consensus on are focused within a one-week period.

- *The Vision Drives the Design* – Consultants rarely live in the communities within which they work, so it comes as no surprise that some proposals are roundly rejected by the community as not "fitting" their vision. By bringing the community to the design table, the team is able to consistently hold to the vision of the community, and develop design alternatives that fit holistically within that vision.
- *Be Inclusive of Ideas* – Some of the best solutions to a given problem come from someone other than an engineer. Be careful not to dismiss any proposal without testing its feasibility within the design team (which includes the stakeholders) and analyzing it thoroughly. Some of the most valuable partnerships are forged by including something in the design that may not be an integral part of the engineering, but is extremely important to the community.
- *Articulate the Vision* – Many stakeholders need assistance in visualizing what a proposal may look like on the ground. Oftentimes, plans and profiles mean little to a stakeholder; for that reason, visualization is important to illustrate various proposals within their contextual setting. The use of hand drawn sketches, renderings, and photo enhancements aid the general public in "seeing the future" of how a proposal fits within the community.

These principles have been used successfully around the nation in developing designs for projects ranging in scope from neighborhood traffic calming initiatives to regional transportation plans.

Success Stories

CSD has been used successfully in designing and constructing many projects throughout the country. From new developments to retrofits of existing urban places, to even rural areas, CSD has allowed for roadway facilities to balance a multitude of roles within their host communities, often at a minimal impact to through travel time or delay. Some recent successes of the application of CSD are summarized as follows:

- *Hernando West (Hernando, Mississippi)* – A 1,100 acre farm adjacent to Downtown Hernando, Mississippi, was proposed for development in late 2005. Katz Builders of Pennsylvania retained the design firm of Looney Ricks Kiss of Memphis to develop a plan for a mixed use traditional community that functioned as an extension of the historic adjacent downtown and its surrounding neighborhoods. The topography of the property included streams, hills, and wooded areas which were carefully integrated into the plan as conservation areas and amenities to the new community. The street network was designed around the principles of context-sensitivity, with multiple connections to the surrounding street network and a system of narrow, two-lane roadways instead of multi-lane facilities. A benefit of the transportation network was that it was able to carry the traffic associated with the development of 4,000 new homes and the retail and office uses (approximately 50% of the existing housing stock in Hernando) without adverse impacts on the existing City's roadway network. The development has already been recognized by the local Sierra Club chapter as a smart growth example, winning an award from the society for its preservation and conservation enhancement efforts.

- *Gilberts Corner Roundabouts (Loudoun County, Virginia)* – The Virginia Department of Transportation had originally proposed a flyover to relieve the traffic pressure and congestion at the intersection of US Highways 50 and 15 in Eastern Loudoun County, Virginia, located only 15 miles west of Dulles International Airport in the Washington, DC metro area. During a four day charrette held in 2000, a concept was developed to replace the proposed flyover with a system of three roundabouts connected by a two-lane piece of additional network. In addition to the ability of the roundabout system to move a similar amount of traffic as the flyover, the roundabouts blended with the surrounding landscape and adhered to the principles of CSD within this historically and aesthetically-rich corridor. VDOT is currently implementing the roundabout system through a design-build process, at a tenth of the cost of the originally-proposed flyover.
- *Towns, Villages, and Countryside (TVC) Comprehensive Plan Element (St. Lucie County, Florida)* - In 2004, the St. Lucie County Board of County Commissioners commissioned a visioning charrette to determine the best pattern of growth for the northern portion of St. Lucie County. The idea of Towns, Villages, and Countryside illustrated this vision characterized by mixed-use development in a traditional form strategically located at crossroads; preservation of open space; an interconnected network of smaller, pedestrian-friendly streets as opposed to multilane, high-speed arterials highways; and the ability to transfer development rights to increase the allowable densities in strategic locations. The Vision is being enabled by an all-new element drafted for the St. Lucie County Comprehensive Plan: the TVC (Towns, Villages, and Countryside) Element. From a transportation perspective, the regional scale of the TVC Element dictates that instead of multi-lane arterial roadways, that mobility would be facilitated by a system of interconnected two-lane roads that provided access between and within the development nodes. Coupled with the form and type of nodal mixed land use specified in the TVC, this system of roadways would facilitate more development density than what was originally allowed by the County's Comprehensive Plan, but in a more sustainable form. The TVC is currently in the process of adoption by St. Lucie County.

These case studies focus on the application of CSD principles in both the local and regional scale, in both urban and rural contexts. Each of these successes was accomplished using the principles of CSD in a collaborative, stakeholder-involved process.

The Next Steps: Mainstreaming Context-Sensitivity

Since the advent of the Smart Growth movement, CSD has gained popularity in its ability to provide projects that are in harmony with the communities in which they exist, and oftentimes can be designed and constructed with less time and cost than conventionally-designed counterparts. More alternatives are available to the roadway designer today to develop balanced facilities while adhering to acceptable and proven design guidelines. Efforts to update design guidelines to address smart growth are constantly ongoing; one such effort is a joint project between the Institute of Transportation Engineers (ITE) and the Congress for New Urbanism (CNU) to address urban street design. The recently-published *Context Sensitive Solutions for Urban Thoroughfares in Walkable Communities* is the result of two years of collaboration between the leading groups in transportation engineering and smart growth. The new document emphasizes the need to balance objectives of transportation and mobility with the livability needed in smart growth areas. The document recognizes the need to consider slower vehicular speeds to achieve design objectives such as accommodation of pedestrians, bicycles, transit vehicles, and parking within a sustainable, mixed-use environment. Other recent efforts include publications such as *Residential Streets, Third Edition* (ITE, American Society of Civil Engineers, and the National Association of Home Builders) to guide street design in slow-speed residential areas, and the ITE document *Traditional Neighborhood Development Street Design Guidelines* that recognizes the need for multi-user streets in traditional neighborhood developments that provide narrower, slower designs to accomplish the goals of smart growth.

As more communities embrace the concept of smart growth, the transportation design profession must adapt to a more in-depth consideration of alternative design strategies to result in roadways that are more consistent with the context in which they are constructed. The inherent flexibility in the accepted design guidelines, the ongoing work of multi-disciplinary groups such as the joint ITE/CNU venture, and the nature of the collaborative design process can insure success in designing livable transportation solutions "within the box" of acceptable and proven design parameters and consistent with the emerging trend toward Smart Growth.

References

American Association of State Highway and Transportation Officials (2004), "A Policy on Geometric Design of Highways and Streets, 5th Edition." American Association of State Highway and Transportation Officials, Washington, D.C.

Federal Highway Administration (1997), "Flexibility in Highway Design." Federal Highway Administration, Washington, D.C.

Institute of Transportation Engineers Transportation Planning Committee 5P-8 (1997), "Traditional Neighborhood Development Street Design Guidelines." Institute of Transportation Engineers, Washington, D.C.

Kulash, W. (2001), "Residential Streets, 3rd Edition." Urban Land Institute, Washington, D.C.

Institute of Transportation Engineers and Congress for New Urbanism (2006), "Context Sensitive Solutions for Urban Thoroughfares in Walkable Communities." Institute of Transportation Engineers, Washington, D.C.

The Impacts of Urban Development on Anthropogenic and Biogenic Emissions and Air Quality: A Case Study in Austin Texas

E. McDonald-Buller,[1] J. Song,[1] A. Webb,[1] G. McGaughey,[1] B. Zhou,[2] K. Kockelman,[2] J. Lemp,[2] B. Parmenter,[3] and D. Allen[1]

[1]Center for Energy and Environmental Resources, The University of Texas at Austin, 10100 Burnet Rd., M/S R7100, Building 133, Austin, TX 78758; *Corresponding author: PH (512) 471-2891; FAX (512) 471-1720; e-mail: ecmb@mail.utexas.edu.
[2]Department of Civil, Architectural, & Environmental Engineering, The University of Texas, 1 University Station, #C1761, Austin, TX 78712-0278; PH (512) 471-0210; FAX: 512-475-8744; e-mail: kkockelm@mail.utexas.edu
[3] Tufts University, Urban and Environmental Policy & Planning, 97 Talbot Ave, Somerville, MA 02144; PH: (617) 627-3394; FAX: (617) 627-3377; e-mail: barbara.parmenter@tufts.edu

Abstract

Future growth due to urban development results in changes to land use and land cover and, consequently, to biogenic and anthropogenic emissions, meteorological processes, and processes such as dry deposition. This on-going project examines the role of urban development, demographic, and technology trends on emissions and air quality using Austin, Texas as a case study. The results contrast the impacts of four different urban growth scenarios on land use/land cover and anthropogenic and biogenic emissions. A community "visioning" project, called Envision Central Texas, has resulted in four pre-determined metropolitan development scenarios based on a projected doubling of the population. The scenarios include: 1) low-density, segregated-use development based on extensive highway provision; 2) concentrated, contiguous regional growth within 1-mile of transportation corridors; 3) concentrated growth in existing and new communities with distinct boundaries; 4) high-density development and balanced-use zoning. Results from these scenarios will be contrasted with those from an integrated transportation-land use model currently under development. A photochemical grid model, the Comprehensive Air Quality Model with extensions, is being used to compare the air quality impacts of urban development from the various scenarios.

Introduction

Urban development results in changes to land use and land cover and, consequently, to biogenic and anthropogenic emissions, meteorological processes, and processes such as dry deposition. This on-going project examines the roles of urban development, demographic, and technology trends on emissions and air quality using Austin, Texas as a case study. Austin was among the first areas to prepare an Early Action Compact or voluntary State Implementation Plan (SIP) under the National Ambient Air Quality Standard (NAAQS) for ozone concentrations averaged over 8-hours. The Austin Metropolitan Statistical Area (MSA), which includes Travis, Williamson, Hays, Bastrop and Caldwell Counties, is located in Central Texas with a population of 1.25 million people and an economic sector based on computer hardware and semiconductor manufacturing, software development, education, and state government.

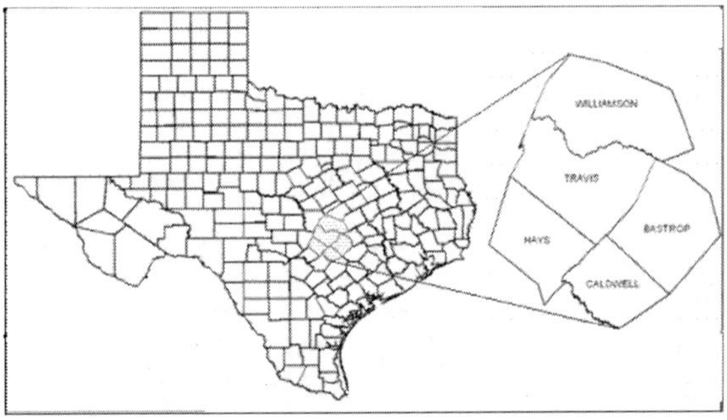

Figure 1. The five-county Austin MSA located in Central Texas.

Objectives

The overall objectives of the project are to:

1. Apply an integrated transportation-land use model (ITLUM) to investigate the impacts of regional development scenarios on the magnitude and spatial distribution of emissions of ozone precursors. Compare ITLUM-based forecasts with four pre-determined metropolitan development scenarios that were developed through a community "visioning" initiative called Envision Central Texas (ECT) and based on a projected doubling of population. These 4 scenarios can be characterized as: A) low-density, segregated-use development based on extensive highway provision; B) concentrated, contiguous regional growth within 1-mile of transportation corridors; C) concentrated growth in existing and new communities with distinct boundaries; D) high-density development and balanced-use zoning.
2. Compare the impacts of the regional development scenarios on predicted ozone concentrations and human exposure patterns using a photochemical grid model.

3. Test the hypothesis that predicted human exposure patterns based on ITLUM emission forecasts will differ from those based on the U.S. EPA's post-Clean Air Act Amendment emission scenario projections.
4. Test the hypothesis that changes in land cover and dry deposition patterns have at least as significant an impact on future air quality as changes in on-road vehicle emission control technologies.

The results presented here contrast the emissions and air quality impacts of different ECT urban growth scenarios. The impacts are specifically examined in the context of changes in biogenic emissions versus anthropogenic emissions from area/non-road mobile sources and on-road mobile sources. The on-going development of the ITLUM is also described.

ECT Model Development and Emissions Estimates

Biogenic and anthropogenic emissions for future development scenarios were developed for use in a regional air quality model. The Comprehensive Air Quality Model with extensions (CAMx), an Eulerian photochemical grid model developed by ENVIRON International Corporation and currently used by the State of Texas for attainment demonstrations, was used to evaluate air quality impacts under the different regional development scenarios. Emissions estimates were developed for a variety of future case scenarios, but all of these scenarios used meteorology from a September 13-20, 1999 historical ozone episode that was used for Austin's Early Action Compact.

It is important to distinguish between land use and land cover (LULC) data and their relative roles in emissions and air quality modeling. Land use categories describe how humans use or intend to use the land, e.g., high or low density, rural or urban. Land cover describes the physical features that occur on the land surface, such as water, vegetation, bare soil, rock and built features, or more specifically, vegetation species. Land uses and land covers affect emissions in a variety of ways, but in eastern Texas, the dominant effect is on biogenic emissions (hydrocarbons originating from vegetation). Accurate predictions of biogenic emissions rely on accurate characterizations of land covers (leaf biomass densities by species). The ECT land development scenarios (10 – 16 land use types) were combined with Texas vegetation maps to arrive at projected changes in land cover for each ECT scenario at a spatial scale of 4 kilometers. The Global Biogenic Emissions and Interactions System version 3.1 (GloBEIS v3.1; Yarwood et al., 1999) was used to develop the biogenic emission inventories for each scenario. Details are described by Song, et al. (2006).

Land use and land cover also affect other processes important in air quality modeling, and one of the most significant of these processes is dry deposition. Dry deposition is the major physical removal mechanism for pollutants during the summer ozone season, and dry deposition rates for each scenario were developed using algorithms based on the work of Wesely (1989) and land cover mappings for each ECT scenario.

On-road mobile source emission inventories were developed for each of the ECT scenarios by combining travel demand model (TDM) output for the 5-county Austin area link network with emission factors from U.S. EPA's MOBILE6 model. Smart Mobility, Inc. developed the Envision Central Texas Transportation Model (ECTTM) with support from the Capital Area Metropolitan Planning Organization (CAMPO). For each scenario, vehicle miles of travel (VMT) by link direction were obtained from the ECTTM and apportioned by hour for each of four day-types: Weekday (average Monday through Thursday), Friday, Saturday, and Sunday. Hourly VMT on each roadway type link was disaggregated into each of the 28 MOBILE6 vehicle types. The MOBILE6 model was then used to calculate emission factors (grams/mile) for volatile organic compounds (VOC), carbon monoxide (CO), and oxides of nitrogen (NOx) based on the year 2030 which is the timeframe for doubling of the region's population. The disaggregated link VMT was matched with the corresponding pollutant-specific MOBILE6 emission type factors tabulated by speed, hour, roadway type, and vehicle type in order to obtain link-level emissions estimates in grams.

Non-road mobile sources include construction equipment, industrial equipment, lawn and garden equipment, and agricultural equipment. Area sources are generally defined as individual sources that are small and numerous, including solvent utilization and industrial processes, and their emissions are estimated collectively. The EPA's NONROAD2005 model was used as the basis for calculating 2030 estimates of non-road emissions for each ECT scenario. Data on single- and multi-family households and average household size for each ECT scenario was obtained from Smart Mobility, Inc. and used to estimate population in the five-county area, shown in Table 1. Population data were used to adjust the state to county-level spatial allocation of equipment in the NONROAD model as well as to develop growth factors for area sources. Land use patterns for each ECT scenario were overlaid on Texas vegetation maps to develop spatial surrogates for allocating county-level emissions to model grid cells at a spatial scale of 4 kilometers.

Summaries of NOx and VOC emissions from biogenic and anthropogenic sources for each ECT scenario are presented in Table 2. Biogenic sources and, because they have been scaled initially with population, area sources are predicted to remain the most significant sources of VOC emissions in the five-county area. Differences in LULC led to 1-6% reductions in daily biogenic emissions in the 5-county area that includes Austin. The regions with reductions in biogenic emissions were located in newly developed areas. Although VMT is predicted to continue increasing, emissions of NOx and VOCs from on-road mobile sources are predicted to decrease through approximately 2025 due to the phase-in of new emission standards. Similarly, NOx emissions from non–road mobile sources are also predicted to decrease due to the phase-in of new emission standards, while VOC emissions are predicted to increase by 5-9%.

Table 1. 2001 population and projected human population for each ECT scenario by county.

County	Average Household Size	Population 2001	ECT A	ECT B	ECT C	ECT D
Bastrop	2.87	61,480	151,023	207,527	264,555	206,950
Caldwell	2.98	33,808	72,029	124,390	177,762	124,261
Hays	2.92	104,514	238,536	247,051	263,590	248,215
Travis	2.53	842,638	1,407,609	1,284,282	1,132,114	1,250,786
Williamson	2.88	276,749	621,035	641,099	679,622	651,846
Total		1,319,189	2,490,231	2,504,349	2,517,644	2,482,059

Table 2. Base Case* and future projected NOx and VOC emissions (tpd) in the 5-county Austin area.

Scenario	Biogenic		On-Road Mobile		Area		Non-road Mobile		Point (no growth)	
	NOx	VOC	NOx	VOC	NOx	VOC	NOx	VOC	NOx	VOC
Base Case*	20.1	210.2	58.0	31.5	10.12	110.66	21.69	22.06	2.8	3.0
ECT A	20.1	197.8	18.6	20.1	20.51	214.28	9.45	23.30	2.8	3.0
ECT B	20.1	204.3	16.2	17.6	22.00	237.64	9.68	23.96	2.8	3.0
ECT C	20.1	204.6	15.8	17.2	23.70	261.62	9.68	23.94	2.8	3.0
ECT D	20.1	206.8	14.7	15.7	21.99	236.13	9.44	23.24	2.8	3.0

*Note that the Base Case included 2007 emission inventories for anthropogenic sources, emission controls adopted for Austin's Early Action Compact, and a biogenic emissions inventory based on a land cover/land use database developed by Wiedinmyer et al. (2000, 2001).

Relative Air Quality Impacts of the ECT Development Scenarios

The ECT development scenarios were compared based on their impacts on air quality due to changes in biogenic emissions and dry deposition and due to changes in anthropogenic emissions. The results presented below focus on the two ECT scenarios that represent the most extreme differences in development patterns: (1) ECT A, which is consistent with Austin's historical pattern of low-density, segregated-use development based on extensive highway provision, and (2) ECT D, which is high-density development and balanced-use zoning. The results for the ECT scenarios are contrasted with a Base Case that included 2007 emission inventories for anthropogenic sources, emission controls adopted for Austin's Early Action Compact, and a biogenic emissions inventory and dry deposition estimates based on a land cover/land use database developed by Wiedinmyer et al. (2000, 2001). The photochemical modeling for the Base Case and all of the ECT scenarios used meteorology from a September 13-20, 1999 historical ozone episode that was used for Austin's Early Action Compact.

Table 3 shows the daily maximum 1-hour averaged ozone concentrations across the week-long air pollution episode for the Base Case and the differences in the daily maximum 1-hour ozone concentration for the ECT scenarios relative to the Base Case; results for the ECT scenarios are segregated by changes due to the impacts of urbanization on biogenic emissions and deposition only, on anthropogenic emissions only, and the combined effects on biogenic emissions, dry deposition, and anthropogenic emissions. With other factors remaining unchanged, future changes in ozone concentrations due to both biogenic emissions and dry deposition (i..e, 'Bio') due to a doubling of population (ECT A) resulted in changes of -0.94 ppb to +0.12 ppb. These concentrations are the differences in maximum daily 1-hour ozone concentrations, compared to the Base Case. The decreases in ozone concentrations were consistent with the loss of vegetative cover in developing areas and reductions in biogenic emissions and dry deposition. Although these impacts appear small, they are comparable in magnitude to some commonly employed air pollution control measures that were adopted as part of Austin's Early Action Compact.

Differences in maximum daily ozone concentrations due to changes in anthropogenic emissions only (i.e., 'Anthro') were far more significant, ranging from -6.97 ppb to -1.43 ppb for ECT A. Area-wide daily maximum ozone concentrations in the Austin region have historically been predicted to be most responsive to NOx emissions reductions, and reductions in ozone concentrations for the ECT scenarios are consistent with decreases in NOx emissions from the phase-in of new standards for mobile sources. The combined impacts due to changes in biogenic emissions, dry deposition, and anthropogenic emissions were quite similar to the 'Anthro' only scenario. Differences in maximum daily ozone concentrations between the ECT scenarios were smaller than the differences between the ECT scenarios and the Base Case; for example, the differences between the ECT A and ECT D development scenarios for the 'Combined' changes ranged from -0.58 ppb to 0.18 ppb (i.e., O_{3ECTD}-O_{3ECTA}).

Table 4 shows the range of differences in ozone concentrations across the Austin area between the ECT scenarios and the Base Case. These results are segregated similarly to those in Table 3 showing changes due to the impacts of urbanization on biogenic emissions and deposition only, on anthropogenic emissions only, and the combined effects on biogenic emissions, dry deposition, and anthropogenic emissions. It is important to note that in contrast to Table 3, the differences in Table 4 are not necessarily associated with the area-wide peak predicted ozone concentration, but instead capture the maximum and minimum differences that occur across the region regardless of time of day or magnitude of the ozone concentrations. Similar to the results in Table 3, ozone concentrations across the area are more influenced by changes in anthropogenic emissions versus changes in biogenic emissions and dry deposition.

Table 3. Daily maximum ozone concentrations for the Base Case and differences in the daily maximum ozone concentrations relative to the Base Case for ECT A and ECT D.

Episode Day	Base Case Daily Max. O_3 Conc. (ppb)	ECT A: Difference in Daily Max O_3 Conc. from Base Case (ppb)			ECT D: Difference in Daily Max O_3 Conc. from Base Case (ppb)		
		Bio	Anthro	Combined	Bio	Anthro	Combined
9/15	80.52	0	-4.61	-4.67	-0.04	-5.27	-5.25
9/16	71.95	0.12	-2.06	-1.75	0.08	-2.18	-2.09
9/17	85.83	-0.11	-6.97	-6.88	-0.03	-6.95	-6.89
9/18	86.17	-0.39	-4.11	-4.09	-0.15	-4.08	-4.08
9/19	90.41	-0.58	-1.43	-1.46	-0.15	-1.34	-1.35
9/20	90.46	-0.94	-5.91	-5.97	-0.12	-5.77	-5.79

Note that: (1) 'Bio' indicates impacts of urbanization due to changes in biogenic emissions and dry deposition only; (2) 'Anthro' indicates impacts of urbanization due to changes in on-road, non-road, and area source emissions only (point source emissions did not change); and (3) 'Combined' indicates impacts due to changes in both 'Bio' and 'Anthro'.

Table 4. Maximum and minimum differences in ozone concentrations between the ECT Scenarios and the Basecase (i.e., [O_{3ECT}-$O_{3BASE\ CASE)}$]) across the Austin area.

ECT Scenario	Range of Differences in Ozone Concentrations Across the Austin Area (ppb)		
	'Bio'- Base Case	'Anthro'-Base Case	Combined – Base Case
A	0.67 at 0900 on 9/15	22.10 at 0600 on 9/20	21.88 at 0600 on 9/20
	-1.37 at 0600 on 9/19	-13.53 at 0800 on 9/18	-13.61 at 0800 on 9/18
D	0.65 at 0900 on 9/15	21.57 at 0600 on 9/20	21.53 at 0600 on 9/20
	-0.81 at 0600 on 9/19	-12.39 at 1400 on 9/20	-12.45 at 1400 on 9/20

Note that; (1) 'Bio' indicates impacts of urbanization due to changes in biogenic emissions and dry deposition only; (2) 'Anthro' indicates impacts of urbanization due to changes in on-road, non-road, and area source emissions only (point source emissions did not change); (3) 'Combined' indicates impacts due to changes in both 'Bio' and 'Anthro'.

ITLUM Development and Comparison with ECT Visions

The 5-county Austin area's year 2030 travel conditions and household and employment distributions are also being predicted using a variation of Steven Putman's gravity-based land use model (LUM) (Putman 1983) as an alternative to the community-oriented, regional visioning process for the ECT development scenarios. This traditional LUM system has three components: a *D*isaggregated *R*esidential *A*llocation *M*odel (DRAM), an *EMP*loyment *Al*location model (EMPAL) and *LAN*d *CON*sumption model (LANCON). The first two borrow very heavily from Putman's specifications, but the third is altogether new, possibly offering better recognition of the land consumption-land intensity relation (e.g., acres of residential use versus number of households). DRAM presumes that current worker (and thus household) distribution is determined by current employment locations (jobs), land usage (thus helping avoid excessive assignment of households to small zones), and travel impedances between all zones. Similarly, the EMPAL model forecasts the spatial distribution of jobs by category (including basic, retail and service). Finally, the LANCON model uses three log-linear multivariate equations to estimate land consumption by type (residential, basic and commercial) in each zone. Variables determining land consumption include the forecasted spatial distributions of households and employment, and prior-year land use conditions. A reasonably standard sequential travel demand model (TDM) (with an upstream step for vehicle ownership forecasting, and an upstream of the motorized mode choice decision) was externally linked to the LUM system in order to update travel conditions and provide a well-defined, series of related steps to all future household and employment forecasts (at five-year intervals). Altogether, the system of equations forms what is called an "integrated transportation-land use model" (ITLUM).

The DRAM, EMPAL and LANCON models were calibrated using 2000 and 2005 demographic and 2005 land use data for Austin's 5-county Metropolitan Statistical Area (MSA) obtained via the Capital Area Council of Governments. The data were refined using the City of Austin's relatively accurate 2003 land use data base, along with year 2004 orthophotos (to fill in over 3,000 parcels that lacked a land use code). Obtaining a second historical land use data set to calibrate the models presently was impractical. Therefore, land use conditions for the year 2000 were backcast in each traffic analysis zone (TAZ), using 2000 and 2005 household and employment counts (along with the 2005 and employment category were assumed to follow an exponential growth pattern, and the intermediate region-wide totals controlled the LUM behavior. The control totals were land use data. In this integrated modeling framework, the EMPAL model runs before the DRAM, followed by LANCON and the TDM. The models were applied every five years. The year-2030 regional households were assumed to be 960,000 thousand versus 476,000 in year 2000 and employment is assumed to be 1.5 million jobs versus 0.67 million in 2000.

As part of this on-going project, emissions and air quality impacts from the development scenario predicted by the ITLUM will eventually be contrasted with those from the four ECT "visions" of Austin's future. In a preliminary analysis, Table 5 compares the ITLUM predictions of households and employment to those of the four ECT scenarios. The county-level household percentages reveal that the ITLUM predicts a relatively concentrated pattern of population. Travis County (home to the City of Austin) is predicted to accommodate more than 62.2% of the regional totals, which is higher than ECT Scenario A's 59.8%. The ITLUM-predicted households in the Bastrop, Caldwell and Williamson counties are noticeably less than what were allocated by the visioning approach; however, the Hays County ITLUM prediction is comparable to the visioning scenarios.

In terms of employment distribution, the ITLUM-predicted pattern is similar to that of ECT A, especially for Bastrop and Caldwell Counties, who have low shares of regional jobs. Assuming regional development follows the 2000-2005 trend, these

Table 5. County-level distributions of households and employment: ITLUM predictions versus ECT visions.

	ITLUM	% of Reg.	ECT A	% of Reg.	ECT B	% of Reg.	ECT C	% of Reg.	ECT D	% of Reg.
House. by County										
Bastrop	46,211	4.8%	52,678	5.7%	72,323	7.8%	92,219	10.0%	72,188	7.8%
Caldwell	22,062	2.3%	24,237	2.6%	41,908	4.5%	59,841	6.5%	41,828	4.5%
Hays	88,008	9.2%	81,713	8.8%	84,619	9.1%	90,293	9.8%	85,023	9.2%
Travis	596,871	62.2%	556,376	59.8%	507,632	54.6%	447,486	48.3%	494,392	53.7%
Williamson	206,930	21.6%	215,666	23.2%	222,658	24.0%	236,118	25.5%	226,398	24.6%
Total	960,082		930,670		929,138		925,957		919,830	
Emp. by County										
Bastrop	43,331	2.9%	29,783	2.2%	68,612	5.1%	107,928	8.0%	68,766	5.1%
Caldwell	19,570	1.3%	18,078	1.3%	48,776	3.6%	80,403	6.0%	48,885	3.6%
Hays	108,407	7.3%	90,552	6.7%	100,720	7.5%	113,600	8.4%	100,900	7.5%
Travis	1,045,094	70.6%	1,043,961	77.4%	890,364	66.0%	726,841	53.9%	889,459	66.0%
Williamson	263,044	17.8%	166,066	12.3%	239,970	17.8%	319,669	23.7%	240,430	17.8%
Total	1,479,446		1,348,441		1,348,441		1,348,441		1,348,441	

two counties will host fewer local jobs than their residents occupy, requiring inter-county commutes (which already is very common to the area). ITLUM-predicted employment opportunities in Williamson and Hays Counties are higher than what is envisioned in ECT Scenario A, but comparable to the community visions (scenarios B, C and D). ITLUM's predicted job count for Travis County lies between the ECT A and other three ECT scenarios.

Visioning is a highly community-oriented planning technique used to create regional land use and transportation goals (FHWA 1996). In contrast, land use models are based on historical trends and attempt to forecast or predict what future land use patterns will look like based on those trends (along with changes in any policy, land use, travel cost or other variable that the analyst has incorporated into the model). While fundamentally different, both the visioning and modeling processes carry benefits (Lemp et al. 2006).

Conclusions

This study is an integrated, inter-disciplinary modeling effort that provides the structure needed for comprehensive modeling of regional land use, transportation, and air quality futures. It addresses key issues in evaluating how regional development trends affect the magnitude and spatial distribution of air pollution and contrasts forecasts based on land use modeling versus community-oriented regional visioning. Integrated modeling efforts, such as the ones described in this study, have the potential to facilitate policy decisions that support balanced growth for U.S. communities.

Acknowledgments

Although the research described above has been funded by the United States Environmental Protection Agency through Science To achieve Results (STAR) Program Grant #RD83183901, it has not been subjected to the Agency's required peer and policy review. It does not necessarily reflect the views of the Agency, and no official endorsement should be inferred.

References

FHWA (1996). "Public Involvement Techniques for Transportation Decision-Making." *Publication No. FHWA-PD-96-031 HEP-30/9-96/(4M)QE.*, U.S. Department of Transportation, Federal Highway Administration and Federal Transit Administration.

Lemp, J., Zhou, B., Kockelman, K. and Parmenter, B. (2006). "Visioning Vs. Modeling: Analyzing the Land Use-Transportation Futures of Urban Regions." Proceedings of the *86th Annual Meeting of the Transportation Research Board (TRB)*, Washington D.C., Jan.

Putman, S. (1983) "Integrated Urban Models: Policy Analysis of Transportation and Land Use." Pion, London.

Song, J., Parmenter, B., Allen, D. and McDonald-Buller, E. (2006). "Impacts of Urbanization on Biogenic Emissions and Air Pollutant Deposition." Presented at the *99th Annual Conference & Exhibition,* AWMA, New Orleans, LA.

Webb, A, Gadda, S. Parmenter, B., Allen, D., Kockelman, K. and McDonald-Buller, E. (2006). "Predicting the Relative Impacts of Regional Development Trends on Air Quality. Presented *at the 99th Annual Conference & Exhibition, AWMA,* New Orleans, LA.

Wesely, M. L. (1989). "Parameterization of Surface Resistances to Gaseous Dry Deposition in Regional-Scale Numerical Models." *Atmos. Environ.*, 23, 1293-1304.

Yarwood, G., G. and Shepard, W. S. (1999). "User's Guide to the Global Biosphere Emissions and Interactions System – Version 2.1." (http://www.globeis.com), 101 Rowland Way, Suite 220, Novato, California.

Linking Conservation and Land Use Planning: Using the State Wildlife Action Plans to protect wildlife from urbanization

Authors: Julia Michalak and Jeff Lerner

Researchers widely recognize habitat loss as the most significant cause of species imperilment in the United States (Wilcove et al. 1998). In recent years, researchers, government agencies, and land use planners have become increasingly concerned about the impacts of urbanization and residential and commercial development on habitats and biodiversity (Brown and Laband 2006, USDA, Forest Service 2006, Ewing and Kostyack 2005, Radeloff et al. 2005, Doyle et al. 2001, Abbitt et al. 2000, Babbitt 1999).

Development has both direct impacts on wildlife through habitat loss and fragmentation and indirect impacts including spreading invasive species, increasing road density, increasing recreation activity, altering hydrologic regimes, increasing pollution, wildfire suppression, noise pollution, and increasing urban and edge predators such as raccoons and cats (Doyle et al. 2001). Numerous studies document changes in species composition across the urban-rural land use gradient (Hansen et al. 2005, McKinney 2002). As humans convert more land from rural and undeveloped uses to residential and commercial developments, so called urban-adaptive species thrive while urban sensitive species, which tend to be of greater conservation concern, decline (Donnelly and Marzluff 2004, Germaine et al. 1998, Germaine and Wakeling 2001, Delis et al. 1996). As a result, numerous authors have called for increased coordination between land use planners and ecologists in an effort to steer development away from important wildlife habitat (Theobald et al. 2005, Broberg 2003, Beatley 2000, Dale et al. 2000, Babbitt 1999).

The most significant potential avenue for reducing development impacts is to integrate conservation and land use planning. As part of the federal State Wildlife Grants Program, every state wildlife agency, in conjunction with numerous partners, recently completed a State Wildlife Action Plan (Action Plans or Plans). The purpose of the Plans is to provide a comprehensive view of wildlife needs in each state and so help focus conservation action on priority issues. All Plans were required to identify declining species, key habitats, conservation threats and actions to prevent further species decline. As such, the Plans have the potential to inform land use planners about wildlife, habitat, and conservation priorities in each state.

We reviewed the Plans from all 50 states and the District of Columbia to see how they discussed development impacts and what strategies the Plans offer to protect wildlife in rapidly urbanizing landscapes. To do this we searched each plan for references to development, urban/suburban growth, sprawl, and land use planning/planners. We then coded each reference into a series of threat and action categories and identified the resulting themes across all Plans.

Threats Assessment

The Plans strongly support assertions from the conservation literature that urbanization and development contribute to species decline. All 51 plans presented development (whether urban, suburban, exurban, residential or commercial) as a concern for wildlife conservation; eight states indicated that development was the *greatest* threat to wildlife statewide; seventeen states indicated that development was a top priority threat to specific regions or habitats; and twelve states emphasized development as a significant issue (see Map 1). These findings indicate that 37 states (73%) consider development an important issue affecting wildlife either regionally or statewide.

Map 1: State Wildlife Action Plans Emphasis on Development Threats
Winter 2007

The spatial distribution of threat emphasis indicates that coastal states, particularly on the East Coast, identified development as a greater concern than states in Great Plains and Midwest. However, we must caution that each state used a different methodology to identify and rank threats and many states did not explicitly prioritize any threats. This variation in methodology makes it difficult to accurately compare development threat levels between states using the Action Plans. We found no relationship between population growth rates between 2000 and 2005 and a state's level of threat emphasis as defined in this report. We did find a slight relationship between population density and threat emphasis with high density states more often prioritizing development as a threat either statewide or for a specific ecoregion or habitat (Figure 1).

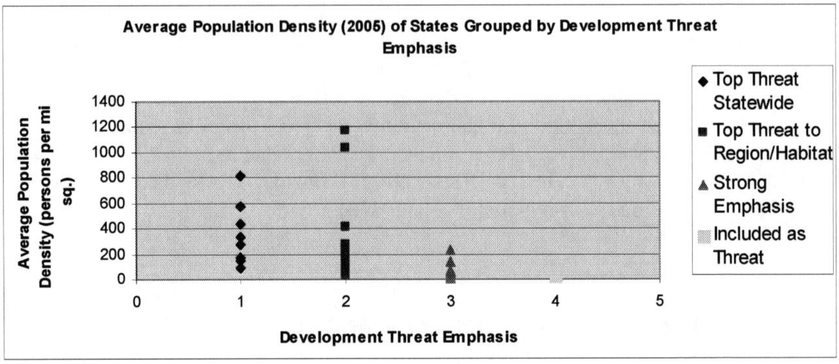

Figure 1: This figure shows the average population density (Calculated using 2005 U.S. Census Bureau population data and land area from U.S. Census Bureau, 2004) grouped by emphasis on development as a threat to wildlife.

Collectively, the threat references revealed a series of underlying mechanisms that link development and species imperilment. These mechanisms coincide with those identified by Doyle and colleagues (2000) in California and by Hansen and colleagues (2005). The threat categories identified include: habitat loss/degradation, habitat fragmentation, altered hydrologic regimes, increased pollution, increased invasive species, increased mesopredators (including native and non-native), increased road density and impacts, direct mortality, noise/light pollution, increased wildlife-human conflict, increased human use of the area (primarily through recreation).

Many states identified particular types of development as being of concern. More than half the Action Plans (28) indicated that a lack of land use planning and increasingly prevalent low density development patterns (frequently referred to as sprawl) exacerbate the habitat loss and fragmentation resulting from development. Almost half the states (22) expressed concern about rural and/or second home/vacation development. Rural development patterns, though frequently low density, have the potential to impact biodiversity patterns and degrade existing protected areas severely (USDA, Forest Service 2006, Hansen et al. 2005).

Actions Assessment

We searched for actions tied directly to development issues and found that all the plans recognized the need to take action to abate development threats. Collectively, the states presented a wide range of actions covering the spectrum from education and incentive based programs to making changes in land use law and policy. We identified the following eleven general categories based on the actions references: coordination with land use planners, incentives programs, regulations, landscape level planning, education, site development design, restoration, mitigation, monitoring, research and capacity building.

Within each of these categories, the states included a diverse array of actions which we coded into action "themes." Some of the most common themes across the plans were:

Support better integration between conservation and land use planning; Increase coordination/communication with land use planners and local decision makers; Provide information, data, support or technical assistance to local planners; Participate in the planning process (through project planning or review); Apply land use planning tools such as zoning, transferable development rights, conservation overlays, and "conservation" subdivision regulations to protect habitat and species.

Both conservation and land use planning are inherently spatial exercises (Margules and Pressey 2000, Daniels and Daniels 2003). As such, one of the best mechanisms for influencing the land use planning process is to identify where important habitat, priority conservation areas and key linkages are on the landscape. By providing spatial information to planners and the public, conservationists form the basis for a transparent conservation planning process. Thirty-three states included maps of either priority conservation areas (25 states) or priority habitat types (8 additional states) (See Map 2). These maps vary greatly in style and detail. Some can be used to prioritize areas for protection through fee simple purchase, less than fee simple, landowner incentives or a variety of land use policy tools. Others will need additional refinement before they are useful to planners.

Thirty-nine states (including DC) indicated that they want to share spatial data with land use planners (see Map 2). Specific actions included creating maps of priority areas, identifing priority areas to protect from development, sharing general spatial data (such as habitat and species locations) and initiating other spatially explicit planning exercises such as watershed planning or Habitat Conservation Planning. To varying degrees, each State Wildlife Agency and/or State Heritage Program develops and maintains spatial data for species and habitat locations throughout each state. As these data sets and maps develop it will be important to coordinate with land use planners to create data that can directly inform land use planning efforts.

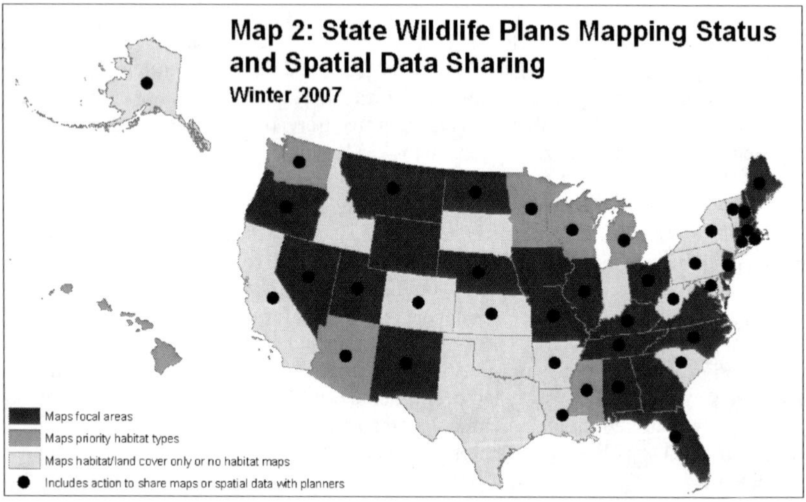

Map 2: State Wildlife Plans Mapping Status and Spatial Data Sharing
Winter 2007

By viewing the collected actions from *all* plans, we were able to identify a number of points of intervention where conservationists can help land use planners reduce the impacts of development on wildlife. These include: 1) integrating conservation priorities into comprehensive or master land use plans; 2) developing model land use ordinance language for zoning regulations, site level development designs, and transferable development rights programs; 3) participating in the permit review process; 4) coordinating residential and commercial development with existing infrastructure capacity; and 5) coordinating with land use decision-makers within and across jurisdictional boundaries. In order to address land use planning comprehensively, an action strategy should include elements from all five of these areas. Very few Plans achieved this goal. Even among those that did, it was not always clear whether the actions were a priority or if the wildlife agency had the capacity or sufficient partnerships in place to implement them.

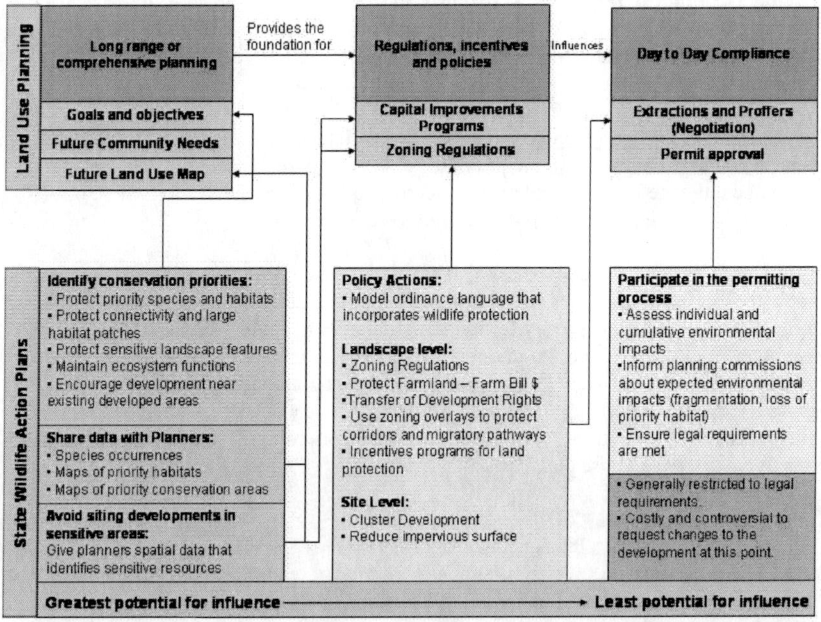

Figure 2: This chart illustrates key points of intervention between the land use planning process (in blue) and the State Wildlife Action Plans (in green). Wildlife agencies and conservation advocates traditionally enter the land use planning process during the permit approval and environmental review stages at the very end of the process. Getting involved early on during comprehensive planning is the best opportunity to create a platform for real habitat and wildlife conservation during land use planning.

By producing this report, we highlight the actions presented by the states that can address development impacts and improve coordination with land use planners. Based on the findings from this report, we have developed a set of implementation recommendations for wildlife agencies interested in initiating a comprehensive program to address development threats and work with land use planners. These include:

Address Land Use Planning Strategically: Given the patchwork nature of land use planning in the U.S., wildlife agencies will not be able to work closely with all local jurisdictions in the state. To narrow the options and gain insight into the scope of development threats, agencies can overlay maps of priority conservation areas with projected future development patterns for the next 50 and 100 years. This exercise will identify both the most biologically important and threatened areas of the state or region. Use this information to develop strategic relationships with land use planners and begin creating more detailed regional conservation plans that can intersect directly with county and city level comprehensive or master plans.

Provide Meaningful Technical Assistance: Provide meaningful technical assistance to land use planners by providing them with maps and data, interpreting those data, and working cooperatively and consistently with planners at multiple levels in the land use planning process.

Target Education Strategically: Target elected officials, planning commissions, and land use planners with educational programs that emphasize the economic, social, and environmental benefits of natural resource conservation. Especially, focus these efforts in areas with high conservation potential and development risks.

Build Capacity: In order to address this challenging issue, Wildlife Agencies will need to devote significant resources and staff time to land use policy. The state wildlife grants funding can provide some money for working on land use planning issues, but it will not be enough. Working with both traditional and non-traditional partners including land use planners, transportation agencies, and non-governmental organizations may help wildlife agencies expand their effective capacity. Some wildlife agencies may have to review their current organizational structure and shift staff from research and inventory to more proactive work on land use planning.

Conclusions

The State Wildlife Action Plans contain information about priority species and habitats that can help land use planners avoid impacts to priority wildlife resources. However, not all Plans included spatial data indicating the location of these priority resources, which will make it more difficult for land use planners to incorporate this information into their planning process. In addition, many of the Action Plans failed to present a strategic and coherent strategy for addressing complex development issues. This report highlights some of the key opportunities for enhanced coordination between land use planners and wildlife biologists.

The State Wildlife Grants program marks the beginning of a new direction for state wildlife agencies. The scope of their efforts is expanding from a focus on game and endangered species to include proactive wildlife conservation of declining species before they are endangered. Protecting these species means protecting their habitat, which necessitates addressing land use planning. Although it is unfamiliar and politically difficult terrain, addressing development patterns is central to protecting and maintaining wildlife populations in the United States today. For many states, land use decisions made over the next decade will permanently determine the fate of the state's wildlife and ecological sustainability. As stewards of the nation's wildlife, the wildlife agencies can provide real leadership in addressing this difficult challenge. The most meaningful step wildlife biologists can take is to initiate long-term, consistent partnerships with land use planners and professionals. A failure to address these issues now will compromise our ability to sustain biodiversity in the future.

References

Abbitt, R. J. F., Scott J. M., and Wilcove D. S. (2000). "The geography of vulnerability: incorporating species geography and human development patterns into conservation planning." *Biological Conservation,* 96(2), 169-175.

Babbitt, B. (1999). "Noah's mandate and the birth of urban bioplanning." *Conservation Biology* 13(3), 677-678.

Beatley T. (2000). "Challenges for Planners." *Journal of American Planning Association,* 66(1), 5-21.

Broberg, L. (2003). "Conserving ecosystems locally: A role for ecologists in land-use planning." *Bioscience,* 53(7), 670-673.

Brown R. M. and Laband D. N. (2006). "Species imperilment and spatial patterns of development in the United States." *Conservation Biology,* 20(3), 239-244.

Dale, V. H., Brown S., Haeuber R. A., Hobbs N. T., Huntly N., Naiman R. J., Riebsame W. E., Turner M. G., and Valone T. J. (2000). "Ecological principles and guidelines for managing the use of land." *Ecological Applications*, 10(3), 639-670.

Daniels, T. and Daniels K. (2003). *The Environmental Planning Handbook.* American Planning Association, Chicago, IL.

Delis, P. R., Mushinsky H. R., and McCoy E. D. (1996). "Decline of some west-central Florida anuran populations in response to habitat degradation." *Biodiversity and Conservation*, 5(12), 1579-1595.

Donnelly, R. and Marzluff J. M. (2004). "Importance of research size and landscape context to urban bird conservation." *Conservation Biology,* 18(3), 733-745.

Doyle, K., Kostyack J., McNitt B., Sugameli G., Whitaker C., Whitcomb-Blaylock K., Byrd J. and Skull G. (2001). *Paving paradise: Sprawl's impact on wildlife and wild places in California.* National Wildlife Federation, Washington D.C.

Ewing, R., Kostyack J., Chen D., Stein B., and Ernst M. (2005). *Endangered by Sprawl: How runaway Development Threatens America's Wildlife*. National Wildlife Federation, Smart Growth America, and NatureServe. Washington, D.C.

Germaine, S. S. and Wakeling B. F. (2001). "Lizard species distributions and habitat occupation along an urban gradient in Tuscon, Arizona, USA." *Biological Conservation*, 97(2), 229-237.

Germaine, S. S., Rosenstock S. S., Schweinsburg R. E., and Richardson W. S. (1998). "Relationships among breeding birds, habitat and residential development in greater Tuscon, Arizona." *Ecological Applications,* 8(3), 680-691.

Hansen A. J., Knight R. L., Marzluff J. M., Powell S., Brown K., Gude P. H., and Jones K. (2005). "Effects of exurban development on biodiversity: Patterns, mechanisms, and research needs." *Ecological Applications,* 15(6), 1893-1905.

Margules C. R. and Pressey R. L. 2000. "Systematic conservation planning." *Nature,* 405(6783), 243-253.

McKinney, M. L. 2002. "Urbanization, Biodiversity, and Conservation." *Bioscience,* 52(10), 883-890.

Radeloff, V. C., R. B. Hammer, S. I. Stewart, J. S. Fried, S. S. Holcomb and J. F. McKeefry. 2005. "The wildland-urban interface in the United States." *Ecological Applications,* 15(3), 799-805.

Theobald, D. M., T. Spies, J. Kline, B. Maxwell, N. T. Hobbs, and V.H. Dale. 2005. "Ecological support for rural land-use planning." *Ecological Applications*, 15(6), 1906-1914.

USDA, Forest Service. 2006. *Cooperating across boundaries: partnerships to conserve rural America.* FS-861. USDA, Forest Service, Washington DC.

Wilcove, D. S., D. Rothstein, J. Dubow, A. Phillips, and E. Losos. 1998. "Assessing the relative importance of habitat destruction, alien species, pollution, overexploitation, and disease." *Bioscience,* 48(8), 607-615.

Implementing Multi-Lane Roundabouts in Urban Areas

S. L. Thieken[1]

[1]P.E., PTOE,Director of Traffic Engineering, Burgess & Niple, Inc., 5085 Reed Road, Columbus, Ohio 43220, PH (614) 459-2050; email: sthieken@burnip.com

Abstract

An explosion of multi-lane roundabout projects has recently occurred in Central Ohio. This paper discusses four transportation improvement studies for which multiple multi-lane roundabouts were chosen as the preferred solution. The projects include: 1) a high volume partial cloverleaf interchange design that will include high-capacity roundabouts at the ramp terminal intersections and other intersections within the corridor; 2) a six-mile extension of a suburban/rural four-lane principal arterial that will include six roundabouts along it's length; 3) two closely spaced, high capacity, roundabouts in a densely developed area, near two schools, 4) roundabouts utilized for access improvements in a future office/commercial growth corridor.

This paper includes a description of these projects and a discussion of how the choice of multi-lane roundabouts in Transportation planning decisions at these locations will reduce delay and vehicle emissions and air pollution while also improving safety, aesthetics, and access management. The analysis presented in this paper is highly abbreviated. Please contact the author for additional information or greater detail of the analysis.

Introduction

In recent years, the benefits of roundabouts have clearly been recognized by many Transportation engineering and urban planning leaders in the Untied States. The benefits of vastly improved traffic safety and aesthetics are relatively easy to describe and illustrate; but it is often less obvious that roundabout result in less traffic delay and fewer stops (resulting in fewer vehicle emissions) and improved access management over traditional traffic signal installations. Governmental agencies and the public have often been tentative to implement multi-lane roundabouts (roundabouts with more than one circulating lane) because of the perceived confusion and complexity of driving them. However, thanks to the knowledge, professionalism, and courage of local engineers and political leaders, all of these benefits will be realized at the four locations described in this paper. It is interesting that while improving air quality was not the driving force behind pursuing roundabouts for any of these multi-lane roundabout projects, each will achieve just that!

Case Study #1 - U.S. 33 and S.R. 161/Post Road Interchange, Dublin, Ohio

The City of Dublin, Ohio has been one of the most progressive advocates of roundabouts in the United States. Dublin opened its first multi-lane roundabout in 2004, which has been successfully in operation ever since. This case study is probably the most aggressive use of roundabouts that the City has pursued to date. Figure 1 illustrates this case study, which will consist of three consecutive roundabouts. This project is expected to begin construction in 2007 and be complete by 2009. Unique features of this proposed project are:

- High volume partial cloverleaf interchange with roundabouts at the exit ramp terminals
- Ramp terminal roundabouts will have three lane entries on most approaches
- High volume roundabout (with two three lane entries) at an adjacent intersection

Since the inception of the project, the City of Dublin had wanted to explore the use of roundabouts at the ramp terminal intersections. However, to simplify and expedite the state and federal approval process on this very time sensitive project, the initial design scope included traditional traffic signals at these intersections. In fact, the first version of the Interchange Modification Study (IMS) submitted to, and approved by, the Ohio Department of Transportation (ODOT) and the Federal Highway Administration (FHWA), included signals at the ramp terminal intersections.

After the approval of the original IMS document, the City approached ODOT about an addendum to the IMS that would evaluate the potential operational benefits of roundabouts at the ramp terminals. ODOT was receptive to the idea of analyzing roundabouts as an alternative intersection traffic control treatment. Frequent and open communication with ODOT was critical in arriving at the appropriate analysis methods, the approval of the addendum, and ultimately the preliminary design of the proposed ramp terminal roundabouts. The addendum to the approved Interchange Modification Study concluded that the roundabout alternative would provide reduced delays and increased safety over the signalized alternative for a comparable cost.

To assess expected traffic operations at the study intersections, a capacity analysis was performed for the 2030 AM and PM peak hours. Several common performance measures, including Level of service (LOS), average vehicle delay, volume to capacity ratios, and 95th percentile queue lengths were calculated as part of the capacity analysis. To analyze the roundabout operation, both the "empirical" *RODEL* program and the "analytical" *aaSIDRA* software were used. Tables 1 and 2 summarize the delay and level of service results from the 2030 roundabout capacity analysis. For comparison purposes, the capacity analysis results for the signalized intersection alternatives are also presented.

Table 1–2030 Delay and LOS for West Ramp Terminal Intersection

Intersection Leg	LOS and Average Vehicle Delay (seconds)											
	RODEL (Roundabout)				aaSIDRA* (Roundabout)				HCS** (Signalized)			
	2030 AM		2030 PM		2030 AM		2030 PM		2030 AM		2030 PM	
North Leg (Off-Ramp)	A	3.0	A	3.6	B	14.0	B	16.3	D	53.2	D	43.0
West Leg (SR 161)	A	3.6	A	3.0	A	9.4	A	6.4	D	53.9	D	48.0
South Leg (University)	A	4.8	A	6.6	B	19.0	C	25.5	D	47.1	D	48.7
East Leg (SR 161)	A	1.8	A	2.4	A	4.6	A	4.4	C	20.7	D	39.4

* 1.2 Environmental Factor used. ** From original IMS.

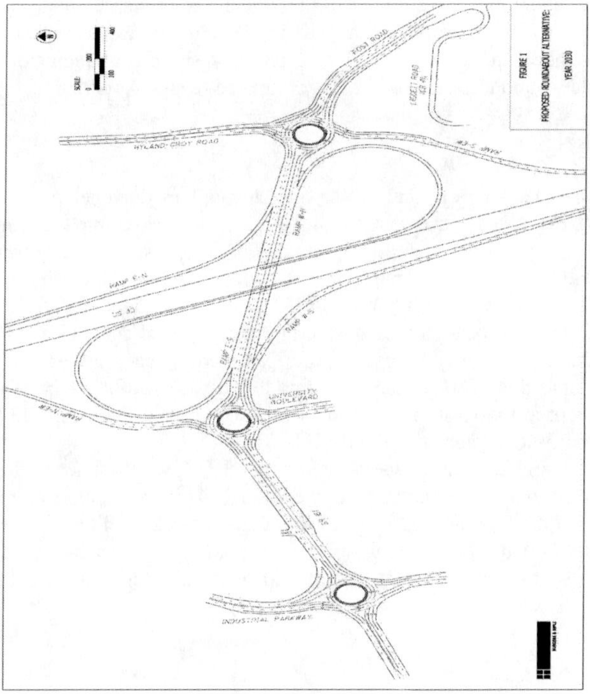

Figure 1 – Case Study #1 Interchange Area Layout

Table 2 – 2030 Delay and LOS for East Ramp Terminal Intersection

Intersection Leg	LOS and Average Vehicle Delay (seconds)											
	RODEL (Roundabout)				aaSIDRA* (Roundabout)				HCS** (Signalized)			
	2030 AM		2030 PM		2030 AM		2030 PM		2030 AM		2030 PM	
North Leg (Hyland Croy)	A	3.0	A	3.0	B	19.7	C	20.8	D	39.9	D	48.1
West Leg (SR 161)	A	3.0	A	2.4	A	7.5	A	6.5	D	35.2	C	32.7
South Leg (Off-Ramp)	A	2.4	A	2.4	B	17.3	B	14.5	D	47.4	D	47.0
East Leg (SR 161)	A	2.4	A	2.4	B	14.6	B	18.6	D	45.8	D	47.3

The analysis shows that during the 2030 AM and PM peak hours, the roundabout intersection alternative will operate with much lower delays than the signalized intersection alternative, resulting in less congestion and vehicle emissions. Additionally, while not directly estimated in the study, it is expected that during non-peak hours the roundabouts will also result in a significant reduction in vehicle emissions due to reduced stops and delay compared to signal operation.

Ramp Metering

The need to install ramp meters at this interchange was documented in the original IMS. ODOT expressed concerns regarding the vehicle backup into the roundabouts as the results of the ramp metering, particularly at the southbound on-ramp from eastbound SR 161. For this ramp, the first IMS stated that, "the ramp meter would need to restrict traffic to a maximum of 853 vehicles per hour to enter the freeway" during the 2030 AM peak hour. Considering the projected demand of 1180 vehicles per hour, this restriction could cause significant vehicle back-up into the west ramp terminal intersection. The concern was that this backup could effectively "lock up" the roundabout at the west ramp terminal intersection, causing traffic to back-up on the southbound exit ramp and spill into U.S. 33.

The proposed solution, which was accepted by ODOT, was to "meter" the eastbound roundabout approach if necessary in the future. A simple single direction signal installation on the west approach could "hold" eastbound traffic prior to entering the roundabout, allowing the traffic within the roundabout to clear. Queue detectors on the U.S. 33 southbound off-ramp would also be installed to actuate the roundabout approach metering.

Safety

The assertion that roundabouts were safer was never challenged. All involved could agree that the roundabout alternative would result in fewer injury and potentially fatal crashes throughout the life of the design, and the alternative would constitute an excellent transportation safety planning solution. In addition to national research, it very was helpful that the City of Dublin could present clear documentation to ODOT of the safety benefits gained at its existing multi-lane roundabout intersections.

Trucks

ODOT required that the side-by-side movement of WB-62 semi-trucks be accommodated in the design. This resulted in roundabout design speeds that are a little higher than desired, but given that U.S. 33 is a freeway and the adjacent land use is industrial in nature, it is clear that large trucks will have to be routinely accommodated at these roundabouts.

Case Study #2 – Sawmill Parkway Extension, Delaware County, Ohio

The overall project consists of the six-mile extension of an existing four-lane divided parkway. This project will be constructed in phases to begin in 2007 and complete by 2010. Figure 2 illustrates the alignment and the proposed access points to the parkway.

The use of roundabouts had been contemplated prior to the design phase of this project, but not included as a part of the original design scope. Roundabouts were formally added as an intersection control alternative after several local residents requested at a public meeting that roundabouts be constructed at public road intersections along the proposed parkway (for safety reasons). A very detailed and comprehensive analysis and report was prepared to document the potential benefits of roundabouts. The final report described the analysis of each type of intersection control (signal, stop sign control, or roundabout control). Factors considered were: functional classification, intersection spacing, traffic delay, corridor travel time, traffic safety, and construction cost.

Roundabouts were found to be the preferred traffic control at seven intersections based on evaluation of the above factors. In terms of reduced delay/congestion and the resulting reduced vehicle emissions, Figures 3 and 4 illustrate the comparison between the calculated cumulative total 24-hour delay at the roundabout and non-roundabout intersection options for 2030 traffic. The only preferred roundabout location that will not be constructed is the U.S. 42 and Sawmill Parkway intersection. This intersection is within the corporation limits of the City of Delaware, thus the decision whether or not to install a roundabout at this intersection lies with the City (state law). The City decided that a traffic signal would be a more appropriate traffic control measure due to the existence of adjacent traffic signals and the large volume of semi-trailer trucks which use U.S. 42.

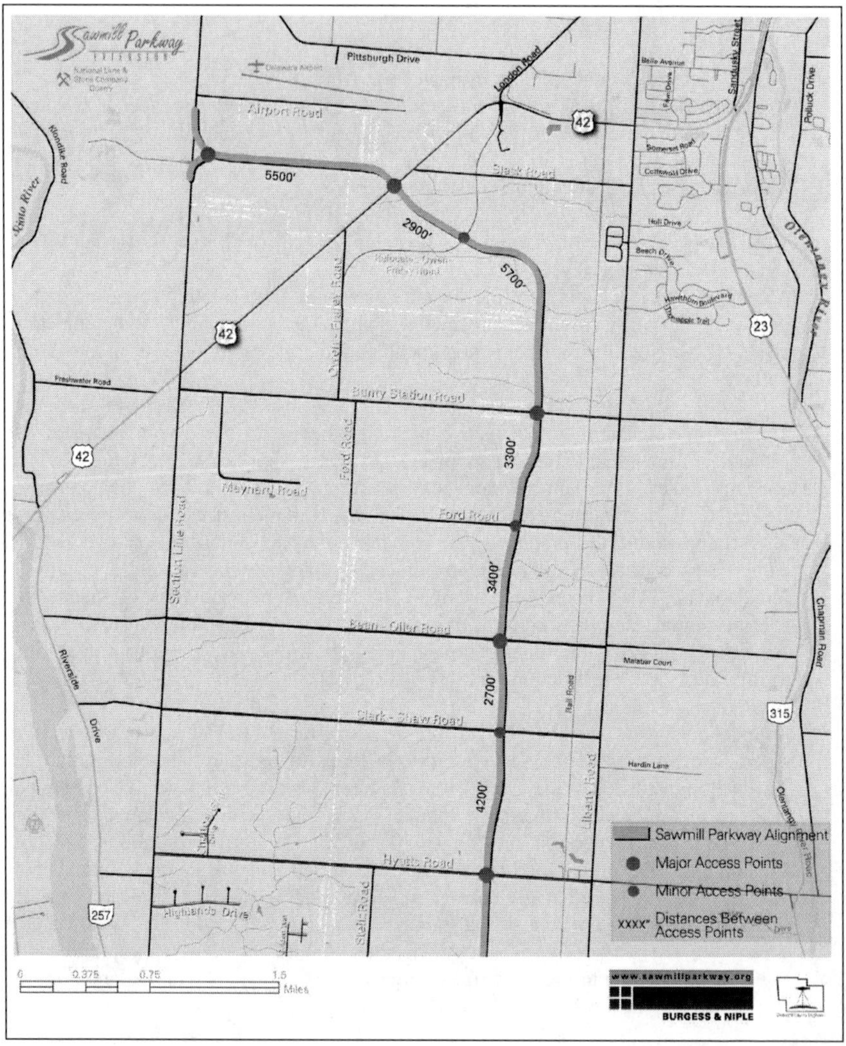

Figure 2–Sawmill Parkway Extension Alignment and Intersections

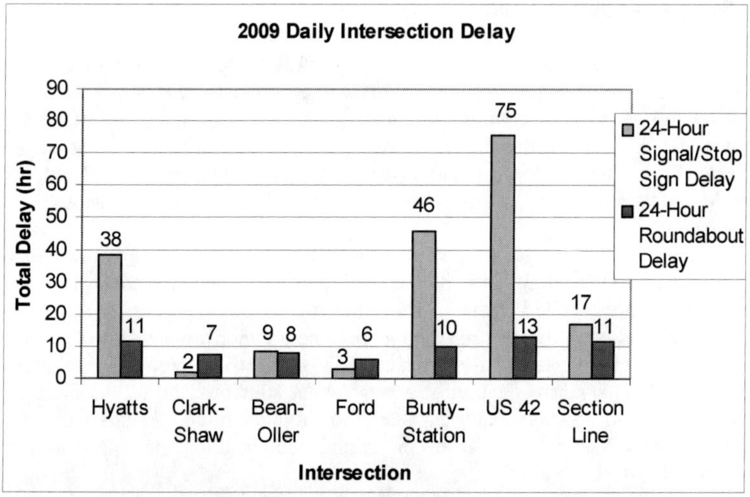

Figure 3–2009 Daily Estimated Intersection Delay Comparison

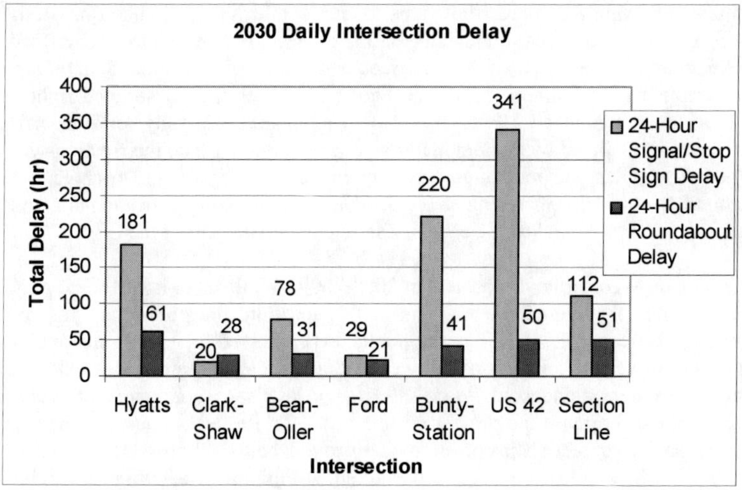

Figure 4–2030 Daily Estimated Intersection Delay Comparison

Case Study #3 – Hilliard Triangle Project, Hilliard, Ohio

The Main Street / Cemetery Road / Scioto Darby Road Improvement Project, often referred to as the "Triangle Project," involves the complete reconstruction of the roadways and intersections within the project limits. Alternatives for pavement widening and expanded traffic signals as well as modern roundabouts were studied at length to determine the best solution for reducing congestion and delay and improving safety and ease of access to homes, businesses, and schools in the area.

Several years ago, the City decided to pursue traffic signal and road widening improvements for the study area. In fact, a set of construction plans was nearly complete for the signalized intersection improvements when access issues due to proposed medians and driveway restrictions became a major problem for businesses in the project area. In search for a better solution the City hired a consultant to perform a preliminary investigation of a roundabout solution. The results indicated that such a solution was potentially feasible. The City then hired the author's firm to perform a more detailed analysis of the roundabout solution, and ultimately to prepare detailed design plans.

The analysis revealed that the roundabout alternatives for the triangle provided reduced traffic delay when compared to the full signalization alternative. Full access to businesses was made possible by the ability of roundabouts to accommodate u-turn maneuvers, allowing for indirect left turns to and from driveways. The final design (ongoing) includes two-lane roundabouts at the intersections of Scioto Darby Road / Main Street and Cemetery Road / Main Street, and a traffic signal at the Scioto-Darby Road / Cemetery Road intersection (see Figure 5). Given the constrained right-of-way, neither the signalized nor the roundabout alternative can fully serve the 2030 traffic demand predicted by the regional travel demand model. The intersections were designed to provide the maximum feasible traffic capacity for the triangle. It is understood by the City that long range land use and thoroughfare planning must address the future potential traffic capacity deficiency in this part of the City.

The City plans to construct this project in 2008-2009. While there is still some public apprehension to the use of roundabouts in this location, the public has generally accepted the concept. The issue that the public has expressed the most concern about is the ability of school children to safely cross the roundabout approaches. Although research has shown roundabouts to be safer for pedestrians, some parents of school aged children still oppose the project on this basis. To further enhance the inherent pedestrian safety characteristics of the modern roundabouts, all crossing locations at roundabouts will be clearly signed and marked. Additionally, at designated school crossing locations, overhead warning flashers will be activated during times that standard school zone flashers and restricted speed limits are in effect. The City also intends to provide school crossing guards at least during the first few weeks of roundabout operation to provide assistance and training to students.

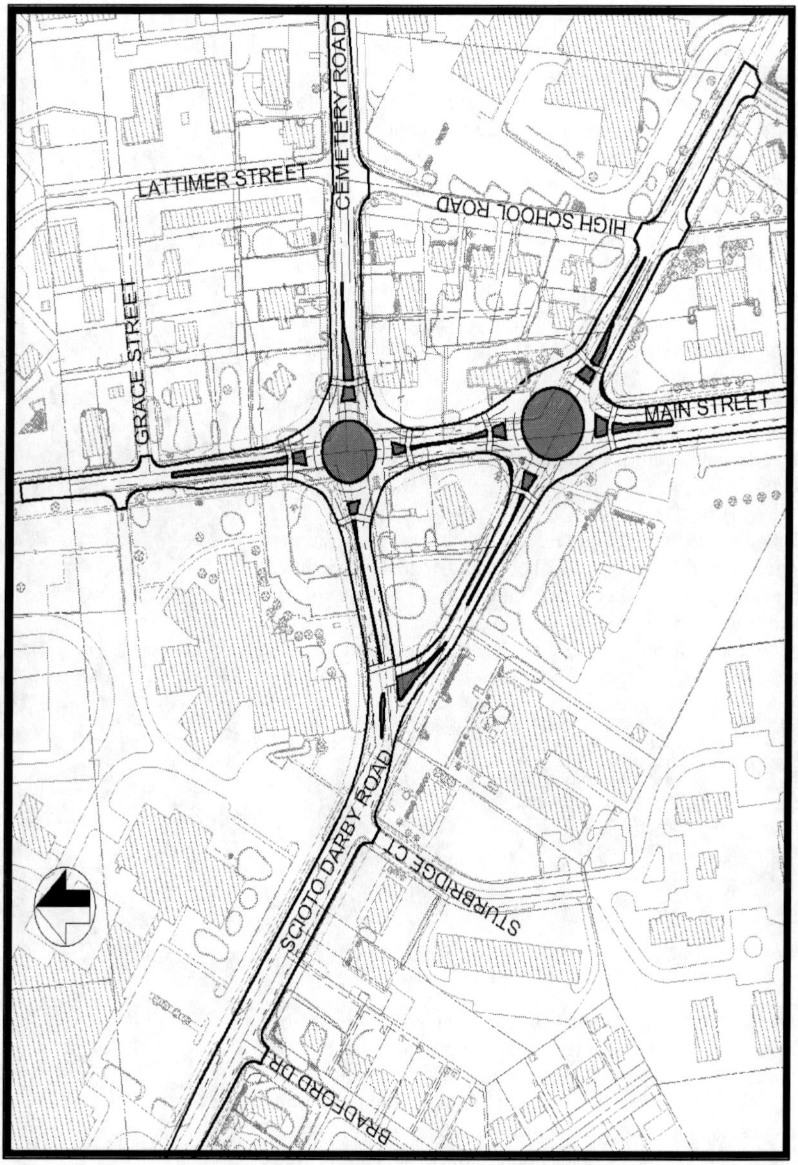

Figure 5 - Hilliard Triangle Project Area

Case Study #4 – Avery Road South Corridor Study, Dublin, Ohio

This project is to address capacity, safety, and land use/access management needs for the section of Avery Road between Shier-Rings Road and Woerner Temple Road. The goals of the study were to identify preferred intersection traffic control, required number of lanes, preferred access locations and treatments, and any additional access improvements necessary. A variety of alternatives was considered, including those with signalized intersections and access roads, and those with roundabouts serving as alternative left turn locations to facilitate the elimination of left turn access at a number of existing driveways. Three potential access management/traffic control scenarios were developed for detailed analysis. An evaluation matrix of characteristics for each scenario was developed. Characteristics evaluated were traffic operations, traffic safety, cost, land use impacts, environmental impacts, community impacts, and economic impacts.

Figure 6 – Preferred Alternativ

The adjacent figure (Figure 6) shows the preferred solution which includes three roundabouts and roadway widening. The roundabout alternative had the advantage of not requiring expensive and complicated cross access roadways, since u-turns could be provided at the roundabouts (see Figure 7 below).

Figure 7 – Indirect Left Turn Using Roundabout

The public involvement process included one-on-one discussions with property owners, public meeting/open house, email and telephone responses to concerned citizens, and a presentation to City Council. Local businesses supported the concept. There is still some public concern over pedestrian safety at one of the intersections that is used by young pedestrians to access a pubic pool. The perception of some local parents is that children cannot safely cross at the roundabout, since there is no traffic signal with pedestrian indications. City is actively working to address these concerns.

Keys to the success of this alternative is the City of Dublin's keen understanding of the importance of access management, and its willingness to, and its aptitude at, acquiring right-of-way and key private properties in support of its City Comprehensive Land Use and Transportation plans, as opportunities present themselves.

Conclusions

While the needs and goals of each of these projects are unique, the result in terms of providing a solution that improves air quality, traffic operations, safety, and community aesthetics is the same. In these case studies project engineers and planners evaluated and promoted all of the benefits of roundabouts. However, through good communications it was necessary to promote the specific benefits that were most likely to "hit home" with those whose support was needed to advance the project; for example: politicians (safety and aesthetics), business owners (access, circulation, and aesthetics), state and federal highway officials (traffic capacity and safety) and the general public (safety and delay). The introduction of roundabouts into the transportation engineering and planning toolbox should ultimately improve the quality of transportation and land use and improve air quality in the United States.

Models to Measure Pedestrian Activity at Intersections

S. S. Pulugurtha[1] and S. Repaka[2]

[1]Assistant Professor of Civil Engineering, Assistant Director of Center for Transportation Policy Studies, The University of North Carolina at Charlotte, 9201 University City Boulevard, Charlotte, NC 28223-0001, USA; PH: (704) 687-6660, FAX: (704) 687-6953, email: sspulugurtha@uncc.edu
[2]Graduate Student of Civil Engineering, The University of North Carolina at Charlotte, 9201 University City Boulevard, Charlotte, NC 28223-0001, USA; PH: (704) 687-2089, FAX: (704) 687-6953, email: srepaka@uncc.edu

Abstract

Increasing congestion and travel delays on urban roads has lead practitioners to look at means to encourage and make alternate modes of transportation more attractive. Allocation of resources to build facilities amicable to such modes of travel is governed by activity at the location of interest. Collecting real world data for several locations is an expensive and time consuming process. On the other hand, unlike trip generation models to estimate vehicle trips, literature documents limited research to model and measure activity pertaining to alternate modes of transportation. This paper focuses on the development of models to measure pedestrian activity at intersections.

Pedestrian activity depends on land-use characteristics (residential, commercial, industrial, etc.), demographic characteristics (population), socio-economic characteristics (income level, employment, etc.), and access to public transportation systems. Data collected at 564 intersections in the Charlotte metropolitan area in North Carolina are used to develop a regression model to measure pedestrian activity at intersections. Pedestrian counts collected at intersections in Charlotte, NC are used as a dependent variable. Factors such as demographic characteristics, socio-economic characteristics, land-use characteristics, and the number of bus-stops are identified using features available in a commercial Geographic Information Systems (GIS) software program. These factors are used as independent variables. Regression analysis through backward elimination of independent variables is used to develop a model to measure pedestrian activity at intersections. This model could be used by practitioners to measure pedestrian activity at a location if data are available. Also, models by the time of day could be developed using the same procedure if pedestrian count data are available by time of the day. The measured pedestrian activity could be used in transportation planning, safety, and operational analyses.

Introduction

Transportation demand modeling has a long history and a complex heritage. The need to estimate the amount, type, and distribution of vehicular traffic in cities is well recognized and traffic models have played an important part in the planning of modern urban growth since the late 1950s. However, the need and ability to model pedestrian movement is a more recent development. Advancements in computational power and understanding have made such modeling approaches feasible giving rise to the emerging field of pedestrian volume modeling.

According to the National Personal Transportation Survey (NPTS), walking and bicycling accounts for nearly 8 percent of all trips. There is a growing need for quantitative traffic estimation tools to improve pedestrian transportation safety, access, and mobility. There has been a significant increase in pedestrian research in the United States during the recent years. This interest is the result of a growing awareness among elected officials, practitioners and the general public that walking is vital to the health of cities and their residents, and that in general, Americans walk far too little. Despite the increased understanding and interest, most urban planners and policymakers charged with making cities safer and more walkable are forced to do so with limited tools and resources. Pedestrian traffic at any place is influenced by psychological, physiological and environmental factors, social relationship to neighborhood pedestrians, purpose of walk, and area topography.

Literature Review

Pedestrian trip generation is defined as the number of trips generated due to pedestrian related activities. Urban land use patterns play an important role in influencing walking trips (Handy, 1996; Loutzenheiser, 1997). People who live in spread out development patterns (sprawl) areas spend less time walking than people who live in mixed-land use, neo traditional neighborhood, or well established neighborhoods, because sprawl requires more frequent and longer trips (Berkovitz, 2001). Other factors such as the number of bus stops, population, travel speed, geometric conditions, time of the day, facilities available at the location, income level, age group, presence of sidewalks, and number of lanes also have an effect on pedestrian movements at the intersections.

Several authors have worked on topics related to modeling pedestrian trips in the past. Examples of such studies on modeling pedestrian trips include examining pedestrian crash data using "time spent walking" and "number of roads crossed" as exposure measures (Keall, 1995), studying the relationship between site design and pedestrian travel in a mixed-use, medium density environment (Hess et al., 1999), forecasting pedestrian volumes in high-density urban areas based on existing land use characteristics and pedestrian volumes for specific locations (Pushkarev et al., 1971), and models to estimate pedestrian counts (Pulugurtha et al. 2006). The main limitations of these studies lies in sites, their selection, and characteristics considered for modeling. This paper attempts to address a few of these concerns by considering a large sample of sites at which pedestrian data were collected.

Methodology

The main objective of this paper is to identify the factors that have a bearing on pedestrian trips and develop a model to measure pedestrian activity at intersections. A Geographic Information Systems (GIS) based methodology was developed to extract data to identify critical factors for modeling pedestrian activity. The methodology involves the following steps.

1. Identify pedestrian sites
2. Extract demographic characteristics for each site
3. Extract landuse characteristics for each site
4. Extract number of transit stops in the vicinity of each site

Each of these steps is discussed in detail next.

Identify Pedestrian Sites

To develop models for a study area, sites should be selected such that they are geographically distributed throughout the study area. Demographic, landuse, and transit characteristics of these sites should represent typical characteristics of the metropolitan area. The number of sites selected should also be large enough to yield statistically significant findings.

Extract Demographic Characteristics for Each Site

Demographic characteristics such as number of household units, population, income, automobile ownership, and total employment may influence a person's decision to walk. To extract these characteristics, census data at block level or planning variables data at traffic analysis zone (TAZ) level are overlaid on a 0.5-mile buffer generated around each site. A 0.5-mile buffer was used based on the assumption that a pedestrian would at most walk for 10 minutes at 4 feet per second. Spatial analysis is conducted to extract population within the polygon area in each buffer.

Extract Landuse Characteristics for Each Site

The number of pedestrian trips could also depend on landuse characteristics. People living in apartments may walk more than residential neighborhoods or vice versa. Likewise, commercial and office type landuses may attract more pedestrians than industrial. To extract these characteristics, landuse coverage is overlaid on a 0.5 mile buffer generated around each site. Spatial analysis is conducted to extract area of landuse by type within the polygon in each buffer.

Extract Number of Transit Stops in the Vicinity of Each Site

People generally walk or bike to access (bus) transit systems. To study its influence, the number of stops within 100 feet of each site are identified and considered for modeling. A 100 feet buffer was used primarily to identify the number of bus-stops near the site.

Statistical Analysis and Discussion

An off-the-shelf software program (Minitab ®) is used for the statistical analysis. Regression analysis through backward elimination technique is used to identify factors that have a bearing on pedestrian activity and develop a regression model to measure the same. Pedestrian counts, demographic characteristics, landuse characteristics, and the number of transit stops collected and extracted at 564 intersections in the Charlotte metropolitan area are used to develop the model. These data were collected between 2002 and 2006 from 7:00 AM to 7:00 PM. Table 1 lists all the independent variables and the dependent variable (pedestrian counts or trips) used to develop the model. Figure 1 shows the spatial distributions of the sites at which pedestrian counts were collected. Variables and output parameters at the end of the statistical analysis are shown in Table 2.

It can be seen from the table that the number of pedestrian trips increases as the number of household units, mixed landuse area in square footage, and the number of transit stops within 0.5-mile from the intersection increase. On the other hand, the number of pedestrian trips decreases as population and urban residential area in square footage within 0.5-mile from the intersection increase. The T-value for all these variables is greater than 2 and the level of significance is greater than 95 percent. However, the coefficient of determination is at about 37 percent.

Conclusions

This paper presents identification of factors and development of a model to measure pedestrian activity at intersections. Data collected at 564 intersections in the Charlotte metropolitan area were used to identify the factors and develop the model. Demographic characteristics, socio-economic characteristics, and landuse characteristics within 0.5-mile from the intersection were used for analysis whereas transit stops within 100 feet were identified and considered. Statistical analyses indicate that the number of household units, population, urban residential area, mixed landuse area, and the number of transit stops are critical independent variables.

The buffer distance could depend on acceptable walking distance/time norms for pedestrians. Sensitivity to changes in buffer distance needs investigation. Aspects such as population by age group and automobile ownership, and network characteristics were not considered in this study. Possible inclusion of these variables may yield better models and results.

Acknowledgments

The authors acknowledge and sincerely thank Charlie Jones, Ashton Watson, Steven Castongia, and Anna Gallup of the Charlotte Department of Transportation for their help with pedestrian counts, land use coverage, and socio-economic characteristics data.

References

Berkovitz, A. (2001). "The Marriage of Safety and Land Use Planning: A Fresh Look at Local Roadways." *Public Roads,* Vol. 65(2).

Handy, S.L. (1996). "Urban Form and Pedestrian Choices: Study of Austin Neighborhoods." *Transportation Research Record: Journal of the Transportation Research Board,* No.1552, TRB, National Research Council, Washington D.C., pp. 135-144.

Hess, P.M., Mouden, A.V., Snyder, M.C., and Stanilov, K. (1999). "Site Design and Pedestrian Travel." *Transportation Research Record: Journal of the Transportation Research Board,* No.1674, TRB, National Research Council, Washington D.C., pp. 9-19.

Keall, M.D. (1995). "Pedestrian Exposure to Risk of Road Accident in New Zealand." *Accident Analysis and Prevention*, Vol.27, 1995, pp. 729-740.

Loutzenheiser, D.R. (1997). "Pedestrian Access to Transit: Model of Walk Trips and their Design and Urban Form Determinants around Bay Area Rapid Transit Stations." *Transportation Research Record: Journal of the Transportation Research Board,* No.1604, TRB, National Research Council, Washington D.C., 1997, pp. 40-49.

Pulugurtha, S. S., Shashi, S. N., and Maheswari P. (2006). "Estimating Pedestrian Counts in Urban Areas for Transportation Planning and Safety Analyses." *American Society of Civil Engineers Proceedings of the Ninth International Conference on Applications of Advanced Technology in Transportation,* Chicago, IL, August 13-16, p (257-262).

Pushkarev, B., and Zupan, J. M. (1971). "Pedestrian Travel Demand." *Highway Research Record*, Vol.355, Highway Research Board, Washington, D.C.

Table 1. Independent and dependent variables.

Independent variables	
Demographic & Socio-economic Characteristic	
1	Household units
2	Population
3	Mean income
4	Area - landuse
5	Total employment
Landuse Categories	
1	Single family residential area
2	Multi-family residential area
3	Urban residential area
4	Research district
5	Institutional
6	Office district
7	Urban residential commerical area
8	Neighborhood business
9	Mixed landuse
10	Business
11	Industrial
12	Historic district
13	Residential - mobile
14	Planned unit development
15	Airport
16	Manufactured house
17	Commercial center
18	Neighborhood service district
19	Hazardous waste district
20	Resort residential
21	Innovative
22	Unknown
Transit system Characteristics	
1	Transit stops
Dependent variable	
Pedestrian counts	

Table 2. Model and output parameters – summary.

Independent Variable	**Coefficient**	**T-value**	**P-value**	
# Household Units	0.71	3.46	0.001	
Population	-0.239	-2.6	0.01	R_2 = 37.12
Urban residential area	-0.00008	-2.65	0.008	R_2(adjusted) = 36.56
Mixed landuse	0.00014	12.01	0	
# Transit stops	196	2.72	0.007	

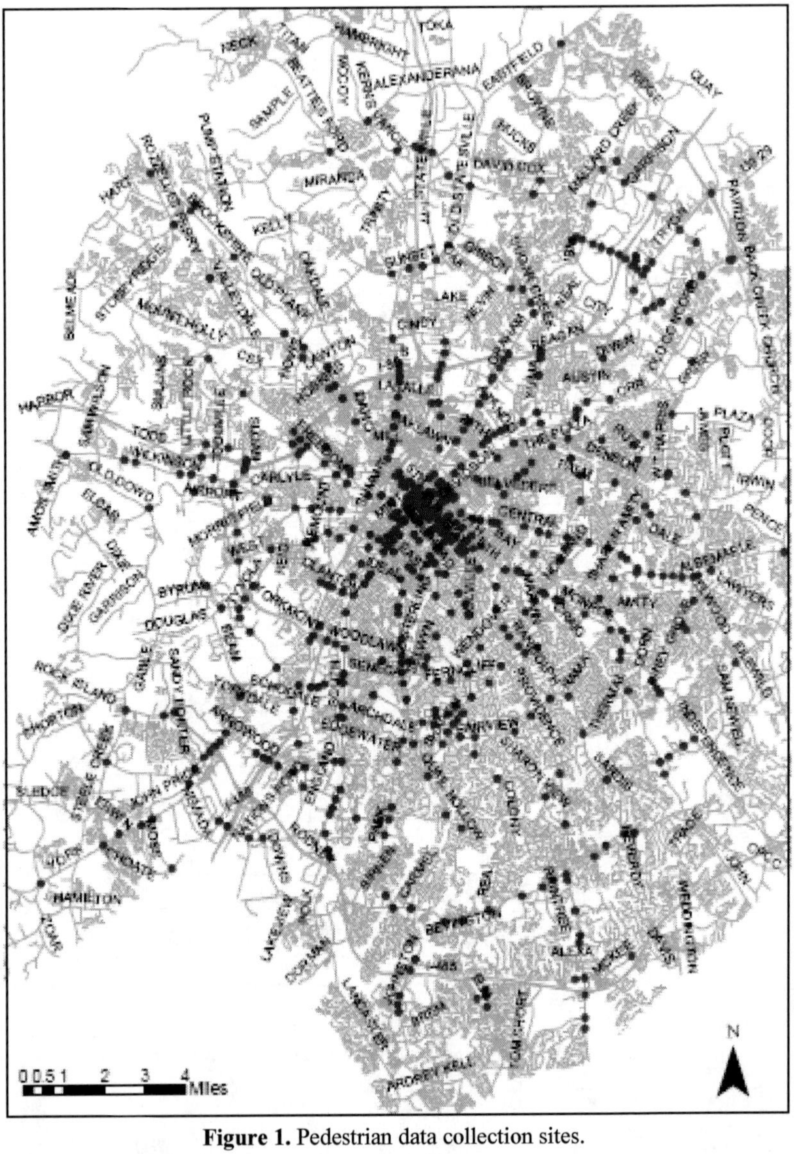

Figure 1. Pedestrian data collection sites.

Innovative Mobility Solutions In Évora: A Historical Medium Sized City In Portugal

A. S. Vasconcelos,[1] S. Taborda,[2] and T. L. Farias[3]
B.
[1]Instituto Superior Técnico, Technical University of Lisbon, Portugal, email: ana.vasconcelos@ist.utl.pt
[2]Instituto Superior Técnico, Technical University of Lisbon, Portugal, email: sofia.taborda@ist.utl.pt
[3]Instituto Superior Técnico, Technical University of Lisbon, Portugal, email: tiago.farias@ist.utl.pt

Introduction

The growth of urban mobility and use of private car leads to congestion, parking and sustainability problems in most cities. Cities around the world face unacceptably high levels of air pollution, noise, congestion, parking problems and accident rates. Urban transport is also one of the main responsible factors for the emission of green house gases, but improvements are being studied and tested all around the world (Newman, P. and Kenworth, J., 1999). The present paper describes innovative mobility systems applied to a city near Lisbon - Évora, with special focus on the interaction between parking policies, traffic management, transit operation and soft modes strategies. The solutions adopted include a new bus network, an improvement in the train shuttle service, a Park & Ride bus line named Blue Line, and an innovative pedagogical parking enforcement strategy adopted in Évora.

The City of Évora

Évora is a mid-sized city, with about 50 thousand inhabitants and located 120 kilometers from Lisbon, the Portuguese capital. The city is classified as UNESCO world heritage, mainly due to its historical center but is also a students' city, derived from the centenary university. This historical center is surrounded by roman walls and has some of the most important monuments of the city within. The urban area is about 1643.5 ha and the historical center is about 103 ha. The county is administratively divided in nineteen sub-regions: 7 urban subregions (3 of them located inside the walls) and 12 rural sub-regions. The inhabitants are the majority of them distributed in the surrounding urban area and less than 6 thousand live in the historical center (Farias, T. L. et al., 2006). Évora was pioneer in what concerns strategic urban planning. The city has programs for revitalizing the historical center, integrated transport and parking projects, a project for green spaces in the whole city and for the construction of other equipments that constitute a base for the future.

The New Bus Grid

The new bus grid is based on three main axes linking for the first time the outer neighborhoods while crossing the inner wall region of the city in two hubs minimizing the impact of heavy vehicle traffic on the historical center while continuing to promote the center as a main gathering point, serving a high number of visitors. The grid was designed by the authors using OD matrices, based on the main travel profiles obtained from a survey using questionnaires sent by mail to all households in the city.

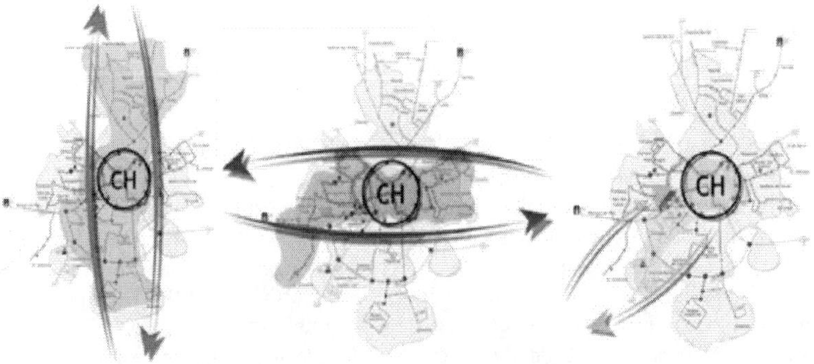

Figure 1 Three axes that structure Évora's urban bus network (from left to right: green lanes' group, red lanes' group and yellow lane) (SITEE-EM, 2007a)

The concept of the new grid consists on the definition of two central hubs, as previously mentioned, where the transshipment point is located since all lanes stop there. The new bus network will be officially launched on the 2nd of April of 2007, accompanied by a massive communication campaign, involving direct actions such as customized information brochures to different areas; a direct telephone line and several media spots and press releases.

LinhAzul - Blue Line

The service LinhAzul (meaning Blue Line) consists in a bus line that connects private car parks outside the fortified city to the inside. Four mini buses connect the peripheral parks to historical center with a continuous circulation, providing a high level of service where people do not wait more than 10 minutes on average for the bus anywhere on the circuit, which takes about 30 minutes to complete it. The fare is also coordinated with the park fee. This new system intends to dissuade the use of private car inside historical center in order to avoid congestion and parking problems inside the city walls, achieve lower emissions and noise levels in the historical center and promote sustainable transport (Simões,

A. M., Farias, T. L. and Quenino, J. H., 2004).

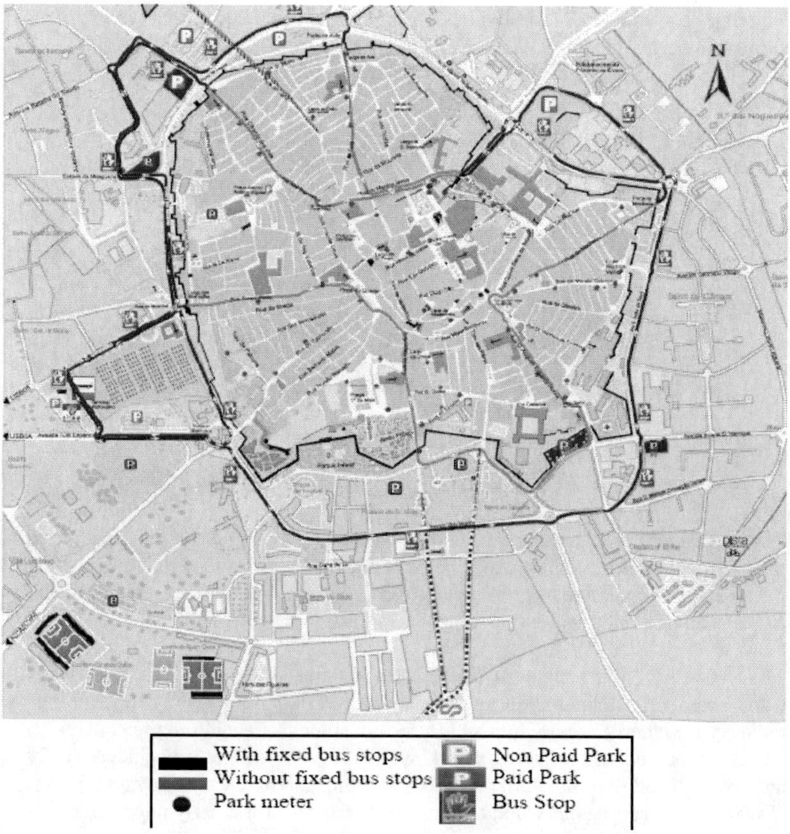

Figure 2 Blue Line Route (SITEE-EM, 2007b)

The mini buses (Mercedes Sprinter mini buses equipped with an Euro 3 motor) circulate continuously in two opposite directions in order to minimize waiting times. There are bus stops outside the city walls, but inside there is a predefined route marked on the street pavement with a blue line and everyone with a valid ticket can ask the bus to stop, Figure 2, just by raising their hand. Since December of 2006, an extra road segment was added to the circuit, to connect the city center to the train station, serving the new train schedule service. This service has been continuously monitored every 6 months since the beginning in order to evaluate customer satisfaction as well as to define possible improvements in the service.

This line fare is 1.00 Euro (ca. 1.25 USD), and is valid for as many trips as needed for one day. If the user leaves the car in one of Blue Line's car parks, the parking fee is also 1.00 Euro and the Blue Line trips are for free. Each bus is equipped with a Euro 3 Motor, automatic gear box and air conditioning. They provide twelve seats and have standing room for nineteen people, see Figure 3. This line is available all working days from 8.00 am to 8.00 pm and on Saturdays from 8.00 am to 2.00 pm. (SITEE-EM, 2007b)

Figure 3 Blue Line buses (SITEE-EM, 2007b)

The motivation for creating the Blue Line was to slowly reduce private cars access to Évora's historical center, except for residents and cargo loading & unloading purposes. This measure was integrated with the closing of some streets to the traffic, and new and renewed parking spaces outside the city walls.

The success of the Line was above expectations with respect to passengers transported. The total for one year reached 150,000 passengers. However, the majority are elderly people followed by young students, leading to the conclusion that its initial goal of promoting a Park & Ride service was not yet achieved. Only around 10% of the Blue Line clients park the car in the surrounding parks. However, this service provided an important transport for elderly population that wouldn't get out of their homes before the Blue Line.

Parking Enforcement Policy

One of the critical mobility issues of the city of Évora was the past parking policy, which until a few years ago was not suitably managed, specially aggravated by the fact that there was an inefficient legal enforcement. (Farias, T. L. et al., 2006). Nowadays Évora is characterized by paid parking spaces within the historical center walls but free of charge spaces outside the walls, except on dedicated parks. The parking policies are currently enforced by the municipal company SITEE-EM (Integrated transport and parking system of Évora), sustained by a real-time enforcement support system (SIAF) which allows parking enforcement officers to precisely know which vehicles have already a non-paying background and all their history even in other cities that work with the same system. A strict enforcement is essential to the success of any parking policy, allowing not only a good control of the policies but also a real knowledge of the situation.

Conclusion

In summary, the success of a mobility management policy depends on an integrated approach, with the combination of several measures, not forgetting the importance of the combination of restrictive measures and positive measures (EUROCITIES - The Network of Major European Cities, 2005).

Consequently, the city of Évora in now one of the leading cities in Portugal concerning sustainable urban mobility management. These facts were decisive in including Évora in the PILOT European research project on the development of SUTP.

References

"EUROCITIES - The Network of Major European Cities (2005)." Literature Review on research and study results relevant to SUTP. PILOT Project.

Farias, T. L., Vasconcelos, A. V., Taborda, S., Brasil, A., Simões, A., Gonçalves, G. and Moura, L. (2006), "Estratégia de Mobilidade da Cidade de Évora." Instituto Superior Técnico, Lisboa.

Newman, P. and Kenworth, J. (1999). "Sustainability and Cities: Overcoming Automobile Dependence." Island Press, California.

SITEE-EM (2007a). Nova Rede de Carreiras Urbanas. Évora.

SITEE-EM (2007b). SITEE-EM - LinhAzul. http://www.evora.net/sitee/linha_azul.asp (>February, 2007)

Simões, A. M., Farias, T. L. and Quenino, J. H. (2004), "Blue Line, a New Bus System: Application to a Case Study in Évora City, Portugal." *Proceedings of the XIII Panamerican Conference, Traffic & Transportation Engineering*. 26 - 29 September 2004, Albany, New York.

Multinomial Modeling of Purpose Driven Trip

M. Penn,[1] F. Vargas,[2] and D. Chimba[3]

[1] Transportation Engineer, Stanley Consultants Inc, 1601 Belvedere Rd, Suite 400 East, West Palm Beach, FL-33406; PH(561)712-2242; FAX (561)689-3003
[2]Senior Project Manager, Stanley Consultants Inc, 1601 Belvedere Rd, Suite 400 East, West Palm Beach, FL-33406; PH(561)712-2242; FAX (561)689-3003
[3]MSCE, EIT, Transportation Engineer, Stanley Consultants Inc, 1601 Belvedere Rd, Suite 400 East, West Palm Beach, FL-33406; PH(561)712-2242; FAX (561)689-3003; email: chimbadeo@stanelygroup.com

Abstract

Activity and purpose based trip making studies are gaining popularity in current transportation research industry. The purpose is determined by activity one is to engage in at the destination of the trip. The trip purpose can be influenced by several factors, in which mainly are social economic. Transportation planners need to know the effect foresee the future origin and destination of certain community based on the household and surrounding social economic data. Based on the importance of determining the effect of trip purpose, this study applies multinomial modeling approach to evaluate various factors which determine purpose based trip. The findings from this study can be used during preliminary feasibility or planning study in determining the expected number of trips. The study utilizes trip data from the 2001 National Household Travel Survey (NHTS). Home base and non-home based work, school, shopping, family, social and religious driven trips are modeled by using different social economic variables. Multinomial logit (MNL) and multinomial probit (MNP) models are used in this study in determining the effect of the variables, due to their capability of modeling multivariate discrete response variables and flexibility of studying effect of the variables for each trip purpose model developed. The modeling result showed the presence of more number of workers in the house to influence each trip purpose mentioned. Lifecycle which classifies each household member as adult or child and retirement status appeared to influence school, family and social related trips. The families with small children were found to make fewer trips compared to those with grown up children and retired people. Individual income in the household affect much trips related to shopping, work, and social. Weekdays and rush hours have positive relation with work and school trips but negatively related to shopping, family and social trip purposes. The MNL and MNP models yield the similar results in terms of the sign and magnitude of the coefficients. However the standard errors in MNP are seen to be tighter than those in MNL. Moreover, many variables appeared to be non-significant in MNP compared to MNL.

Overview and Literature Review

Trip generation is and essential stage in transportation planning. Failure to identify the accurate trip production and attraction will end up in faulted or over estimated or underestimated number of trips from one zone to another. In the process of determining number of trips from one zone to another, the trip purpose and activity are the main determinants. In other words all trips are for certain activity or purpose. Planners will have advantage if they will know the characteristics of the community they are planning for in terms of their social economic factors. By knowing or foreseeing the social economic characters of the people living or visiting the community will help in planning adequate facilities for the community.

The effect of the social economic factors to the transportation planning can be determined by using statistical approach. One of the approaches is to determine if presence of one variable within the household will increase or reduce the number of trips. For instance, planner will like to know what type of trip purposes in which the community in household with more than 4 children under the age of 6 will make. Another planner will be interested in knowing the main trip purpose in elderly community or the community in which many households have few vehicles than drivers and vice versa.

To answer of these kinds of planning questions can be done by looking in the coefficient of these variables from the statistical models. The planning engineers can develop these models and interpret the effect of these variables and eventually use them in planning. Modeling development in estimating household production rates to support transportation planning process has a long decade of history. Data driven analysis, such as correlation analysis and cross-categorization, may well serve the purpose identifying basic, linear relationships among the underlying variable and various explanatory variables. Exploration of the possible nonlinear relationship might require further modeling efforts and insights. For total trip rates, various estimation methods have been developed, including linear regression analysis, cross classification (category) analysis (FHWA 1975, Caldwell and Demetsky 1978), adaptive tree based classification (Washington and Wolf 1997, Strambi and Bilt 1998), and generalized linear model (Said and Young 1990, Lan and Hu 2000).

Activity-based planning approach (Kitamura 1988, Kitamura et al. 1997, Goulias 1999, Ma and Goulias 1999, Bowman and Ben-Akiva 2001) has activated new avenue and opportunity for renewing traditional planning process whose results have been criticized far from perfection. In particular, Kitamura et al. (1997) and Goulias (1999) employed simulation-based methods to reproduce the multitude of factors underlying individual travel behavior and collective activity pattern. Micro-simulation approach was implemented by the former to achieve activity associated travel pattern. The later used Markov chain model to carry out Monte-Carlo simulation for realizing non-stationary switching of activity participation from year to year, and stationary switching in day-to-day activity participation pattern. These simulation methods represent innovative vein of approach dealing with activity-based planning process and provide a mean of tracking not only the mean evolution but also variability of the travel pattern. Yet, at present time they are not meant for and suitable for practitioners and planning agents such as MPO to use for planning purposes. It is still essentially practical to use deterministic equation for prediction application. Ma and Goulias (Ma and Goulias 1999) used Poisson regression and its variants to estimate individuals' daily activity frequencies by activity type (purpose). Simultaneous estimation of trip rate breakdown by multiple dimensions of trip-making characteristics will give more insights but might require substantially richer data.

This study performs a joint estimation of purpose and mode specific trip estimation using the 2001 National Household Travel Survey data (NHTS). The paper attempts to answer questions like which are the factors that affect the activity and purpose based trip making, the finding which can help transportation planners in trip generation. The modeling structure is intended to build in a way consistent with trip decision-making process, i.e., either hierarchically deciding what to do (purpose) and followed by what mode to use or making decisions simultaneously. The rest of the paper is organized in the following fashion. First, description of the trip data used in this study, then concept of multinomial models is briefly discussed. The model results are presented with discussion on the effect of each variable used in the model. The conclusion is given at the end of the paper.

Trip Data

The data used in this paper was obtained from the 2001 National Household Travel Survey (NHTS, 2001). The data dated from April 2001 through May 2002 was collected from a national sample of the civilian, non-institutionalized population of the United States. The survey includes demographic characteristics of households, people, vehicles, and detailed information on daily and longer-distance travel for all purposes by all modes. According to the NHTS website, the data collected for daily trips taken in a 24-hour period includes: purpose of the trip (work, shopping, etc.), means of transportation used (car, bus, subway, walk, etc.), day of week when the trip took place, number of people in the vehicle i.e., vehicle occupancy, and driver characteristics (age, sex, worker status, education level, etc).

In addition, the 2001 NHTS also collected additional data on trips to a destination 50 miles or more from home (long-distance travel) that started from home and ended at home during a four-week travel period. Data collected on long-distance trips included all the data mentioned above for daily trips with the exception of travel time and the time of day the trip took place. Although the NHTS was divided into four hierarchical data files to facilitate analysis, the authors focused on the travel day trip file since it contains the data about each trip the person made on the household's randomly assigned travel day. The travel day file contained a sample size of 642,292 trips but not all variables were applicable for this particular study.

The 2001 NHTS travel day trips file contains more than 140 variables. Some of them were not useful for the purpose of this study. By using Matlab data mining program and Stata statistical software, authors were able to retrieve the whole travel day trips file. From the literature reviews and prior-knowledge, fifteen variables were chosen for the study. These variables as described below are the ones which mainly determine the purpose of the trip. After pruning those unwanted observations, 453,446 trips (data points) remained as disaggregate data. The data has been appropriately weighted according to the specific sampling strategy used in data collection. The purpose of weighting was to expand the sample data to estimates for the whole nation and adjust for non-response.

Total income of household members was one of the variables retained for analysis. This variable represents the derived total income for the household. The variable is coded as HHINCTTL in the NHTS database. The original data categorize the income into 18 different groups starting 1 to 18 in which each number represents the range of the household income. This study utilized the numerical data for income by taking the midpoint of the ranges from each category. In order to facilitate trip purpose decision of the members in the household, the household income was divided by the total number of persons (house size) in order to get approximately average income for individual members in the house. This created variable "individual income" used in analysis.

The count of household members coded as HHSIZE in NHTS database was also retained. The variable is used to study effect of household size to trip purpose making. Other countable variables in the household included in this study were number of vehicles in household coded as HHVEHCNT, number of household members with jobs coded as WRKCOUNT and number of adults in household coded as NUMADLT. In order to facilitate the study, number of children in the household was calculated by subtracting the number of adults from the number of household members. In this case the numbers of children together with number of adults were used in the models instead of number of household size. Number of drivers in household coded as 'DRVRCNT' in NHTS data file was used too. The variable was used as a numerical variable in the modeling process in order to test how more or less number of drivers in the household can affect the trip purpose. In order to evaluate if excess or less number of vehicles with respect to the number of drivers in the household has effect in trip purpose making, the ratio of number of vehicles to the drivers in the household was evaluated. The household with fewer vehicles than the drivers was coded as 0 and those in which there were more vehicles than the driver was coded 1.

Life cycle coded as LIF_CYC in NHTS data was also used. This variable classifies each household member as adult or child and retirement status for adults. From 10 categories in this variable, five groups for the purpose of modeling were created as follows;

- 0 for household with one adult, no children and 2+ adults, no children
- 1 for household with one adult, youngest child 0-5 and 2+ adults, youngest child 0-5
- 2 for household with one adult, youngest child 6-15 and 2+ adults, youngest child 6-15
- 3 for household with one adult, youngest child 16-21 and 2+ adults, youngest child 16-21
- 4 for household with one adult, retired, no children and 2+ adults, retired, no children

The presence of the household in urban or rural area coded as URBRUR was also treated as categorical variable with rural coded 0 while urban coded 1. The authors created two groups based on travel day, weekends and weekdays. The interaction between day of the week and rush hour was also investigated as one of the variables. The morning hours between 7:00-9:00 AM and evening hours between 4:00-7:00 PM were identified as rush hours while the remaining hours including weekends where termed as non-rush hours. The day of the week was interacted with rush hours in which three variables were categorized; weekdays rush hour, weekdays non-rush hour and weekends coded as 2, 1 and 0 respectively. Also included in the dataset was the gender of the subject (R_SEX) with male coded as 0 while female as 1.
Travel day trip purpose (WHYTRP01) was retained as the response variable for this study. Through this variable, different trip purposes were derived and grouped according to similarities, then used as a response variable in the models. In NHTS data file the variable is coded from 0 to 83 with each code representing different trip purpose. Authors divided the purposes into eight different groups according to similarities of the trip purposes. The groups include home based work trips, school trips, shopping trips, social and recreational trips, church trips, family and personal trips and none-home based work trips and others coded as 1, 2, 3, 4, 5, 6, 7 and 8 respectively. From these eight groups the MNL and MNP models were developed and discussed. The activity based trips purposes created are defined as follows (NHTS);

- Home Based Work Trip: Include working at home and going to work from home.
- None Home Based Work: Include returning to work, attending business and other work related.
- School: Include go to school, go to school as student, and go to library or school related.
- Shopping: Include shopping/errands, buy goods in groceries/clothing/hardware store, buy services like video rentals/dry cleaner/post office/car service/bank and buy gas.
- Social and Recreational: include social/recreational, go to gym/exercise/play sports, rest or relaxation/vacation, visit friends/relatives, go out/hang out in entertainment/theater/sports event/go to the bar and visiting public place like historical site/museum/park/library
- Family and personal: Include, family personal business/obligations, use professional services: attorney/accountant, attend funeral/wedding, use personal

services like grooming haircut/nails, pet care: walk the dog/vet visits and attend meeting like PTA/home owners association/local government and medical/dental services
- Church/Religious: Include, going to religious services
- Others: Include transport someone, pick up someone, take and wait, drop someone off, meals, social event, get/eat meal, coffee/ice cream/snacks and medical/dental services.

Multinomial Logit (MNL) and Multinomial Probit (MNP) Concepts

Based on the fact that there were eight different trip purposes for modeling, several distributions which take into account multi-response variables seemed to be applicable. Some of the distributions which consider multi-response variable include multinomial logit, multinomial probit and ordered logit or probit. These distributions consider dependent variables which can take more than two categorical levels or indicators. The multi-level response variable is distinguished of whether the variable has an ordered or unordered level. One disadvantage of ordered distributions is that each category of the dependent variable is evaluated simultaneously with respect to the independent variables. In other words, it is not possible to produce different sets of estimators for the different dependent variable values (Drucker and Khattak 2000, unpublished). For instance one variable can be relevant to use in work trip model but not relevant in family trip model. In order to assess the categories of the dependent variable separately, the modeling approach which could have considered discrete nature of trip purpose at the same time flexible in using appropriate explanatory variables for different trip purposes was considered.

Alternatively authors applied Multinomial Logit (MNL) and Multinomial Probit (MNP) based on their ability to consider multi-response dependent variables. For MNP a discrete outcome is derived assuming that the disturbances are multivariate normally distributed (Washington et al. 2002). Multinomial probit model is an alternative to MNL logit regression analysis. These two models are very similar to one another; the difference is based on the log odds. Multinomial logit model is based on logistic function while probit uses the cumulative normal probability distribution. The reader must keep in mind that the two procedures are so similar that they can easily be confused with one another. The logistic regression and probit analysis produce predicted probabilities that are very similar (http://www.gseis.ucla.edu/courses/ed231c/231c.html, Accessed 2006).

The principle of MNL and MNP are built on the assumption that the choice between any pair of alternatives of the response variable is independent of the availability of other alternatives. It implies that the random part of utility function is independent between the alternatives. The multi-level response variable is distinguished of whether the variable has an ordered or unordered level.

One problem of MNL is independence of irrelevant alternatives (IIA). This is due to the fact that the error terms are assumed to be independent distributed from each other. Although this independence has the advantage that the likelihood function is quite easy to compute, in most cases the IIA assumption leads to unrealistic predictions. At this stage is where MNP come into effectiveness. Multinomial probit breaks down the IIA assumption and allow the error term to be correlated with each other. While MNL follows logistic function, the MNP assumes the error terms follow the multivariate normal distribution and are correlated across the choices.

In understanding how the two models work, suppose there are J categories of purpose based trips as the response variable. This implies that one needs to construct J-1 equations from MNL. Each of these J-1 equations is logistic regression comparing a group with the reference category (base) or comparison group. By using maximum likelihood, MNL simultaneously estimates the J-1 logit functions. The probabilities of other members in other categories are compared to the probability of membership in reference category. Suppose the utility function is written as:

$$U_{ki} = X_k \beta_i + \varepsilon_{ki}$$

Where X_k denotes the individual independent variable, β_i denotes the coefficient associated with each independent variable and ε_{ki} is the error term. Suppose the response variable (trip purpose), q is subjected to different levels or categories denoted as 0…i. Then,

$$q_k = j, \text{ if } U_{kj} \geq U_{ki} \text{ for } j \neq i$$

In this study i=0, 1, 2, 3, 4, 5, 6 and 7 where U_{k0} represent other trip purposes, U_{k1} represent home based work trips and U_{k2} represent school trips, Uk3 represent shopping related trips, Uk4 representing social and recreational trips, U_{k5} represent church and religion services trips, U_{k6} represent family and personal trips n U_{k7} representing non-home based work trips. For these eight categories, it requires seven equations, one for each category in relation to the reference category which in this case is Uk0. The general logistic equation for MNL is given by,

$$P(q_k = j) = \frac{e^{X_k \beta_i}}{1 + \sum_{i=1}^{J} e^{(X_k \beta_i)}}$$

As one sets reference category to zero, the equations for the probabilities become:

The estimation can be performed by using maximum likelihood (ML) in which the log likelihood function is given as:

$$\log L = \sum_{k=1}^{K}\sum_{j=1}^{J} q_{kj} \log\left(P_{kj}\right),$$

With $q_{kj} = 1$ if the trip record k falls into category j and $q_{kj} = 0$ otherwise.

For MNP, the estimation problem would in principle be similar to the MNL, except the logit function being replaced by the normal density function. The practical obstacle is that it is highly time consuming to evaluate higher order multivariate normal integrals, even for a sample of moderate size (Ma and Goulias 1999). Several researchers as well as the software STATA (Stata Press Publication 2005) use the random sampling technique to approximate multivariate normal probabilities.

Modeling Results

The MNL and MNP model results for each activity based trip purposes are shown in Table 1. For each model type, the coefficients, standard error and Z-values for each variable is presented. The following are the descriptions of the model results.

Home and Non-Home Based Work Trip Model (HBW and NHBW)

Home based work refer to trips involving individual working at home or going to work from home while none home based work refer to trips involving returning to work, attending business meeting and other work related travels. Except for number of children in the household, households with youngest kid between 0-5 and weekends, all other variables were positively related to home based work trip purposes. The coefficients in NHBW are restricted the same as those in HBW except for weekday rush-hour in which the coefficient is negative, indicating that most of NHBW trips occur in mid-day. It is obvious that the household with more number of workers will have more frequent trips to the work, the reality which is supported by the model output where the work count has positive coefficient and very powerful z-value. Work trip is also strongly associated with average individual income of the household with positive coefficient explaining that the higher the average individual income in the house, the higher we have to expect more work trips. The same effect of income on work related trips was also found in previous studies (Goulias et al. 1992, Druker and Khattak 2000 unpublished). The number of adults in household has positive coefficient indicating more number of adults in household will work related trips.

Having more number of vehicles than drivers in the household will increase work trips. Weekdays and urban areas have more work trips compared to weekends and rural areas. The interaction between week day and rush hour is seen to be very powerful in NHBW. This means somebody who is involved in NHBW has a room to choose the time interval of the day to go to work without being constrained during rush hour times.

School Trip Model

School trips were referred to trips going to school or library as a student. As expected weekdays have more school trips compared to weekends in the model. Most schools colleges and Universities hold classes during the weekdays rather than weekends. At the same time many parents and school buses take students to school during morning rush hour and return them back during evening rush hour time, the reality which accepts the positive coefficient of weekday rush hour. More number of children in the household will favor school trips.

Lifecycle groups with children are shown to be the primary determinant to school related trips. The households with one or two adults and youngest child ranging from 6-15 years old are seen to strongly increasing school trips followed by household with youngest child 16-21 and finally those with youngest child 0-5. The retired family with one or two adults but without a child has negative coefficient indicating they don't go to school because of their ages after retiring. The result from school model agrees with the previous one found in 1990 (Goulias et al. 1992).

Individual income of the household members have negative coefficient in the model, the outcomes which might be contrary to the expectations. As the individual income increase in the household which increase the income, presence of the activity can cause the household member to ignore going to school. This means the effect of higher income goes hand to hand with the source of income which can influence school trips. It is also observed that urban areas have more school trips than rural areas.

Table 1: Model Results

	Multinomial Logit (MNL)			**Multinomial Probit (MNP)**		
Home and Non-Home Work Based Trips	Coef.	Std. Err.	Z-Value	Coef.	Std. Err.	Z-Value
Number of Adults in Household	0.183	0.01	17.48	-0.022	0.009	-2.46
Number of Children in Household	-0.038	0.008	-5.07	-0.076	0.005	-14.19
Number of Workers in Household	0.106	0.01	10.38	0.198	0.007	28.88
Female	-0.152	0.015	-10	-0.192	0.011	-17.94
Urban Areas	0.174	0.013	12.87	0.012	0.010	1.27
Individual Income	8.9E-06	5.80E-07	15.43			
Excess Vehicles than Drives	0.246	0.023	10.65	0.093	0.015	6.05
One and 2+adults with youngest child 0-5	-0.084	0.02	-4.22	-0.074	0.014	-5.42
Weekday Rush-Hour	0.199	0.018	11.1	0.213	0.013	16.41
(Weekday Rush-Hour(NHBW))	-0.946	0.035	-26.68	-0.842	0.018	-47.55
Weekends	-0.042	0.021	-2.05	-0.035	0.015	-2.42
Weekends(NHBW)	-1.769	0.057	-31.25	-1.223	0.026	-46.87

Constant (NHBW)	-1.927	0.028	-69.56	0.381	0.023	16.84
School Trip						
Number of Children in Household	0.097	0.016	6.14	0.035	0.009	3.7
Female	-0.092	0.03	-3.12	-0.079	0.017	-4.52
Urban Areas	0.1	0.032	3.13	0.022	0.018	1.18
Individual Income	-1.9E-05	1.90E-06	-10.13	-1.3E-05	1.1E-06	-11.87
One and 2+adults with youngest child 0-5	0.797	0.07	11.45	0.347	0.037	9.37
One and 2+adults with youngest child 6-15	1.266	0.065	19.55	0.626	0.034	18.42
One and 2+adults with youngest child 16-21	1.205	0.072	16.79	0.566	0.038	14.91
One and 2+adults retired no child	-0.472	0.087	-5.42	-0.164	0.042	-3.94
Weekday Rush-Hour	1.721	0.04	43.14	0.882	0.021	41.15
Weekends	-0.406	0.063	-6.45	-0.193	0.030	-6.34
Constant	-2.886	0.087	-33	-1.381	0.046	-30.29
Shopping Trips						
Number of Adults in Household	0.139	0.013	10.48	0.077	0.009	8.84
Number of Children in Household	-0.052	0.012	-4.29	-0.044	0.008	-5.75
Number of Workers in Household	-0.086	0.014	-6.35	-0.043	0.009	-4.85
Female	0.053	0.018	3.02	0.026	0.012	2.14
Individual Income	2.6E-06	6.5E-07	3.93	1.4E-06	4.1E-07	3.37
Excess Vehicles than Drives	0.137	0.027	5	0.092	0.018	5.15
One and 2+adults with youngest child 0-5	-0.19	0.032	-5.9	-0.077	0.021	-3.67
One and 2+adults with youngest child 6-15	-0.185	0.028	-6.64	-0.060	0.018	-3.24
One and 2+adults retired no child	0.197	0.023	8.49	0.236	0.016	14.5
Weekday Rush-Hour	-0.274	0.021	-12.92	-0.188	0.014	-13.31
Weekends	0.171	0.022	7.66	0.105	0.015	6.92

	Multinomial Logit (MNL)			Multinomial Probit (MNP)		
Social and Recreational Trips	Coef.	Std. Err.	Z-Value	Coef.	Std. Err.	Z-Value
Number of Adults in Household	0.18	0.018	10.21	0.045	0.011	3.95
Number of Children in Household	0.036	0.01	3.64			
Number of Workers in Household	-0.063	0.015	-4.12	-0.023	0.010	-2.41
Female	-0.108	0.02	-5.41	-0.109	0.013	-8.41
Urban Areas	0.143	0.02	7.06	0.047	0.013	3.65
Individual Income	7.8E-06	8.80E-07	8.76	1.6E-06	5.6E-07	2.88
Excess Vehicles than Drives	0.176	0.031	5.66	0.078	0.019	4.06
One and 2+adults with youngest child 6-15	0.162	0.023	6.91	0.121	0.015	7.97
One and 2+adults with youngest child 16-21	0.237	0.034	6.97	0.151	0.022	6.96
One and 2+adults retired no child	0.253	0.03	8.54	0.199	0.019	10.51
Weekday Rush-Hour	0.198	0.024	8.14	0.085	0.015	5.51
Weekends	0.601	0.025	23.72	0.344	0.016	20.87
Constant	-1.154	0.046	-25.01	-0.433	0.031	-14.07
Religion Related Trips						
Number of Adults in Household	0.22	0.036	6.03	0.075	0.018	4.08
Number of Children in Household	0.042	0.019	2.19			
Number of Workers in Household	-0.085	0.031	-2.74			
Female	0.221	0.043	5.12	0.068	0.022	3.12
One and 2+adults with youngest child 6-15	0.184	0.053	3.46	0.129	0.027	4.87
One and 2+adults retired no child	0.686	0.064	10.65	0.413	0.033	12.56
Weekday Rush-Hour	1.807	0.104	17.41	0.703	0.040	17.41
Weekends	3.161	0.099	31.82	1.347	0.039	34.57
Constant	-5.075	0.125	-40.73	-2.352	0.055	-42.98
Family and Personals Trips						
Number of Adults in Household	0.205	0.024	8.58	0.057	0.013	4.2
Number of Workers in Household	-0.108	0.022	-4.94	-0.043	0.012	-3.58
Female	0.137	0.027	5.06			
Individual Income	4.9E-06	1.2E-06	4.21			
Excess Vehicles than Drives	0.293	0.041	7.24	0.149	0.023	6.54
One and 2+adults with youngest child 0-5	-0.235	0.04	-5.9	-0.184	0.022	-8.32
One and 2+adults with youngest child 6-15	-0.098	0.037	-2.63	-0.071	0.021	-3.45
One and 2+adults retired no child	0.455	0.041	11.09	0.309	0.023	13.36
Weekday Rush-Hour	-0.042	0.031	-1.38	-0.053	0.018	-3.03
Weekends	-0.298	0.036	-8.27	-0.192	0.020	-9.46
Constant	-1.567	0.063	-24.99	-0.691	0.036	-19.29

Shopping Trip Model

Shopping trip was modeled as a collection of trips related to shopping (errands), buying goods in groceries or store, buying services like video rentals or post office, car service, bank and buying gas. An increase in the number of adults in the household is shown to increase the number of shopping trips. Adults are always responsible for buying their own belongings, food and other requirements for their kids. More number of workers in the house will reduce shopping trips especially on weekdays. During the weekdays the workers make trip to working places and most time the shopping trips are made on weekends. Individual income also increases shopping trips. The more the income the more desire to buy various commodities, the theory which can increases driving to shopping centers (Goulias et al. 1992).

Social and Recreational Trips

These trip types were those which included going to social areas and recreational, going to gym or exercise or to play sports, resting or vacation, visiting friends or relatives, going to hang out in entertainment or to the bar and visiting public place like historical sites. Social and recreational trips are seen to be linked much by more number of adults in the household, number of children, individual income and weekends. It must be known that there are some social services which are open during weekdays and not only in weekends. That's why both weekdays and weekends have positive coefficients, but weekends is strong than the weekdays. Contradictory to expectation, females has negative coefficient to social and recreational trips, the result which have no accurate explanation yet. Both lifecycles have positive coefficient meaning all age groups in one way or another are associated in social activities or attending recreational areas.

Family and Personal Trips

Family and personal trips represented the trips related to personal business obligations; use professional services like attorney, attending funeral or wedding and attending meeting medical or dental services. It is observed from the model results that increase in number of adults in the household, number of children increase family and personal trips. At the same time households with retired people are likely to make more family and personal trips. Families with children aging 0-15 have lower trips to the family and personal issues compared to those with grown up children or retired people. Households with good income and extra vehicles than the drivers influence increase in family and personal trips.

Religious Trips

As expected, weekends have very strong coefficient to religious trips. Christians go to church on Sundays and some of them on Saturdays; the reason which might has caused the coefficient to be such strong. Though not documented, but it is the opinion of the authors that senior people and children aged 6-15 are the ones mostly found in the churches or mosques compared to other age groups the situation which increase number of trips if these lifecycle groups are present in the household. Female has positive coefficient indicating they have higher frequency of going to church compared to men, the reality which is true as personal opinion of the authors based on the experience.

Comparison between MNL and MNP

As it can be seen in Table 2, the MNL and MNP are clearly similar in terms of the sign of the coefficients and magnitude of Z-values. One observation seen during analysis is that MNP probabilities require very complicated multivariate integral computations compared to MNL which utilize logistic function. In running the models, MNL was very quick to reach final optimization while MNP was very time consuming, indicating how complex it was. In terms of the results, the standard errors in MNP were relatively tighter than those in MNL, implying that MNP is a more efficient estimator. But few variables were significant in MNP compared to MNL, the case which is linked to good and more accurate estimation. In general, both models reveal similar results and they can be applicable for modeling multivariate response variables like activity based trip purposes. Planning engineers can utilize either of them in determining the coefficient of the variable during planning stage.

Practical Application of the Model Results

Planning engineer will need to study the coefficient of the variable in the model to see if it is negative or positive. The negative coefficient will indicate that presence of factor in the household lower the probability of making such kind of trip purpose. In case the coefficient is positive, that will indicate that the presence of that variable in the household will increase the probability of making trip for that trip purpose. If the aggregate variables households within the community are know, then by using the model, the planner will have preliminary result of what to be expected in terms of capacity of the transportation facility, parking, parks and so forth.

Conclusions

This paper modeled relationship between purpose-driven household trip productions to various household variables. The 2001 NHTS trip data was used in the study. Eight different categories of purpose-driven trip were identified including home and non-home based work trips, school trips, shopping trips, family and personal trips, social and recreational trips and religion related trips. Multinomial logit (MNL) and multinomial probit (MNP) models were used to estimate how different variables affect those specified trip purposes. Both models (MNL and MNP) allow the use of one purpose-based trip purpose as a reference category while analyzing other categories. Comparison of the results from MNL and MNP models gave similar results in terms of the sign of the coefficients in general. The standard deviations in MNP were relatively tighter than those in MNL, implying that MNP is a more efficient estimator.

A number of variables were identified as having significant effects on activity-specific trip productions. In the work models (HBW and NHBW) the significant positive coefficient variables in trip production were found be number of adults, number of workers and individual income in the household. Other positive related variables to work trips included urban areas, presence of more vehicles than drivers in the household and weekday rush hours. The negatively related variables to work trip included households with children aged 0-5, number of children in the house and weekends. The signs for weekday rush-hour indicator variables indicate that HBW trips occur during rush hours of weekdays but NHBW trips occur during non-rush hours of weekdays.

For school trips, number of children, the families with children aging 0-21, weekday rush hours and urban areas were found to be significant with positive coefficients while female, individual income, retired people and weekends were found to have negative coefficients. Shopping trips were seen to be positively supported by more adults in the household, individual income, presence of females, more vehicles in the household, retired people and weekends. The household with more number of children and more number of workers have to expect fewer trips for shopping based on the model results.

In terms of recreational and social trips, all factors seemed to be positively correlation except number of workers in the house which have a negative coefficient. Weekends have the strongest coefficient followed by number of adults and individual income and retired people in affecting social trips. Religious trips also have positive coefficient with number of adults, number of children, females, family with children and weekends. More number of workers in the household has shown to decrease religion related trips. Family and personal trips are seen to be positively related to weekends, retired people, number of adults in the household, excess vehicles in the household and females. Presence of more workers in the household, children aged 0-16 weekends and weekday rush hours decrease family and personal trips.

References

"Federal Highway Administration Trip Generation Analysis." *U.S. Department of Transportation*, Washington, D.C., 1975.

Caldwell III, L. C. and Demetsky, M. J. (1978). "An Evaluation of the Transferability of Cross Classification Trip Generation Models." *Virginia Highway and Transportation Research Council*, VA, VHTRC 78-R39.

Washington, S. and Wolf, J. H. (1977). "Tree-Based Versus Ordinary Least Squares Linear Regression Models - Theory and Example Applied to Trip Generation." *Transportation Research Record* 1581, TRB, pp. 82-88.

Strambi, O. and Van De Bilt, K.A. (1998). "Trip Generation Modeling Using CHAID, A Criterion-Based Segmentation Modeling Tool." *Transportation Research Record* 1645, TRB, pp. 24-31.

Said, G. M. and Young, D. H. (1990). "A general linear model framework for estimating work trip rates for households in Kuwait." *Transportation Research* - A, Vol. 24A, No. 3, pp. 187-200.

Lan, C-J and Hu, P. (2000). “A Mixed Generalized Linear Model for Estimating Household Trip Production.” *Transportation Research Record* 1718, pp. 61-67.

Kitamura, R. (1988). “An Evaluation of Activity-based Travel Analysis.” *Transportation*, Vol. 15, 1-2, pp. 9-34.

Washington. S. P., Karlaftis, M. G, and Mannering, F. L. (2002). “Statistical and Econometric Methods for Transportation Data Analysis.” *Chapman & Hall/CRC.*

Kitamura, R., Chen, C., and Pendyala, R. M. (1997). “Generation of Synthetic Daily Activity-Travel Patterns.” *Transportation Research Record* 1607, pp. 154-162.

Goulias, K. G. (1999). “Longitudinal Analysis of Activity and Travel Pattern Dynamics Using Generalized Mixed Markov Latent Class Models.” *Transportation Research*, Part B: 33-8, pp. 535-557.

Ma J. and Goulias, K. G. (1999). “Application of Poisson Regression Models to Activity Frequency Analysis and Prediction.” *Transportation Research Record* 1676, pp. 86-94.

Phil, E. Applied Categorical & Non-normal Data Analysis, http://www.gseis.ucla.edu/courses/ed231c/231c.html>(July 18, 2005)

Bowman, J.L. and Ben-Akiva, M.E. (2001) “Activity-Based Disaggregate Travel Demand Model System With Activity Schedules.” *Transportation Research*, Part A: Policy and Practice, Vol. 35, 1, pp. 1-28.

Drucker, J. and Khattak A, J. “The Propensity to Work from Home: Modeling Results from the 1995 Nationwide Personal Transportation Survey.” *Department of City and Regional Planning, University of North Carolina at Chapel Hill.*

Goulias, K. G., Pendyala, P. M. and Kitamura, R.(1992). “Practical method for the estimation of trip generation and trip chaining.” *Transportation Research Record 1285*, TRB, pp. 47-56.

“2001 National Household Travel Survey (NHTS) User’s Guide. January 2004 (Version 3)”, Accessed July 19, 2005.

A Stata Press Publication: Stata Base Reference Manual, Vol 2 K-Q, Release 9. (2005). *Stata Corp.LP*, College Station, Texas.

How Reliable are our Roads?
A Case Study Using Charlotte, NC Data

S. S. Pulugurtha[1] and N. Pasupuleti[2]

[1]Assistant Professor of Civil Engineering, Assistant Director of Center for Transportation Policy Studies, The University of North Carolina at Charlotte, 9201 University City Boulevard, Charlotte, NC 28223-0001, USA; PH: (704) 687-6660, FAX: (704) 687-6953, email: sspulugurtha@uncc.edu
[2]Graduate Student of Civil Engineering, The University of North Carolina at Charlotte, 9201 University City Boulevard, Charlotte, NC 28223-0001, USA; PH: (704) 687-2089, FAX: (704) 687-6953, email: nspasup1@uncc.edu

Abstract

Rapid growth in travel demand during the last decade has led to increased delays and recurring congestion costs on urban roads in many areas. In addition, crashes, construction and rehabilitation activities, weather, and special events contribute to significant levels of non-recurring congestion cost. Travelers' perception of reliability of road network is based on factors contributing to both types of congestion. However, literature documents little to no research to combine and measure congestion using both the disparate factors. Thus, there is a need to research, define and measure congestion based on factors contributing to both recurring and non-recurring congestion.

Travel time is a measure which is simple and easy to understand. Reliability can be measured either in terms of travel time or in terms of variation in travel time. Using measures such as travel time reliability helps address questions such as "how reliable are our roads?" and "which path is more reliable to reach a destination from an origin within an "on-time window" during a certain time of the day?" The focus of this paper is to develop and illustrate the working of a Geographic Information Systems (GIS) based methodology to estimate travel time reliability for each link on major roads. The methodology involves 1) estimating travel time using Bureau of Public Roads equation for each link, 2) identifying the number of crashes by severity type on each link, 3) estimating travel delays due to crashes for each link, 4) estimating total travel time based on 1 and 3 for each link, and, 5) estimating travel time reliability for each link using results from Step 4. Factors such as link volume, link capacity, the travel speed, crashes and their types, and time to restore normal conditions after a crash are collected and used in the computations. As travel demand and crash occurrence depend on the time of the day, temporal variations in travel time for each link are also addressed. Data collected for the Charlotte metropolitan area in North Carolina are used to demonstrate the working of the methodology. The outcomes will help practitioners and researchers identify congested segments and suitable mitigation strategies.

Introduction

Increasing congestion has been a problem of concern to transportation system managers, general public, and elected officials due to its impact on mobility and economy. It is reported that an average American travels 78 minutes a day, over 80 percent of which is by automobile (Kockelman, 2003). The Texas Transportation Institute in their recent urban mobility study estimates that over 45 percent of peak period travel or roughly one-third of the total vehicle miles traveled (VMT) occur under congested conditions in many United States metropolitan areas.

Congestion occurs when demand exceeds capacity. Congestion during peak periods is typically referred to as a recurring congestion whereas congestion due to crashes, inclement weather conditions, or due to mechanical breakdown of vehicles is referred to as non-recurring congestion. To a user, recurring congestion during peak periods may be acceptable but not non-recurring congestion during off peak periods. Travel time, delays, and related indices are typically used to indicate recurring congestion whereas reliability may be used for both recurring and non-recurring congestion.

Congestion could arise anywhere along the roadway in the transportation network. When present, it reduces the capacity of the roadway and makes the traffic condition unstable. Unfortunately, transport networks and roadways are not 100 percent reliable, and as society's reliance on transport links and its expectations of the infrastructure performance grow so do the consequences of network failure. Thus, estimating reliability becomes an increasingly important attribute of road networks. A recent survey showed that travel time reliability is one of the most important factors for route choices indicating that travel time reliability is either the most or second most important reason for choosing a primary commute route (Abdel-Aty et al., 1997).

This paper makes an attempt to define and measure travel time reliability of each link in the transportation network by combining travel time and crash data. A Geographic Information Systems (GIS) based methodology is developed and used. Data for the Charlotte metropolitan area are used to illustrate the working on the methodology.

Literature Review

Reliability, in general, is defined as the ability of an item to perform a required function under given environmental and operational conditions and for a stated period of time. Reliability of road networks can be measured either in terms of travel time or in terms of a variation in travel time. Using measures such as travel time reliability helps address questions such as "how reliable are our roads?" and "which path is more reliable to reach a destination from an origin within an "on-time window" during a certain time of the day?"

The network reliability has been a growing area of interest to the transportation community. The definition of network reliability that is of interest here pertains to connectivity. Specifically, the network reliability is the probability that the origin and destination are connected due to the probabilities of link existence. Difficulties exist in directly applying the definitions and methodologies to the transportation area.

The operational conditions of road networks could be normal or abnormal (after some exceptional disasters). The reliability analysis of road networks could be divided into two dimensions: 'pure network' analysis and 'flow network' analysis. The 'pure network' reliability analysis is applicable to the situation after major exceptional events, comprising connectivity analysis and capacity reliability analysis.

Iida and Wakabayashi (1989) proposed two approximation methods for determining the connectivity reliability between a pair of nodes in a transportation network. These methods were based on reliability graph analysis using minimal path sets and cut sets. Iida and Wakabayashi state that complete enumeration of the minimal path sets and/or minimum cut sets was necessary to find an exact value for the reliability. In another study, Iida (1999) present basic equations to determine reliability based on connectivity when a system is in a series or in parallel. However, to apply this concept one must be able to predict the probability and extent to which a link would be damaged.

The measure of network performance reliability is more complex issue. Chen et al. (1999) defined network performance reliability as the probability that the road network can accommodate a certain level of traffic demand. This capacity-related reliability was estimated by investigating the network reserve capacity (a maximal origin-destination matrix multiplier subjected to link capacity constraints) in a Monte Carlo simulation procedure.

Bell (1999, 2000) defined another measure of network performance reliability as the utility of users when they are extremely pessimistic about the state of the network. The utility was obtained in a mixed strategy Nash equilibrium. Another set of measures of reliability was proposed by Asakura (1999). Their definitions could be viewed as a straightforward extension of the connectivity reliability, where reliability was estimated as a combination of network performance index in each possible network status.

Methodology

The objective of this paper is to estimate the reliability of the Charlotte road network considering factors related to recurring and non-recurring congestion. Reliability in this study is expressed in terms of travel time and its variation. The methodology to estimate travel time and its variation for each link includes the following steps.

1) Define study area
2) Data Collection
3) Estimate travel time under recurring conditions
4) Estimate travel time under non-recurring conditions
5) Integrate travel time components to study its variation

Define study area

The core metropolitan comprising sub-urban areas were congestion prevail is considered as the study area. Only major roads in this study area are considered as congestion primarily occurs on these roads. Also, data available for these roads is relatively more detailed and accurate at present compared to minor roads such as collector roads and local roads.

Data Collection

Data required to calculate travel time for each link include free flow speeds during peak and off peak periods, capacity during peak and off peak periods, and traffic volumes during peak and off peak periods. Summing traffic volumes for peak periods and off peak periods results in daily traffic volumes. Travel demand forecasting model outputs could be used as collecting data in the field for a large network is an expensive and time consuming process.

Crash data and other data pertaining to temporary changes in network conditions (due to weather, construction activities, etc.) could be used to calculate increase (or variations) in the travel time. However, collecting such data and studying their impact on travel times other than historical crash data is tedious though not impossible.

Estimate Travel Time under Recurring Congestion Conditions

In this step, travel time during morning peak period (6:30 AM to 9:30 AM), evening peak period (3.30 PM to 6:30 PM), mid-day (9:30 AM to 3:30 PM), and night (6:30 PM to 6:30 AM) were calculated using the Bureau of Public Road equation (FHWA, 1979). The equation is shown below.

$$t \equiv t_0(1+\alpha[\frac{v}{c}]^{\beta}) \quad \ldots$$

Equation (1)

Where,
t_0 is free flow travel time,
v is link volume, and,
c is link capacity.

The α and β values for each link may change one region to another. In this paper, these were obtained from the regional agency as used in the regional travel demand forecasting model.

The percent difference between the travel time during peak period and off peak period (night) of a link is defined as reliability of the link. These values can be combined for a path to obtain reliability value for a path. Thus, variations between peak period and off peak period travel times were used to explain variations in reliability of roads during peak hours. Also, links could be compared to determine how reliable our roads are when compared amongst themselves.

Estimate Travel Time under Non-recurring Congestion Conditions

Variation in travel time is a function of the number of crashes during the considered time period. It also depends on the type (severity) of crashes during that period. To extract these characteristics, crash data are overlaid on a 100 feet buffer generated around each link in the road network. The crash data used in this paper was obtained from the responsible local agency for analyses and illustration. Spatial analysis is conducted to extract the number of crashes by severity in each buffer. These crashes are converted into their equivalent travel time value by multiplying the number of fatal and severe injury crashes by 120, number of less severe injury crashes by 45, and property damage only (PDO) crashes by 15 and then adding them together. These numbers or weights are based on anecdotal evidence that fatal and severe injury crashes influence travel time for about 2 hours, less severe injury crashes influence travel time for about 45 minutes, and PDO crashes influence travel time for about 15 minutes. The equivalent travel time value is divided by 365 as the number of crashes is for one year.

Integrate Travel Time Components to Study Its Variation

In this step, data from the previous two steps are combined (integrated) to study its variation. Combining the two estimated travel times by adding them together do not make sense as they are disparate components in a way. Also, in general, their impact can be felt for several miles. Due to this, scores were calculated for travel time under recurring condition and travel time during non-recurring congestion conditions. As an example, to calculate score under recurring conditions, the travel time variation for a link based on link volumes during the day is divided by the maximum value for all the links and then multiplied by 50. Likewise, to calculate score under non-recurring conditions, the travel time variation for a link based on crash data during the day is divided by the maximum value for all the links and then multiplied by 50. Thus, the maximum possible total score for a link is 100.

Analysis & Discussion

The working of the methodology is illustrated using data obtained for the Charlotte metropolitan area in North Carolina. Only major and minor thoroughfare roads (principal and minor arterials) in the network were considered for analysis. Travel times variations due to link volumes and crashes were calculated. Figures 1, 2, and 3 show percent variations in travel times during the morning peak period, the evening peak period, and using daily link volumes when compared to night time link volumes. Links were categorized based on percent variation in travel times (less than 0, 0 to 50, 50 to 100, greater than 100). It can clearly be seen that network is comparatively more unreliable during the evening peak period when compared to the morning peak period.

Figure 4 shows considered links by score (which includes impact of both link volumes and crashes). The links were categorized into less than 15, 15 to 30, 30 to 45, and 45 to 60. The figure clearly helps identify the impact of crashes on travel time. It also helps identify those links that have high number of crashes.

Conclusions

This paper presents a methodology and discussion on reliability of the Charlotte road network. Travel time for each link were calculated considering both traffic and crash data. Reliability is explained 1) based on percent variation in travel time, and, 2) score based on percent variation in travel time and the impact of crashes on travel times. The methodology helps identify links which are unreliable. It also helps identify those links with significant crash problems.

Acknowledgments

The authors acknowledge and sincerely thank Charlie Jones and Anna Gallup of the Charlotte Department of Transportation for their help with crash data and travel demand forecasting model output data.

References

Abdel – Aty, M.A., Kitamura, R. and Jovanis, P.P. (1997). "Using stated preference data for studying the effect of advanced traffic information on drivers' route choice." *Transportation Research,* Vol.5C, pp 39 – 50.

Asakura, Y. (1999). "Evaluation of Network Reliability Using Stochastic User Equilibrium." *Journal o f Advanced Transportation*, Vol. 33, No. 2, pp. 147-158.

Bell, M. G. H. (1999). "Measuring Network Reliability: A Game Theoretic Approach." *Journal of Advanced Transportation*, Vol.33, No.2, pp 135 – 146.

Bell, M. G. H. (2000). "A Game Theory Approach to Measuring the Performance Reliability of Transport Networks." *Transportation Research*, Part B, Vol. 34, pp 533- 545.

Chen A., Yang, H., Lo, H. K., and Tang, W. H. (1999). "A Capacity Related Reliability for Transportation Networks." *Journal of Advanced Transportation*, Vol.33, No.2, pp 533-545.

Iida, Y. (1999). "Basic Concepts and Future Directions of Road Network Reliability Analysis." *Journal of Advanced Transportation.* Vol. 33, No. 2, pp. 125-134.

Iida, Y. and Wakabayashi, H. (1989). "An Approximation Method of Terminal Reliability of Road Network Using Partial Minimum Path and Cut Sets." *Transport Policy, Management, and Technology Towards 2001: Selected Proceedings of the Fifth World Conference on Transport Research.* Ventura, California: Western Periodicals Co.

Kockelman, K. (2003). "Chapter 12. Traffic Congestion, In Handbook of Transportation Engineering." *Editor: Myer Kutz, McGraw-Hill Publications.*

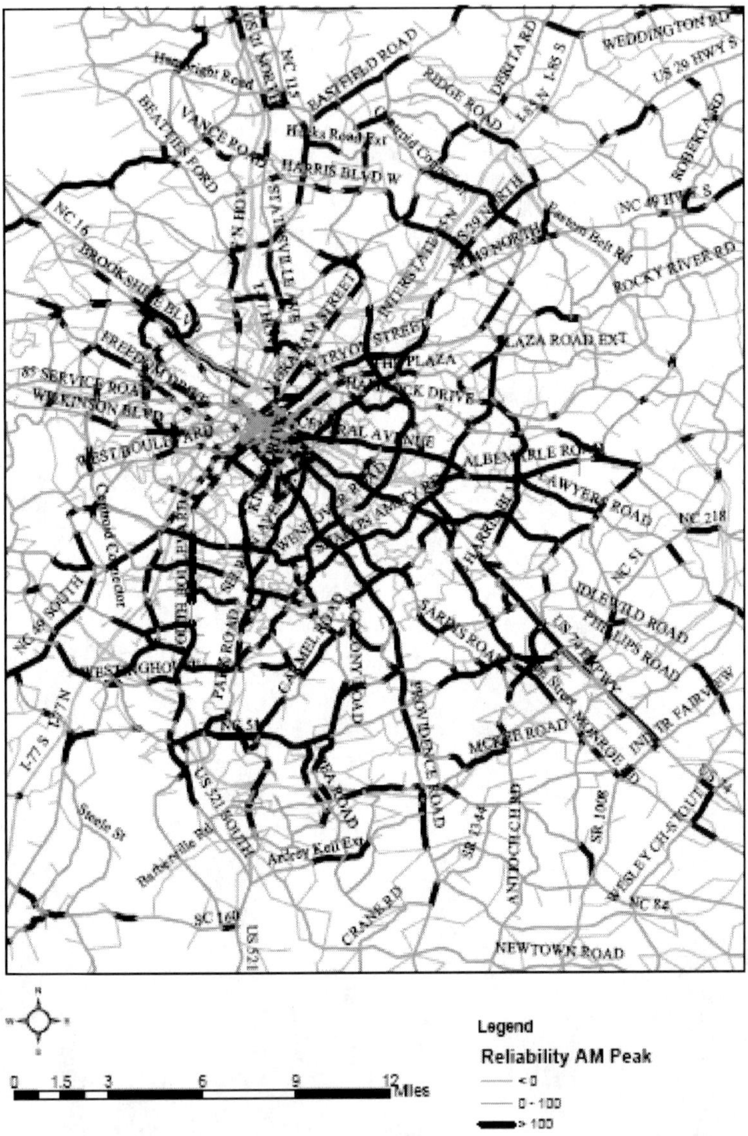

Figure 1. Reliability in terms percent variation in travel time during morning peak period.

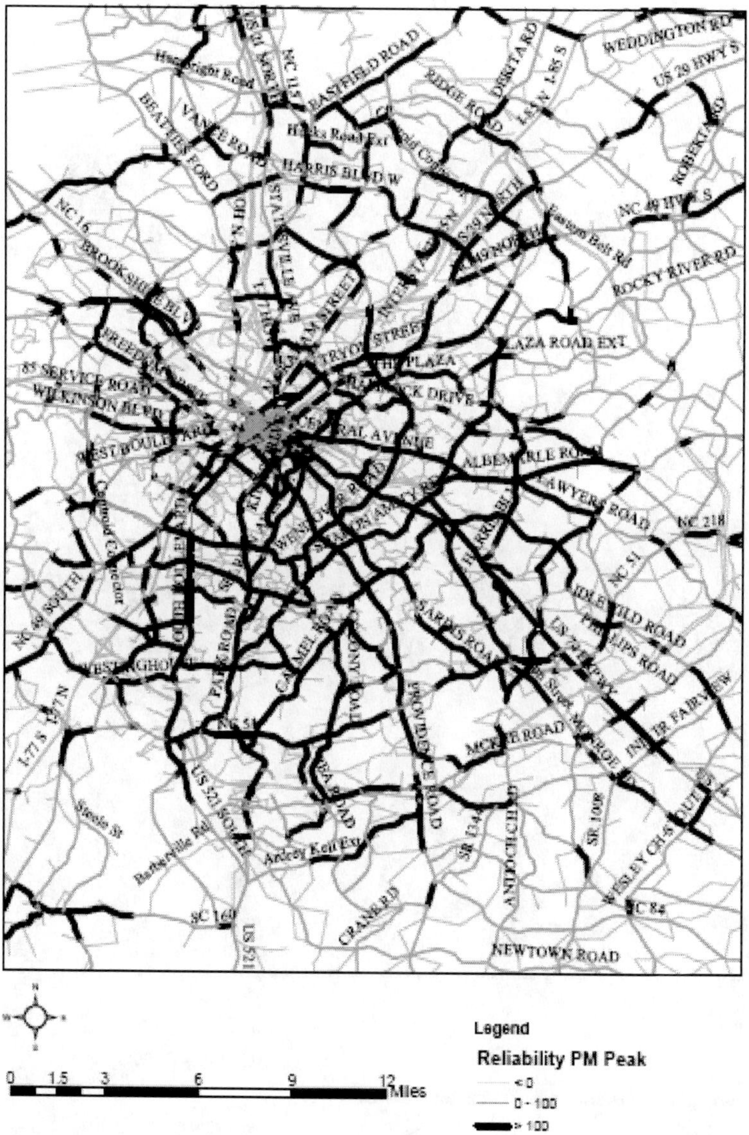

Figure 2. Reliability in terms percent variation in travel time during evening peak period.

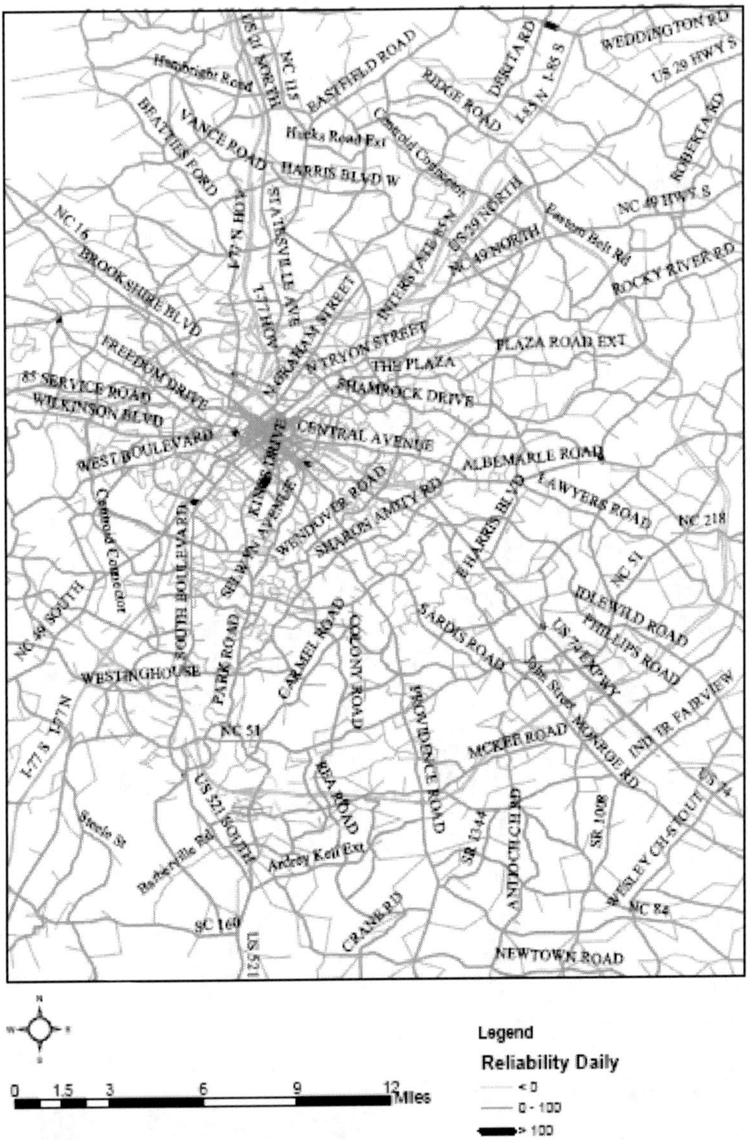

Figure 3. Reliability in terms percent variation in travel time based on daily link volumes.

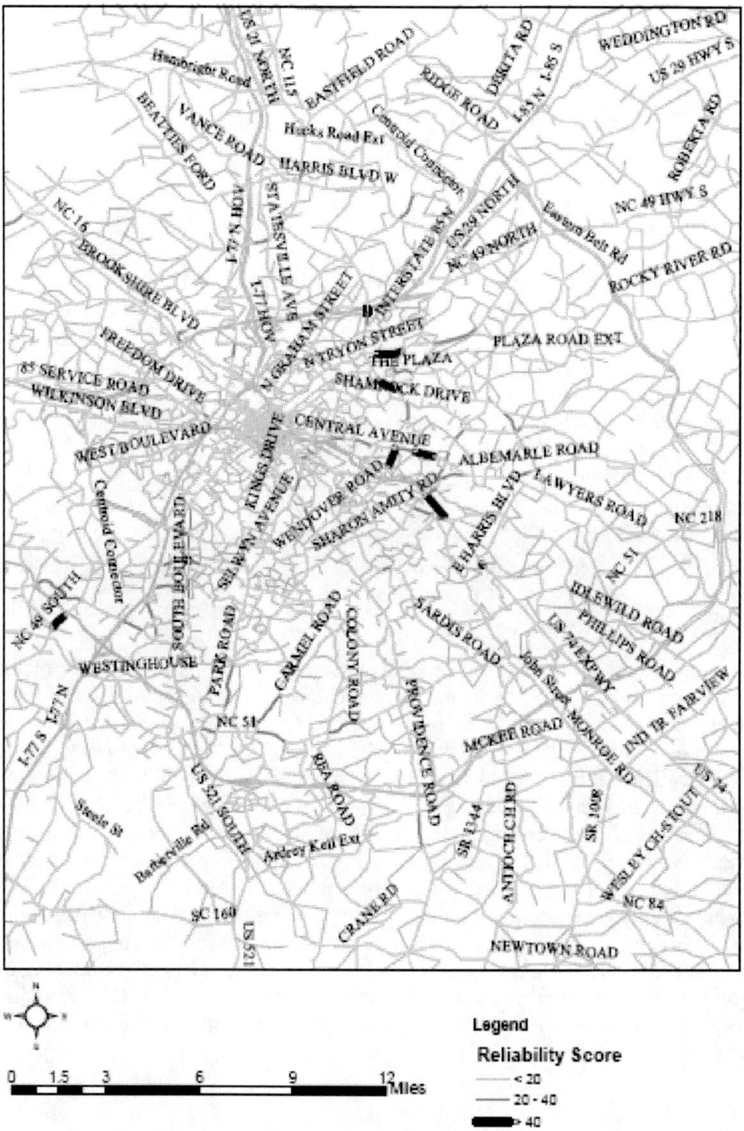

Figure 4. Reliability in terms score based on daily link volumes and crashes.

Variability of Vehicle Emissions and Congestion Forecasting

Michael Claggett, Ph.D.,[1] and H. Sarah Sun[2]

[1]Air Quality Modeling Specialist, Federal Highway Administration Resource Center, 4001 Office Court Drive, Suite 800, Santa Fe, New Mexico 87507;
PH (505) 820-2047; email: Michael.Claggett@dot.gov
[2]Transportation Specialist, Federal Highway Administration, Planning and Environment, HEPI-30, 1200 New Jersey Avenue, SE; Washington, DC 20590;
PH (202) 493-0071; email: Sarah.Sun@dot.gov

Abstract

There is an increasing reliance on mobile source emissions analyses as part of the transportation planning process, including regional and hot-spot analyses to fulfill the requirements of the transportation conformity rule and project-level highway air quality assessments in response to the requirements of the National Environmental Policy Act. The general procedure of an emissions analysis is to employ the U.S. Environmental Protection Agency's MOBILE6.2 model to obtain on-road mobile source emission factors, and multiply by the vehicle miles of travel to construct emission inventories. Critical to the outcome of such studies are future predictions of vehicle activity. This is especially true in evaluating transportation alternatives to assess how well a proposed project may or may not mitigate vehicle congestion and reduce emissions. This paper investigates one aspect of vehicle congestion and emissions forecasting by applying different vehicle speed estimating methodologies used for sketch planning purposes and shows the variability of MOBILE6.2 emission factors as a function of travel demand and the associated capacity (i.e., volume-to-capacity ratio).

Introduction

In the current regulatory framework, potential changes in mobile source emissions among transportation alternatives are evaluated using the U.S. Environmental Protection Agency's (EPA) MOBILE6.2 model (U.S. EPA, 2003), along with forecasts of vehicle miles of travel (VMT). Emission factor predictions obtained from the MOBILE6.2 model are also a function of numerous vehicle activity parameters such as VMT fraction by vehicle class, by hour of the day, by highway functional system, by vehicle speed, as well as others. While additional inputs such as external conditions (e.g., ambient temperature); vehicle fleet characteristics; vehicle fuel specifications; and state programs (e.g., implementation of an inspection maintenance program and/or use of reformulated gasoline) can affect the magnitude of emissions specific to a locale; these would be common to all alternatives under review. The most important factors affecting emission differences among the available transportation options are distinctions in vehicle activity. An essential facet of a mobile source emissions analysis is to determine the extent to which a project alternative is expected to mitigate congestion on links of a transportation network operating at or over capacity. Anticipated changes in MOBILE6.2 emission factors as a function of congestion mitigation is investigated herein along with the variability of results produced among several vehicle speed forecasting methodologies.

Vehicle Congestion Forecasting

The sensitivity of MOBILE6.2 emission factors to changes in vehicle speeds has been documented by the U.S. EPA (2002a) and the Federal Highway Administration (FHWA) (Tang, 2003; Tang, 2005). Surveys of vehicle speed forecasting techniques are also available from a variety of sources (NCHRP, 1997; Dowling Associates, 1997; NHI, 2003). The de facto standard for computing travel speeds is the Highway Capacity Manual (HCM) (TRB, 2000). However, as acknowledged by the U.S. EPA (2001), the HCM techniques require detailed, facility-specific information that is unlikely to be available at the planning level. Because of this practical consideration, the U.S. EPA (2001) recommends that the Bureau of Public Roads (BPR) formula be applied instead to forecast vehicle speeds on a regional basis for typical urban areas.

In its most recent update in 2000, the Highway Capacity Manual provides recommended procedures for forecasting highway performance measures for area-wide planning applications, including speed estimation. Because of the recognized practical considerations, these procedures are simplifications of the more elaborate techniques provided elsewhere in the HCM. An alternative technique is also provided in the HCM based on the traditional BPR formula. A third technique, based on methodology developed by the Texas Transportation Institute (TTI) for the National Highway Institute (2003) is examined as well.

The data required to compute vehicle speeds on a link vary by facility type or functional class and generally consist of demand data (daily traffic, peak hour volume); data for estimating free-flow speed (facility type, speed limit); and data for estimating capacity (number of lanes, percent trucks, terrain). Facility types are used in the HCM; the TTI method uses FHWA's Highway Performance Monitoring System (HPMS) area types and functional classes; and MOBILE6.2 uses its own facility driving cycle definitions. Table 1 provides a suggested association of HPMS area type/functional class to HCM facility type to MOBILE6.2 facility type.

Table 1. HPMS functional class – HCM facility type – MOBILE6.2 facility type mapping.

HPMS Area Type	HPMS Functional Class	HCM Facility Type	MOBILE6.2 Facility Type
Rural	Interstate	Basic Freeway	Freeway/Freeway Ramp
	Other Principal Arterial	Rural Multilane Highway	Freeway
	Minor Arterial	Rural Two-Lane Highway	Arterial/Collector
	Major Collector	Rural Two-Lane Highway	Arterial/Collector
	Minor Collector	Rural Two-Lane Highway	Arterial/Collector
	Local	Rural Two-Lane Highway	Arterial/Collector or Local
Small Urban (Population 5,000 to 49,999)	Interstate	Basic Freeway	Freeway/Freeway Ramp
	Other Freeways	Basic Freeway	Freeway/Freeway Ramp
	Other Principal Arterial	Class I	Arterial/Collector
	Minor Arterial	Class II	Arterial/Collector
	Collector	Class III	Arterial/Collector
	Local	Class IV	Local
Small Urbanized (Population 50,000 to 199,999)	Interstate	Basic Freeway	Freeway/Freeway Ramp
	Other Freeways	Basic Freeway	Freeway/Freeway Ramp
	Other Principal Arterial	Class II	Arterial/Collector
	Minor Arterial	Class II	Arterial/Collector
	Collector	Class III	Arterial/Collector
	Local	Class IV	Local
Large Urbanized (Population 200,000 or more)	Interstate	Basic Freeway	Freeway/Freeway Ramp
	Other Freeways	Basic Freeway	Freeway/Freeway Ramp
	Other Principal Arterial	Class III	Arterial/Collector
	Minor Arterial	Class III	Arterial/Collector
	Collector	Class III	Arterial/Collector
	Local	Class IV	Local

The HCM approach for computing vehicle speed on a link is to divide the link length (in mi) by the link travel time (in h). The link travel time in turn is a function of the link travel time at the free-flow link speed; the zero-flow control delay at signalized intersection; the expected duration of demand; the link volume to capacity ratio (V/C); a calibration factor; and the link length as provided in Equation 30-5 of the Highway Capacity Manual (TRB, 2000). An alternative computation of the link travel time is provided in Equation C30-1 of the Highway Capacity Manual, based on the BPR curve. In this equation, the link travel time is a function of the link travel time at the free-flow link speed; the link V/C; and BPR parameters. The TTI method is to characterize vehicle link speed as the harmonic mean of two components: the free-flow speed and delay speed (NHI, 2003; Claggett, 2006a). The delay speed is a function of the link V/C above a minimum amount of delay.

Congested vehicle speeds were computed using the three methods for conditions representative of the HPMS small urbanized area type for interstate and other principal arterial functional classifications. Speeds were computed for volume-to-capacity ratios ranging from 0.05 to 1.25 in increments of 0.05. Emission factors provided by the MOBILE6.2 model are for an average trip, the basis of which is the Federal Test Procedure adopted to simulate a typical urban trip of 7.5 mi in length. Consequently, a link length of 7.5 mi was used in the speed computations. The other assumptions adopted are provided in Table 2.

Emission Factor Forecasting

The EPA's MOBILE6.2 model (dated 24-Sep-2003) was used to forecast episodic emission factors of carbon monoxide (CO) and the precursors to ozone formation (volatile organic compounds, VOC and nitrogen oxides, NOx); as well as, annual average emission factors of selected mobile source air toxic (MSAT) compounds (benzene and diesel particulate matter, DPM). MOBILE6.2 was run using EPA's corrected external file for computing particulate matter emission factors (dated March 2006). FHWA's Easy Mobile Inventory Tool (EMIT) (Claggett, 2006b) was employed as an interface to MOBILE6.2 to facilitate the creation of speed look-up tables.

The evaluation months and temperatures utilized in the analysis depend on the pollutant evaluated. Typically, highest CO levels are observed in the wintertime and highest ozone levels are observed in the summertime. The appropriate context for evaluating MSAT compounds is the long term – an annual basis at a minimum. For CO, the MOBILE6.2 model was run for January assuming the median of the normal daily minimum and maximum temperatures measured in the U.S. during the coldest month of the year (which happens to be January) – 23.5 °F and 41.0 °F, respectively. For the precursors to ozone formation (VOC and NOx), an evaluation month of July and the median of the normal daily minimum and maximum temperatures measured in the U.S. during the warmest month of the year (which happens to be July) were used – 64.0 °F and 86.1 °F, respectively. For the selected MSAT compounds (benzene and DPM), seasonal emission factors were developed as the basis for

Table 2. Speed calculation parameters.

Technique	Parameter	Interstate	Other Principal Arterial
HCM Approach	Free-flow speed, mph [a]	65	40
	Link length, mi	7.5	7.5
	Calibration parameter, h^2/mi^2 [a]	1.48E-05	5.02E-04
	Signals per mile [a]	0	0.6
	Average effective green time per cycle for all signals on link [b]	0	0.45
	Average cycle length for all signals on link, s [c]	0	100
	Speed at capacity, mph [a]	52	18
BPR Formula	Free-flow speed, mph [a]	65	40
	Link length, mi	7.5	7.5
	a [d]	0.25	0.38
	b [d]	9.0	5.0
TTI Method	Free-flow speed, mph [a]	65	40
	A [e]	0.015	0.050
	B [e]	4.2	3.9
	M [e]	5.0	6.0

[a] Highway Capacity Manual (TRB, 2000), Exhibit 30-4
[b] Highway Capacity Manual (TRB, 2000), page 10-23
[c] Highway Capacity Manual (TRB, 2000), Exhibit 10-16
[d] Highway Capacity Manual (TRB, 2000), Exhibit C30-2
[e] National Highway Institute (2003)

computing annual averages. For the January through March winter season, the evaluation month is January; the calendar year is 2010; and the median of the average daily minimum and maximum temperatures measured in the U.S. are 27.8 °F and 47.0 °F, respectively. For the April through September summer season, the evaluation month is July; the calendar year is 2010; and the median of the average daily minimum and maximum temperatures measured in the U.S. are 55.5 °F and 77.7 °F, respectively. For the October through December winter season, the evaluation month is January; the calendar year is 2011; and the median of the average daily minimum and maximum temperatures measured in the U.S. are 35.2 °F and 54.3 °F, respectively.

Emission factors of DPM include the organic carbon, elemental carbon, and sulfate portions of diesel exhausts for the maximum particle size cutoff of 10 μm that can be considered in the MOBILE6.2 model. The diesel fuel sulfur level used is consistent with the 49-state average value reflecting more stringent federal controls (i.e., 11 ppm for 2010). The analysis was based on the 2007/2020 30 ppm fuel specifications for the northeastern states with no reformulated gasoline program (U.S. EPA, 2002b). Emission reductions that may be realized from a local inspection/ maintenance program or anti-tampering program were not taken into account.

The emissions analysis was conducted by accounting for the vehicle emission types specific to the operation on a link, e.g., the exhaust running and evaporative running loss emissions components. The evaluation relied on the VMT fractions by vehicle type calculated internally by the MOBILE6.2 model based on the specified calendar year of evaluation and from national average default data for: 1) vehicle population for the 16 composite vehicle classes; 2) vehicle registration by age distribution; 3) diesel fractions; and 4) mileage accumulation rates. A single distribution is computed to represent the fraction of total highway VMT accumulated by each of 16 combined vehicle types for a day. Speed look-up tables of emission factors were constructed by employing the AVERAGE SPEED command and specifying vehicle speeds from 5 mph to 65 mph in 1 mph increments for freeway and arterial roadway scenarios. The key assumptions employed in the MOBILE6.2 modeling are summarized in Table 3.

Table 3. Emission factor calculation parameters.

Parameter	Season		
	Winter	Summer	Winter
Pollutant: CO			
Calendar Year	2010	–	–
Evaluation Month	January	–	–
Min/Max Temp (°F)	23.5 / 41.0	–	–
Reid Vapor Pressure (psi)	13.2	–	–
Pollutant: VOC / NOx			
Calendar Year	–	2010	–
Evaluation Month	–	July	–
Min/Max Temp (°F)	–	64.0 / 86.1	–
Reid Vapor Pressure (psi)	–	8.6	–
Pollutant: BENZ / DPM			
Calendar Year	2010	2010	2011
Evaluation Month	January	July	January
Min/Max Temp (°F)	27.8 / 47.0	55.5 / 77.7	35.2 / 54.3
Reid Vapor Pressure (psi)	13.2	8.6	13.2
Aromatic Content (%)	23.1	27.1	23.1
Olefin Content (%)	14.1	9.9	14.1
Benzene Content (%)	0.73	1.03	0.73
E200 (%) [a]	51.8	44.4	51.8
E300 (%) [b]	83.3	81.1	83.3
MTBE Content (%)	0.6	3.4	0.6
Diesel Sulfur (ppm)	11.0	11.0	11.0

[a] Vapor percent of gasoline at 200 °F
[b] Vapor percent of gasoline at 300 °F

Results

Congested vehicle speeds characteristic of HPMS small urbanized area interstates and other principal arterials are presented in Figures 1 and 2, respectively, as a function of changes in the volume-to-capacity ratio. A graph of the freeway mobile source emission factors contained in the speed look-up tables are provided in Figure 3. Emission factors for arterials exhibit nearly identical speed profiles. Figures 4 through 9 relate the information on traffic congestion and emission factors to offer insights into how vehicle emissions are expected to vary from an over-capacity to under-capacity operating condition and vice-versa. Emission factors for CO, VOC/NOx, and benzene/DPM are given for a wide range of volume-to-capacity ratios for small urbanized area interstates and other principal arterials. To facilitate a ready comparison of the pollutant emission factors, the appropriate figures are also grouped on a single page for small urbanized area interstates (refer to Attachment 1) and other principal arterials (refer to Attachment 2).

What's notable in the comparison of the different vehicle speed estimating methodologies is how well the HCM approach and BPR formula track each other for small urbanized area interstates – at least for the calculation parameter set adopted. Different results are likely if different assumptions are made. For small urbanized area other principal arterials, the HCM approach and TTI method give similar results for $V/C \leq 0.75$. Substantially lower speeds are predicted with the TTI method for overcapacity conditions on small urbanized area interstates and other principal arterials compared to the HCM approach and BPR formula. Next comes the question of which speed forecasting technique provides reasonable estimates for highly congested traffic conditions. One indicator is to contrast speed estimates to EPA's test cycles used for developing speed correction factors in MOBILE6.2. Table 4 provides an evaluation of vehicle speeds obtained from each of the forecasting techniques for overcapacity conditions to the average speeds of EPA's highly congested test cycles for freeways and arterials. The EPA's level of service (LOS) G for freeways is interpreted as an extreme breakdown or forced-flow condition. Forecasts from the TTI method compare favorably to EPA's congested speed test cycles.

The profile of MOBILE6.2 emission factors with speed is typified by high emissions at slow vehicle speeds for all pollutants with the exception of particulate matter (refer to Figure 3). Consequently, there is no sensitivity to changes in speed for DPM. For CO and NOx, emission factors increase again for speeds above 30 mph, while for VOC and benzene, emissions decrease throughout the entire range from slow to fast speeds.

The implication of this for congestion mitigation projects is the prospect of diminishing emissions by improving persistently slow vehicle speeds. But since some emission factors also increase with increasing speeds above roughly 30 mph, a mixed result can occur for freeway congestion mitigation projects. According to the HCM approach and BPR formula, speeds for exceedingly overcapacity conditions on

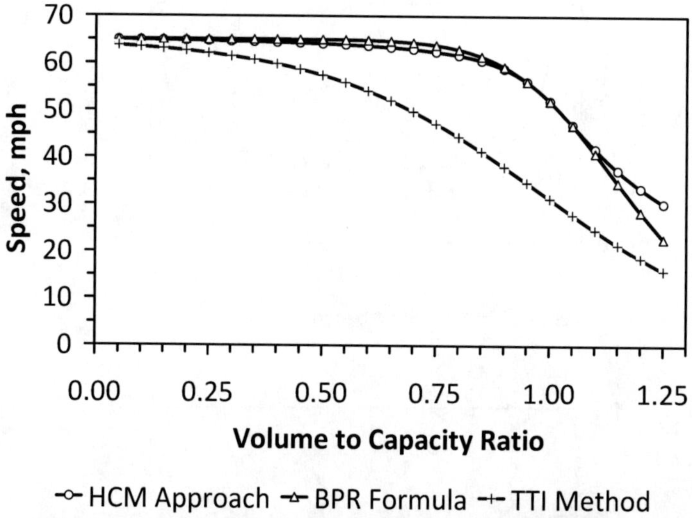

Figure 1. Forecast travel speeds for a small urbanized area interstate.

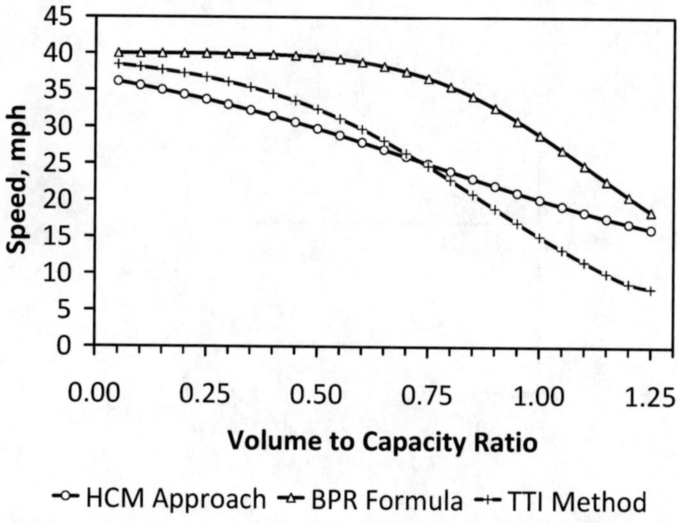

Figure 2. Forecast travel speeds for a small urbanized area other principal arterial.

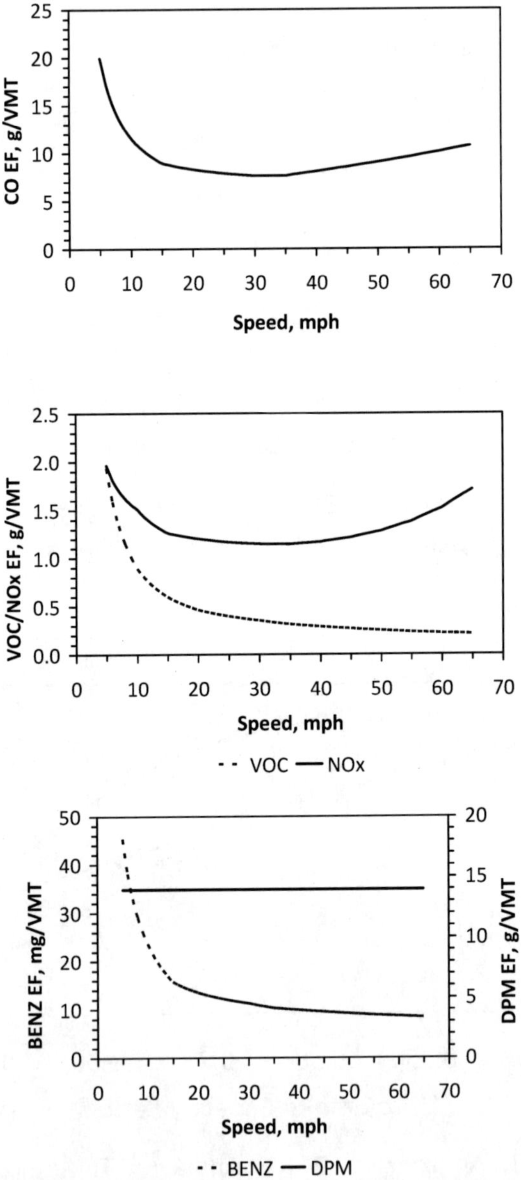

Figure 3. Forecast MOBILE6.2 emission factors for a small urbanized area interstate.

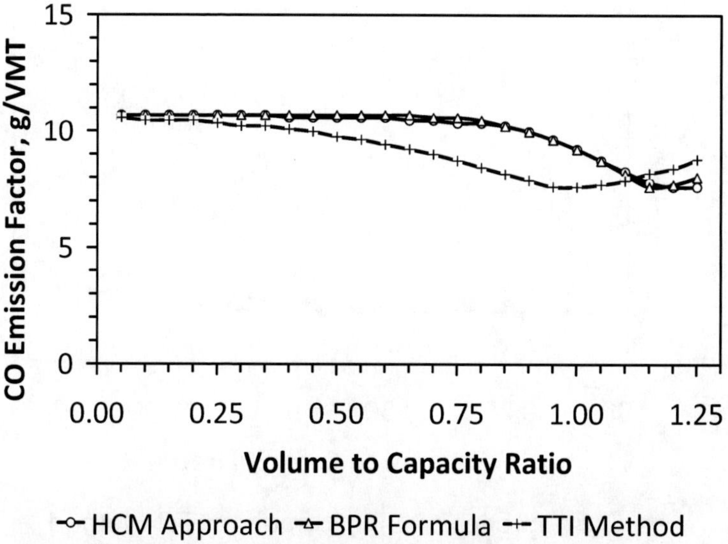

Figure 4. CO emission factors versus level of congestion for a small urbanized area interstate.

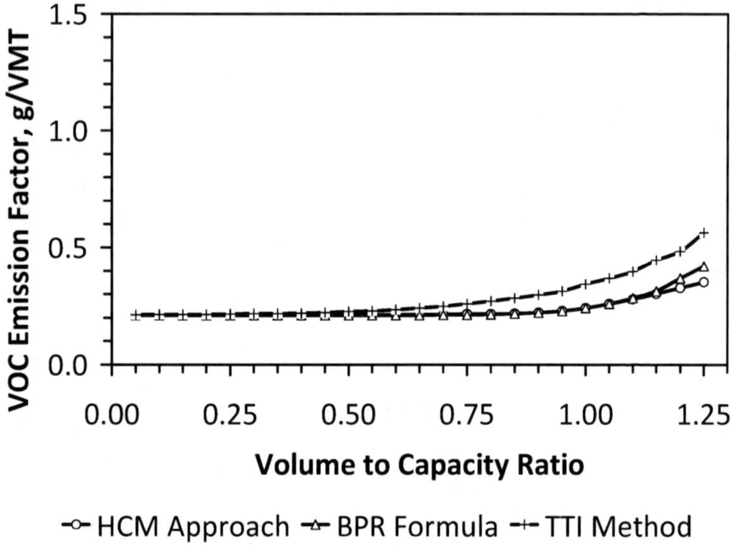

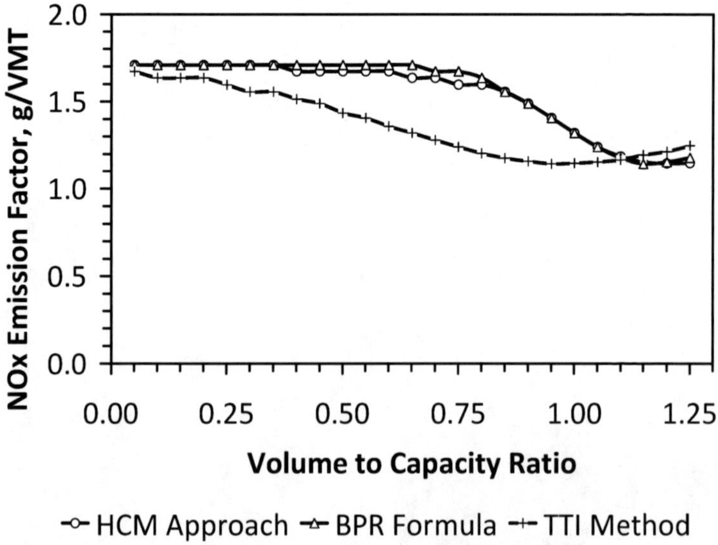

Figure 5. VOC and NOx emission factors versus level of congestion for a small urbanized area interstate.

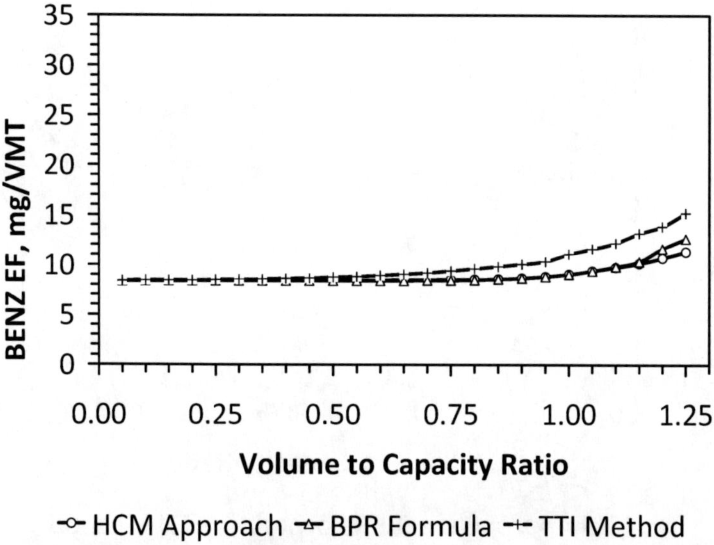

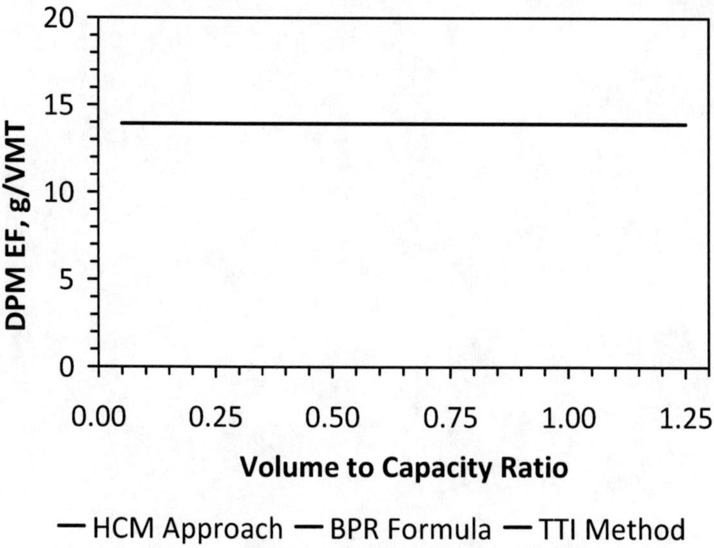

Figure 6. Benzene and DPM emission factors versus level of congestion for a small urbanized area interstate.

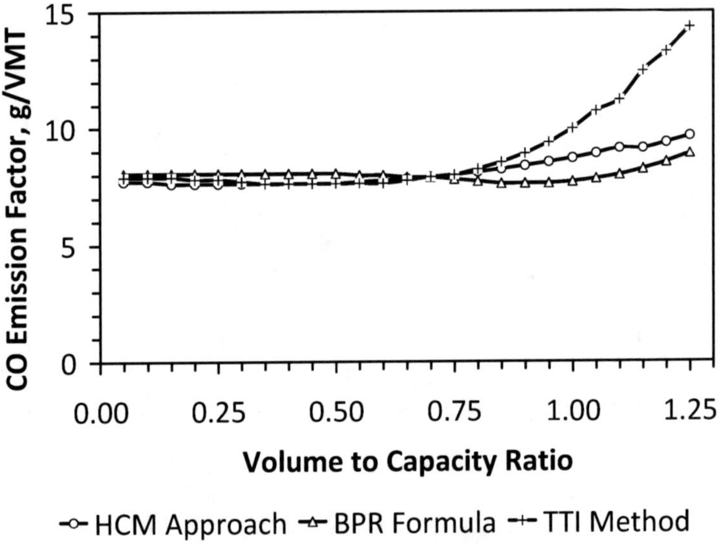

Figure 7. CO emission factors versus level of congestion for a small urbanized area other principal arterial.

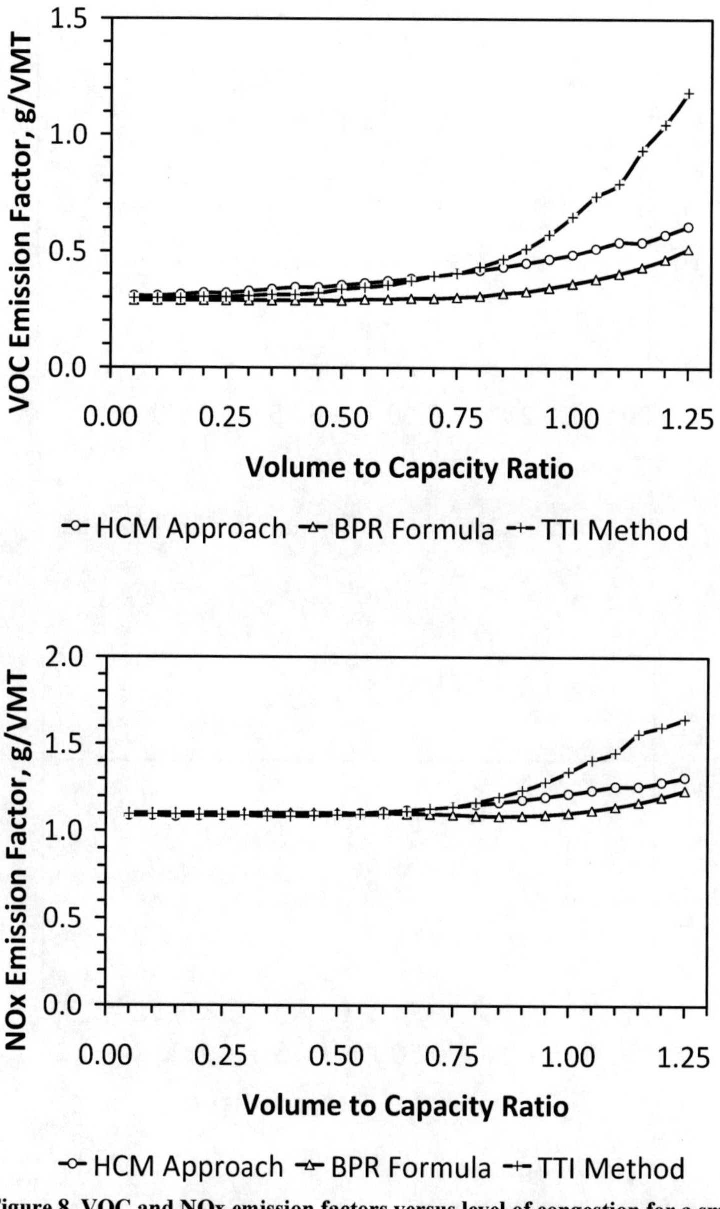

Figure 8. VOC and NOx emission factors versus level of congestion for a small urbanized area other principal arterial.

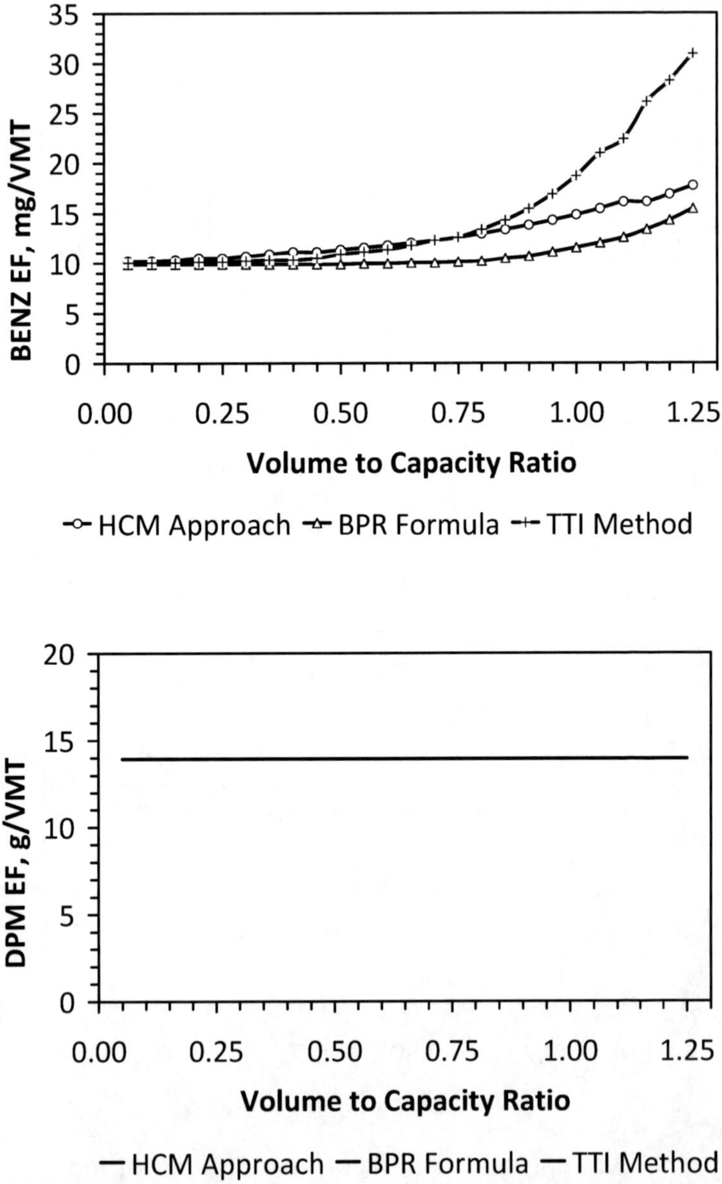

Figure 9. Benzene and DPM emission factors versus level of congestion for a small urbanized area other principal arterial.

Table 4. Reasonableness check of congested vehicle speeds.

Interstate				
Speed Methodology	Small Urbanized		Large Urbanized	
	V/C = 1.0	V/C = 1.25	V/C = 1.0	V/C = 1.25
HCM 2000	52	30	51	29
BPR Formula	52	23	51	27
TTI Method	31	16	30	16
EPA – LOS F	19			
EPA – LOS G	13			

Other Principal Arterial				
Speed Methodology	Small Urbanized		Large Urbanized	
	V/C = 1.0	V/C = 1.25	V/C = 1.0	V/C = 1.25
HCM 2000	20	16	13	11
BPR Formula	29	19	18	9
TTI Method	15	8	14	8
EPA – LOS E-F	12			

small urbanized area interstates are a minimum of 28 to 31 mph – right around the infection point on the emission factor versus speed curves for CO and NOx. Much slower speeds are forecast with the TTI method (i.e., a minimum of 16 mph). Thus, as indicated by the HCM approach or BPR formula, relieving this congested condition will increase CO and NOx emissions. This is somewhat contradicted by the TTI method, whereby CO and NOx emissions are predicted to decrease if congestion is improved to a volume-to-capacity ratios approaching 1.0. Traffic flow improvements beyond this point will increase CO and NOx emissions.

Any congestion relief will reduce CO, VOC/NOx, and benzene emissions for small urbanized area other principal arterials. The degree of the reduction depends on the speed estimating approach with the TTI method projecting the largest. Because of limitations inherent in the MOBILE6.2 model, no emission changes for DPM (or particulate matter) are affected by mitigating congestion on roadways.

Conclusions

Mitigating persistent congestion on highways reduces emissions from motor vehicles on a unit vehicle-mile of travel basis. In forecasting potential emission benefits, the degree to which emissions are expected to decrease depends largely on the speed estimating methodology implemented and the manner in which it is employed. For the assumptions used in this investigation, the TTI method predicted the greatest congestion intensity and consequently, the largest emission benefits associated with congestion relief. For small urbanized area interstates, the HCM approach and the BPR formula produced similar results. For small urbanized area other principal arterials, the BPR formula predicted the least amount of congestion and the smallest benefits associated with congestion relief. Admittedly, different result may be obtained with different assumptions.

However, any real world analysis should apply the speed estimating technique that most aptly reflects local conditions. A selection should be made objectively, adopting a transparent process of investigation, testing via reasonableness checks, documentation, and review. Subjective criteria, such as which methodology affords the largest or smallest benefit associated with congestion relief, should not be considered. It is critical to determine the speed estimating methodology most representative of vehicle congestion in a locale, especially for situations when travel demand exceeds capacity.

Disclaimer

The content of this paper solely represents the work of the authors and does not reflect the policy, guidance, or procedures adopted or recommended by the U.S. Department of Transportation and the Federal Highway Administration. This document is disseminated in the interest of information exchange and the U.S. Government assumes no liability for use of the information. This paper does not constitute a standard specification or regulation.

References

Claggett, M. and T. L. Miller (2006), "Variability of Mobile Source Air Toxic Emission Factors using MOBILE6.2". In *Proceedings of the 2006 TRB 85th Annual Meeting*. CD-ROM. Transportation Research Board, National Research Council, Washington, DC.

Claggett, M. and J. Houk (2006), "An Emission Inventory Tool for Transportation Conformity Applications in Rural and Small Urban Areas". *Proceedings of the 2006 TRB 85th Annual Meeting*. CD-ROM. Transportation Research Board, National Research Council, Washington, DC.

Dowling Associates (1997), "Travel Model Speed Estimation and Post Processing Methods for Air Quality Analysis". DOT-T-98-5. Federal Highway Administration.

National Cooperative Highway Research Program (NCHRP) (1997), "Planning Techniques to Estimate Speeds and Service Volumes for Planning Applications". *NCHRP Report 387*. Transportation Research Board.

National Highway Institute (NHI) (2003), "Estimating Regional Mobile Source Emissions". *NHI Course Number 152071*. U.S. Department of Transportation, Federal Highway Administration.

Tang, T., M. Roberts, and C. Ho (2003), "Sensitivity Analysis of MOBILE6 Motor Vehicle Emission Factor Model". ww.fhwa.dot.gov/resourcecenter/teams/airquality/mobile6_v2.pdf (>February, 2007.

Tang, T., M. Claggett, J. Byun, M. Roberts, J. Granell, and D. Aspy (2005), "MOBILE6.2 Modeling of Exhaust Air Toxic Emission Factors: Trend and Sensitivity Analysis". *Transportation Research Record: Journal of the Transportation Research Board*, TRB, National Research Council, Washington, D.C., 1941, 99-106.

Transportation Research Board (TRB) (2000), "Highway Capacity Manual 2000". Committee on Highway Capacity and Quality of Service.

U.S. Environmental Protection Agency (EPA) (2001), "Development of Methodology for Estimating VMT Weighting by Facility Type". *Publication EPA420-R-01-009. M6.SPD.003*. Office of Transportation and Air Quality.

U.S. Environmental Protection Agency (EPA) (2002), "Sensitivity Analysis of MOBILE6.0". *Publication EPA420-R-02-035*. Office of Transportation and Air Quality.

U.S. Environmental Protection Agency (EPA) (2002), "Technical Description of the Toxics Module for MOBILE6.2 and Guidance on Its Use for Emission Inventory Preparation". *Publication EPA420-R-02-029*. Office of Transportation and Air Quality.

U.S. Environmental Protection Agency (EPA) (2003), "User's Guide to MOBILE6.1 and MOBILE6.2, Mobile Source Emission Factor Model". *Publication EPA420-R-03-010*. Office of Transportation and Air Quality.

U.S. Environmental Protection Agency (EPA) (2004), "Technical Guidance on the Use of MOBILE6.2 for Emission Inventory Preparation". *Publication EPA420-R-04-013*. Office of Transportation and Air Quality, August 2004.

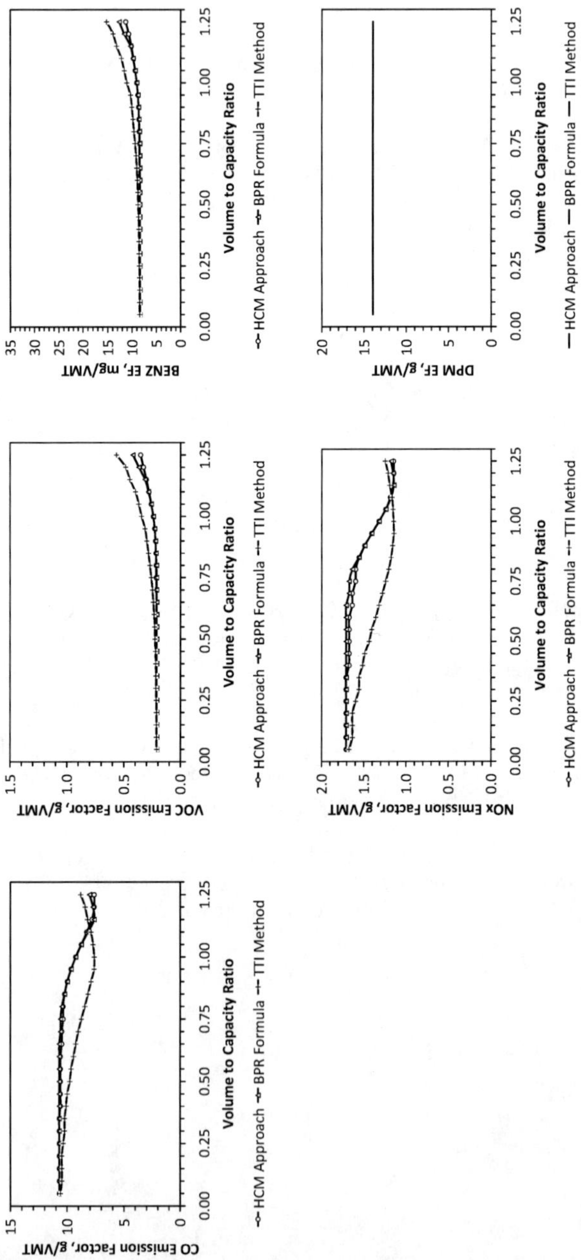

Attachment A-1.
Emission factors versus level of congestion for a small urbanized area interstate.

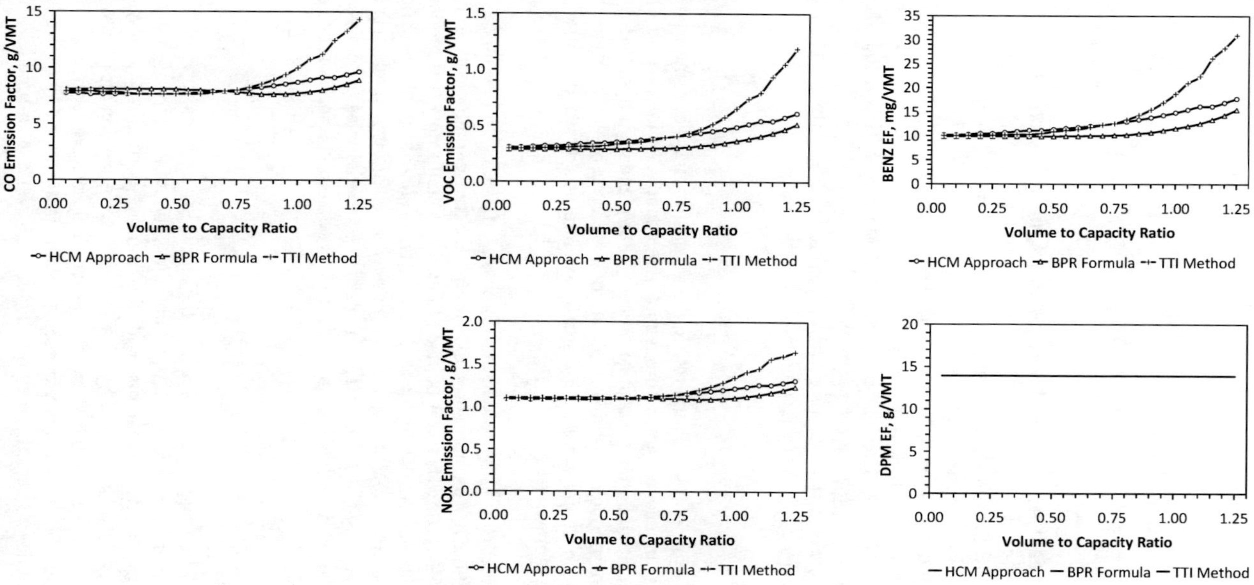

Attachment A-2.
Emission factors versus level of congestion for a small urbanized area other principal arterial.

VMT Forecasting Alternatives for Air Quality Analysis

D. Szekeres[1], N. Koppula[2], and J. Frazier[3]

[1]Michael Baker Jr., Inc., 4431 North Front Street, Harrisburg, PA 17110, PH (717) 221-2019; email: dszekeres@mbakercorp.com
[2]Michael Baker Jr., Inc., 1304 Concourse Drive, Linthicum, MD 21090, PH (410) 689-3461; email: nkoppula@mbakercorp.com
[3]Michael Baker Jr., Inc., 1304 Concourse Drive, Linthicum, MD 21090, PH (410) 689-3464; email: jfrazier@mbakercorp.com

Abstract

Traffic growth forecasting plays a pivotal role in the air quality analyses conducted in nonattainment and maintenance areas throughout the country. However, traffic forecasting is a difficult and often inaccurate process. For planning purposes, many agencies have begun producing estimates of future development and traffic with ranges. This approach demonstrates the disparity of forecasts based on different sources of data and assumptions.

There are opportunities for using alternative forecasts of vehicle miles of travel (VMT) within the air quality planning process. Within the transportation conformity process, alternative VMT forecasts can be used to evaluate conformity assumptions or to adjust travel demand model forecasts. Within the statewide implementation plan (SIP) process, alternative VMT forecasts can play an even greater role. They may be used to develop consistent statewide forecasting methodologies for emission inventories or may be used to develop and assess risks when preparing conformity emission budgets.

Introduction

Traffic growth forecasting plays a pivotal role in the air quality analyses conducted in nonattainment and maintenance areas throughout the country. Analyses include transportation conformity determinations and the mobile portion of statewide implementation plans (SIPs) as required by the Clean Air Act Amendments (CAAA). Various methodologies are available to perform such analyses; however, they are often dependent on the data, tools, and resources available to each area. Urban areas often have more complex forecasting methods including travel demand models, but many of those tools have not replicated the growth of vehicle miles of travel (VMT) reported through historical Highway Performance Monitoring System (HPMS) data. Small urban and rural areas often depend on less sophisticated methods, typically based on the historical trends of traffic data used for reporting HPMS.

VMT forecasting is a difficult and often inaccurate process. VMT forecasts are impacted by many factors including socio-economic and demographic growth, changes in the cost of travel, urban sprawl, technological innovation, social change, and legislative factors. More faith is often held for VMT forecasts conducted for larger geographical scopes (e.g. statewide) or for shorter rather than longer horizons. A recent emphasis has been focused on addressing these issues within the planning process, especially during the development of the long-range plans by Metropolitan Planning Organizations (MPOs). Many agencies have begun producing estimates of future development and traffic with ranges. This approach demonstrates the disparity of forecasts based on different sources of data and assumptions. The approach has led to the concept of scenario planning, a process that considers the various factors that shape the future, and then facilitates consensus on how to deal with the growth and accommodate future transportation needs.

The application of scenario planning concepts to the federally-mandated air quality analyses may serve to be even more difficult. Producing a range of potential emission estimates does not connect with current air quality regulations that require conformity tests and often require comparisons to emission budgets developed by other agencies. This paper addresses how potential VMT forecasting alternatives may be used within the air quality analysis process. The paper highlights the tools and methodologies used in Pennsylvania and highlights their role within the state's air quality planning process.

Alternatives for Forecasting VMT

Several tools and methodologies are used to forecast VMT for air quality purposes. These range from sophisticated travel demand models in large urban areas to simple historical trending of HPMS VMT in smaller urban or rural areas. In Pennsylvania, a range of VMT forecasting methodologies have been used and are available throughout the state.

Travel Demand Models

Since 1990, the CAAA and accompanying regulations have significantly increased the role of regional travel forecasting models in the estimation of pollutant emission levels in non-attainment and maintenance areas. Traffic model applications include the calculation of emission estimates for TIP and long-range plan conformity analyses.

Travel demand models provide greater sensitivity to changes in transportation investments or policies as compared to many manual calculation procedures.

However, the forecasts produced by travel models often have significant inaccuracies. Addressing long-term changes to social, legislative, or technological factors are often outside the scope of most travel demand models. Many regional models, especially in smaller urban areas, lack robust demographic and land use forecasting components and feedback mechanisms. As a result, the models often lack the methodologies to account for mode and route shifts associated with induced travel. In smaller regions where external travel is a significant portion of total VMT, the travel model may not account for significant changes to thru or truck travel within the region. All of these issues can have significant impacts on a travel demand model's estimate of VMT and congestion.

In Pennsylvania, multiple regions use travel demand models for air quality conformity. Pittsburgh and Philadelphia are the two largest regional travel models each containing land use and demographic allocation components. Six other smaller urban travel models are also used within the state. These regional models rely on MPO-prepared demographic forecasts and allocations as input to the travel demand model. Practices for producing forecasts vary by MPO and often include the use of population and employment forecasts from other agencies or private sources. In the future, a statewide model being developed by the Pennsylvania Department of Transportation (PennDOT) may also be used to provide external travel and regional truck and freight forecast data as input to the regional travel models.

Historical Trending of HPMS

For areas without travel demand models, estimates of VMT based on FHWA's HPMS count data are often used to generate regional emissions estimates for each analysis year. However, HPMS is based on collected count data and does not provide estimates of future travel. An option used by a number of regions, including past efforts for non-travel model areas in Pennsylvania, is to estimate future roadway traffic volumes based on historic trends of HPMS VMT. In Pennsylvania, growth factors have been developed for each county and functional class grouping and applied to the current year roadway volumes in each county.

There are significant limitations and inaccuracies with this method. Without a travel demand model, traffic diversions are not accounted for and the forecasting methodology does not provide for future demographic changes that may occur in the region. However, in Pennsylvania, these trended forecasts have often provided better short term traffic projections than the regional travel demand models, as compared to reported HPMS VMT totals.

Other Statistical Methods

Several states including Pennsylvania and Kentucky have recently developed alternative statistical methodologies to estimate future regional VMT. In 2005, PennDOT undertook a study to develop growth rate forecasts for county and functional classes across the state. The forecasts were established as a potential source of information for traffic, corridor, or air quality studies. As part of that study, a statewide traffic growth forecasting system was developed that incorporates historic traffic data and socioeconomic forecasts (PennDOT, March 2005). The forecasting system contains the following improvements over simple trending analyses as used in the past:

- Strong statistical basis and consistent with state of the art
- Thorough documentation of approach vs. alternatives
- Increased robustness through the inclusion of both county-level historical traffic trends and county-based demographic projections from an independent data sources
- Incorporates socio-economic data (households, mean household income) and a relative measure of transportation capacity (lane miles per capita)
- Expedient and inexpensive update process

The resulting forecasting system includes the development of VMT forecasts and growth rates for four functional classifications in each Pennsylvania county: urban interstate, urban non-interstate, rural interstate, and rural non-interstate. The forecasts use statistical relationships based on historic HPMS VMT trends and future county socioeconomic projections from Woods and Poole Economics, Inc. Several sets of statistical models provide a range of growth rates.

Application of Alternative VMT Forecasts to Air Quality Analyses

The concepts of scenario planning have been applied to the long-range plan development process for a number of regions across the country. These concepts recognize that a single set of future land use and travel patterns cannot be accurately predicted. As a result, a range of scenarios are often developed and shared with the public, which are then used to develop a future plan and vision for each region. Although the scenario planning concepts may not be directly applicable to air quality regulations, there are opportunities for using alternative VMT forecasts within the air quality planning process. This section discusses several approaches as related to transportation conformity and SIP emissions inventory development.

Application of Alternative VMT Forecasts to Transportation Conformity

The CAAA requires transportation planners in nonattainment and maintenance areas to consider the air quality impacts of their proposed plans, programs, and projects. These activities, if subject to federal involvement, must be shown to conform based on the requirements for each pollutant.

The application and examination of alternative VMT growth forecasts, or ranges, may be limited for transportation conformity. More flexibility in evaluating and choosing appropriate VMT forecasts may occur for areas without emission budgets that rely on the interim conformity emission tests. For areas with transportation emission budgets, alternative VMT forecasts must be evaluated and applied carefully.

Transportation budgets established within a SIP utilize VMT forecasts deemed appropriate at that time. Future changes to those assumptions or the use of alternative VMT forecasts could jeopardize conformity determinations for a region.

In Pennsylvania, alternative VMT forecasts have been used for several purposes within the transportation conformity process. They include the following:

Evaluate Conformity VMT Forecast Assumptions

Within each region, available VMT forecasts are examined to determine the most appropriate for transportation conformity. If a travel demand model is available, those areas rely on the model forecasts and the associated input demographic forecasts. Although only one VMT forecast option is used for transportation conformity, other alternatives may be examined to evaluate their potential impact on a positive conformity determination. The analysis and air quality results may also serve as a performance measure to evaluate long-range plan alternatives and for related public involvement activities. Figure 1 illustrates the 2025 plan analysis for the Delaware Valley Regional Planning Commission (DVRPC). The impacts of various forecasts were used to evaluate potential air quality impacts as part of the long-range plan development process.

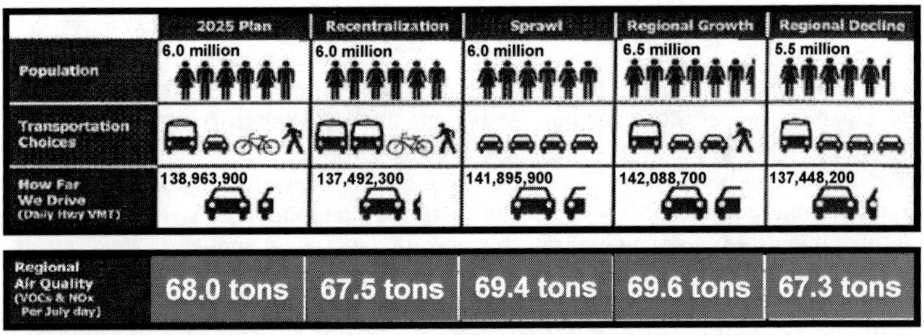

Figure 1: "Regional Analysis of What-If Transportation Scenarios" (DVRPC, July 2003)

Historical HPMS trends and results from Pennsylvania's statewide forecasting study have also been used to evaluate travel demand model growth for smaller urban areas across the state. Large discrepancies in growth assumptions between the alternative sources have initiated review of travel model and land use assumptions to identify the reasons for significant discrepancies. Thus, alternative forecasts serve as an important quality assurance component of the emission analysis process.

Adjust Travel Model Forecasts

For areas using travel demand models, the models are often calibrated and validated to HPMS count data and VMT totals. However, periodic adjustments can also be made to the model outputs, without a formal calibration/validation process, to ensure that the reported values and forecasts are consistent with actual data.

In Pennsylvania, many of the less-robust small urban model outputs are adjusted during a post processing phase to match HPMS VMT on a triennial basis. Although the forecasted growth rates may not change from a previous analysis, the starting point of that growth may be shifted up or down to reflect the current HPMS VMT. Thus, if a 2000 validated travel model under-produces 2005 VMT, as compared to HPMS, a factor is provided to adjust the model 2005 VMT to HPMS, and that factor is carried forth to future years. This process can be repeated as new baseline VMT becomes available from HPMS.

Application of Alternative VMT Forecasts to SIP Mobile Emission Inventories

The SIP contains the regulations and control strategies for meeting clean air standards and associated federal CAAA requirements. The SIP's emission estimates are used to determine whether attainment will be reached or maintained, control strategies that may be needed to reach the goals, and the emission budgets that will be used for future transportation conformity runs. An important feature of SIPs, as compared to transportation conformity, is that they are updated less frequently. Per SAFETEA-LU legislation, transportation conformity must be updated every 4 years; however, in many states, conformity is updated even more often due to changes or amendments to the TIP. SIPs are updated on a less timely fashion and require a significant effort to prepare. In many cases, SIPs are only updated due to inadequate conformity budgets or new federal regulations or standards. This creates an important and complex issue when considering emission budgets since they may have to serve transportation conformity for many years. With the latest planning requirement for transportation conformity, land use and VMT forecasts may evolve significantly from those used for past SIP calculations.

In Pennsylvania, the evaluation and application of alternative VMT forecasts have played a key role in the development of mobile emission inventories. A review of key applications includes:

Develop Consistent Statewide Forecasting Methodology

Within Pennsylvania there are 23 urban and rural planning organizations. Nine of the planning organizations have travel demand models, which are maintained at varying levels, and each have different methods for preparing the land use and demographic forecasts into the model. Over the past decade, many of the travel models have not predicted the trends that have occurred in the reported HPMS VMT. Further complicating matters have been model enhancements, changing land use forecasts, and other changing model parameters. Areas without travel demand models have relied on state count databases and VMT growth rates for air quality analyses.

The state environmental agency, DOT, and consultant staff decided to develop a consistent statewide approach for producing VMT forecasts and emissions to support statewide implementation plans, rate-of-progress plans, maintenance plans, and EPA's periodic national emissions inventory (NEI). The approach uses state count data updated on a triennial basis to reflect base-year conditions. VMT is forecasted using growth rates for each county and functional class combination. The growth rates are determined by examining multiple sources including:

- Travel models – regional, statewide when available
- Historical HPMS trends
- Other forecasting models (e.g. PA VMT forecasting methodology – 3 scenarios of growth)
- MPO scenario planning efforts

Each VMT forecasting methodology is evaluated and compared. Figure 2 illustrates a typical comparison of growth rates that is produced. The choice of appropriate growth rates is an important decision. The decision reflects the need to develop adequate transportation conformity budgets (as discussed in next section), to determine if attainment can be demonstrated, to identify what control strategies may be needed to offset growth, and to address and consider the potential for worst-case growth.

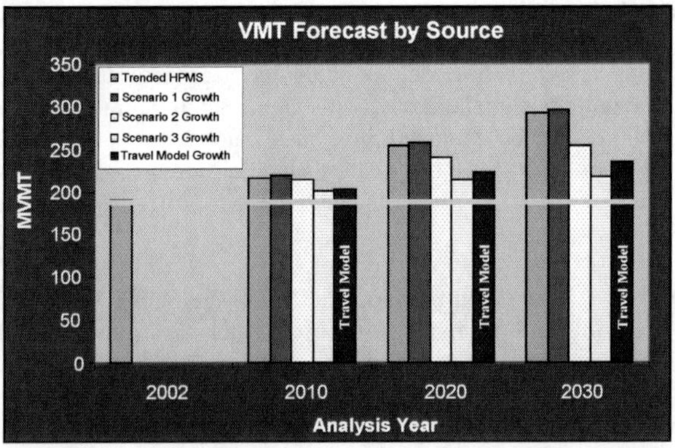

Figure 2: Alternative Growth Sources in Pennsylvania

Risk Management for Conformity Budgets

In Pennsylvania, the development of SIP motor vehicle emission budgets involves a complete review of available VMT forecasts and an assessment of potential impacts on future conformity analyses. The budgets should meet the SIP purpose and provide for future transportation goals. As a result, the budgets are developed to:

- Anticipate likely future developments
- Allow for uncertainty
- Safeguard conformity for extended time frame

Many states are now realizing the importance of having state DOT and MPO staff actively involved in and aware of SIP development activities. Having attainable emission budgets is critical to meeting conformity requirements. The emission budgets based on VMT forecasts that do not provide adequate VMT growth can create significant conformity problems. VMT forecasts are reviewed to identify worst-case growth conditions. Such growth can be used to determine future budgets directly or in the development of allowable safety margins. In Pennsylvania, the creation of motor vehicle budgets have utilized the most conservative estimates of VMT growth from the available scenarios, in balance with the need to demonstrate air quality goals and control strategies to implement.

Conclusion

Understanding and accounting for the uncertainty of VMT forecasts can play an important role in federally-mandated air quality analyses. Addressing alternative VMT forecasting methodologies and results can be used to evaluate and adjust current forecasting practices and as an important quality assurance tool. In addition, accounting for potential growth is a key issue in developing appropriate transportation conformity budgets as part of the SIP emission inventory process.

References

Delaware Valley Regional Planning Commission (DVRPC). (2003). "Regional Analysis of What-If Transportation Scenarios." http://www.dvrpc.org/LongRange.htm

Pennsylvania Department of Transportation (PennDOT). (2005). "Statistical Evaluation of Projected Traffic Growth, Traffic Growth Forecasting System: Final Report." *prepared by Michael Baker Jr., Inc.*

Impact of Metrorail Stations in Washington, DC on Land Use and Development: Shady Grove Metro Station Case Study

V. O. Davis[1]

[1]Malcolm Pirnie Inc, 3101 Wilson Blvd, Suite 550, Arlington, VA, 22201; PH (703) 465-4233; FAX (703) 351-1305; email: vdavis@pirnie.com

Abstract

The Metrorail (Metro) in Washington, DC commenced operation in 1979. The red line extension to the Shady Grove Metro Station in Montgomery County, Maryland opened in 1984. Seven years prior to the opening of the Shady Grove transit station, the Maryland National Capital Parks and Planning Commission (MNCPPC) adopted and approved a Shady Grove Sector Plan to provide guidance for development around the transit station and to reduce the potential negative effects of a Metro station complex. In 1990, the MNCPPC adopted and approved another sector plan that recognized that transit should form the basis for land use and zoning recommendations. In the twenty-two years the station has been in operation, there have been significant land use changes within a one-mile radius of the station. The land use has transformed from predominately agriculture and forest to a mixture of low and medium-density residential, institutional, and industrial, with small pockets of commercial and open space.

Introduction

Historically, transportation has been a foundation for the growth and land use patterns of cities. Public transit has had a more profound influence on the location and intensity of growth within its principal service corridors of a metropolitan area (Huag, 1995). To control and manage the imminent development in the service corridors and around fixed transit stations, local jurisdictions have used a variety of zoning and planning tools (TCRP, 2003). Robert Cervero completed a comprehensive study in 1995 on the impact of the Bay Area Rapid Transit (BART) on land use and development conducted twenty years after BART began operations. This paper attempts to replicate this study using the Shady Grove Metro Station as the point of emphasis.

Cervero analyzed the land use changes near transit facilities to determine the role BART played in emergence of new development. Since land around BART was consciously zoned for higher density, the analysis showed single-family housing increased at a negligible rate. The new housing structures within 1/4 mile of the transit stations were multi-family dwellings. Cervero found BART only influenced land uses changes where there were supportive conditions like incentive zoning, local citizen support, and a buoyant local economy. BART had little impact when those supportive conditions did not exist (Cervero, 1995). Cervero's findings underscore the important role local jurisdictions have in the intensity and rate of development around transit stations.

The Shady Grove Metro Station was chosen for analysis for several reasons. One, it is located approximately twenty-four miles from downtown Washington, DC, which suggests it is less likely to be influenced by urban development as the case would be in the central business district. Secondly, due to the presence of major county facilities, it has been identified as an area of regional importance. Thirdly, Shady Grove Metro station has been in operation for over twenty years, which is an adequate amount of time to observe the land use changes. Lastly, the local government played an instrumental role in guiding development around the station.

Methodology

The analysis was broken into four steps: (1) selection of analysis area round the Shady Grove Metro station, (2) analysis of demographic and economic changes, (3) analysis of land use changes, and (4) analysis of land use policies.

(1) Selection of an analysis area around the Shady Grove Metro Station

In order to determine the radii for analysis, other studies were examined. The BART study analyzed changes within ½-mile radius of the BART stations (Cevero, 1995). In 1984 and 1992, the Metropolitan Washington Council of Governments (MWCOG) conducted two studies to examine employment trends before and after Metrorail. For these studies, MWCOG utilized a 0.7-mile radius around the metro stations, which they define as a 15-minute walking travel time. This translates into a travel speed of about 2.8 miles per hour. For this analysis, a 0.5 and 1-mile radii were chosen as the Shady Grove Metro station analysis, which will be referred to as the 'station impact zone' for the remainder of the paper. This translates into approximately an 11 and 21-minute walking travel time.

(2) Analysis of demographic and economic changes

The BART study compared population and employment trends in the BART areas to the non-BART areas (Cevero, 1995). This analysis examines the population and economic trends in the Shady Grove Metro station impact zone to Montgomery County using 1980, 1990, and 2000 U.S. Census data. The 1980 census, which was four years prior to the opening of the Shady Grove Metro Station, represents the pre-metro station conditions. The 1990 and 2000, represent the six and sixteen-year post-metro station conditions. In order to simplify the analysis, the data was analyzed based on the census tracts where any portion of the tract fell inside the station impact zone.

The population analysis examined the change in population and number of housing units. The economic indicators selected for this analysis were median housing value, median monthly rent, and median household income. In order to determine the percent change over the decades, these values were all translated into 2005 dollars using the Consumer Price Index (CPI) conversion factors developed at Oregon State University (Sahr, 2006).

(3) Analysis of land use changes

The Maryland Department of Planning (MDP) has a data depository for geographic information systems (GIS) data. The land use/land cover shape file was available for 1973 and 2002. MDP has over twenty types of land uses. These land uses were simplified into the following categories:

- Low-density residential
- Medium-density residential
- High-density residential
- Commercial/retail
- Industrial
- Institutional[1]
- Agriculture
- Forest
- Water
- Open Urban Land[2]
- Barren land

(4) Analysis of Land Use Policies

In order to determine the effectiveness of the land use policies, the 1977 and 1990 sector plans were reviewed for goals and policy recommendations as it related specifically to the Shady Grove transit station and its impact zone. The purpose of this analysis was to measure how well the policies and principles of the sector plans were able to control the impending development.

Findings

Based on the analysis, there are four conclusions regarding development and land use within a one-mile radius of the Shady Grove Metro Station. They are:

1. Shady Grove Metro Station experienced the most significant growth in population and housing within the first six years
2. Economic growth within the station impact zone is not affected by the economic condition of the county or Metropolitan area.
3. Shady Grove Metro Station attracted higher density residential development.
4. Policy was able to provide a framework, but not control development

[1] Institutional is public facilities, such as the metro station complex.
[2] Open urban land includes, but not limited to, golf courses, parks, recreational areas, cemeteries, and undeveloped land within urban areas. Open urban land did not fall within the station impact zone.

Shady Grove Metro Station experienced the most significant growth in population and housing within the first six years

The Shady Grove Metro Station impact zone experienced an increase of 41% in population and 43% in housing between the period of four year prior and six years after the metro station was in operation. Between 1990 and 2000, the impact zone still experienced an increase in both population and housing; however, it was at a significantly lower rate. It is possible that the development occurred during the construction of the metro station in anticipation of the station opening in 1984. However, given the analysis years of pre- and post metro, this study can only conclude that the most significant population and housing growth occurs within a short timeframe of the opening metro station.

Economic growth within the station impact zone is not affected by the economic growth of the county

When Montgomery County experienced economic growth between 1980 and 1990, the station impact zone experienced economic growth at a higher rate. For example, the housing value increased in the county by 20%, where housing value in the station impact zone increased by 33%. During this same period, the Washington, DC Metropolitan area experienced a significantly higher economic growth than the county or the station impact zone. Between 1990 and 2000, median house value and median rent decreased in Montgomery County and the DC Metropolitan area. In contrast, the median rent and median house value increased in the station impact zone. Based on the findings of this study, it is possible that economic growth will occur in a station impact zone, regardless of the economic conditions of the county or the metropolitan area.

Shady Grove Metro Station attracted higher density residential development.

Within 0.5-mile of the Shady Grove Metro Station, the prevalent residential land use is high-density at 15% of the total land area. Medium density was the next highest residential land use at 11% of the total land area. When evaluating the 1-mile radius, the prevalent residential land use is medium-density at 31% of the total land area. High-density residential is only 6% of the total land area. As expected, high-density residential occurs within 0.5-miles of the transit station.

Policy was able to guide, but not control development

In the BART study, Cervero surmised that land-use impacts of the BART system were localized due to local land use control (Cervero, 1995). The Maryland National Parks and Planning Commission adopted and approved the two sector plans that attempted to guide the development around the Shady Grove Metro Station. Based on the policy analysis, this study shows that the sector plans were able to guide development, but were unable to control development. The 1977 plan created policy to have low-density residential and commercial/industry around the metro station. The land area of low-density residential decreased by 74% within a 1-mile radius of the metro station. The prevalent land use within the impact zone is medium-density residential. The 1977 plan also recommended medium-density housing to serve as a buffer between non-residential and residential land uses. The 2002 land use map of the station impact zone displays the low-density residential land use immediately adjacent to the metro station on the eastern side. The 1990 land recommends higher density land use around the transit stations. While there is high-density residential within a 0.5-mile radius of the metro station, it only accounts for 15% of the total land area.

Consistent with the findings of BART, the Shady Grove Metro Station impact zone experienced significant land use and development changes in its twenty-two years of operation.

References

Cater, J. (1984). "Employment Change in Metrorail Station Areas 1976-1980". Washington, DC: Metropolitan Washington Council of Governments.

Cervero, R.. (1995). "Land Use and Development Impacts." Berkeley: Institute of Urban and Regional Development, Working Paper No 308.

Green, R. and James, D. M. (1993). "Rail Station Area Development: Small Area Modeling in Washington." *DC,* New York: M.E. Sharp.

Huag, W. S. (1995). "Transit and Regional Economic Growth: A Review of the Literature." Berkeley: Institute of Urban and Regional Development, Working Paper No 310.

Maryland National Park and Planning Commission. (1977). "Shady Grove Sector Plan Transit Station Area." Silver Spring, MD: Montgomery County Planning Board.

Maryland National Park and Planning Commission. (1990). "Shady Grove Sector Plan Transit Station Area." Silver Spring, MD: Montgomery County Planning Board.

Griffiths, R. E. (1982). "Employment Trends before Metrorail." Washington, DC: Metropolitan Washington Council of Governments.

Sahr, R. C. (2006). "Consumer Price Index (CPI) Conversion Factors 1800 to Estimated 2016 to Convert to Dollars of 2005". Oregon: Oregon State University.

Transportation Research Board. (2003). "Land Use and Site Design." Washington, DC: Transit Cooperative Research Program Report 95, Chapter 15.

U.S. Census Bureau. (2006). "Census 1980 of Population and Housing." Gathered by Veronica o. Davis. November.

U.S. Census Bureau. (2006). "Census 1990 of Population and Housing, Summary File 3." Generated by Veronica O. Davis; using American FactFinder; http://factfinder.census.gov; (>October 2006).

U.S. Census Bureau. (2006). "Census 2000 of Population and Housing, Summary File 3." Generated by Veronica O. Davis; using American FactFinder; http://factfinder.census.gov; (>October 2006).

Use of GPS Probe data and Passenger data for Prediction of Bus Transit Travel Time

Y. Ramakrishna, [1] P. Ramakrishna, [1] V. Lakshmanan, [2] and R. Sivanandan[3]

[1]Graduate Student, Transportation Engineering Division, Department of Civil Engineering, Indian Institute of Technology Madras, Chennai – 600036, India.
[2]Project Associate, Transportation Engineering Division, Department of Civil Engineering, Indian Institute of Technology Madras, Chennai – 600036, India.
[3]Professor, Transportation Engineering Division, Department of Civil Engineering, Indian Institute of Technology Madras, Chennai – 600036, India.

Abstract

Intelligent Transportation Systems (ITS) are gaining popularity in developing countries like India. Two major components of ITS are Advanced Traveler Information System (ATIS) and Advanced Public Transportation Systems (APTS). A combined application of these two components is the estimation of bus arrival times. Accurate bus arrival time information will enhance the credibility of the transit system, thereby leading to higher patronage. Significant research has already been conducted using Automatic Vehicle Location (AVL) system for predicting bus arrival times. However, there is limited literature available on the application of probe vehicle speed data and passenger data for predicting bus transit travel time.

It is believed that passenger data influences bus travel time in cities of developing countries such as India. Thus, this paper focuses on the application of probe vehicle speed data and passenger data for predicting bus transit travel time. The improvement brought about by the probe vehicle speed data and passenger data in prediction of travel time over the use of only GPS data is also studied. Along with passenger data, GPS data was collected using probe buses on one of the busiest bus routes on weekday evening peak hours in Chennai city, India. Preliminary data analysis revealed that similar traffic conditions prevail over the route during the evening peak hours on all weekdays. Thus, Multiple Linear Regression (MLR) models which do well in such recurrent traffic conditions were developed. Results conclude that: use of passenger data and speed data from probe buses helped improve the performance of the model.

Introduction

Public transport in most developing countries is perceived less desirable than personal vehicle. To bring about the shift from private vehicle to public transportation usage, public transport must be more desirable than private vehicles. Public transportation can be promoted by providing Real-Time transit information. Real time information displays at bus stops would be highly valued by passengers, as it enables them to take better decision on their bus route choice.

Significant research has already been conducted using AVL for predicting bus arrival times. **Lin and Zeng (1999)** used current time and location data provided by an AVL system for predicting the bus arrival time at a downstream location. These were applicable to rural transit routes only and did not consider congestion and headway. **Ojili (1999)** used one-minute time zones along the bus route, which implicitly ignored the effect of congestion and variation in dwell times. However, for larger cities that have wide variations in travel times and in dwell time, the model would be inaccurate. **Wall and Dailey (1999)** used the AVL and historical data, to estimate the bus travel time. The travel time on a particular route includes delays due to traffic control and at bus stops. However, delays were not explicitly considered.

Chen and Chien (2001) have hypothesized that direct measurement of link travel times is preferable over loop detector data for freeway travel time studies. However, for congested urban arterials traffic congestion and dwell time could be important inputs to prediction of bus arrival times.

Jeong and Rilett (2003) considered traffic congestion, in terms of schedule adherence, and dwell time at stops as input variables for bus travel time prediction. It was concluded that congestion and dwell time at bus stop improved the travel time prediction models in urban congested areas.

Shalaby and Farhan (2002) used AVL and Automatic Passenger Counter (APC) dynamic data for real time bus arrival time prediction. APC cannot be used in overcrowded conditions, which are generally prevalent in urban cities of developing countries.

It is observed that, there is limited literature available on the application of probe vehicle data and passenger data for predicting bus transit travel time in congested urban arterials. It is believed that there is a need to study bus travel time prediction, using probe vehicle data and passenger data.

Problem Definition and Scope of Work

The paper focuses on the application of probe vehicle data and passenger data for predicting bus transit travel time. A congested bus corridor (Bus Route No.21G) connecting Tambaram (suburb) to Parrys (CBD) in the city of Chennai is chosen as a case study for this purpose.

Description of Test Route

The road network in Chennai Metropolitan Area (CMA) is of radial and circular/orbital pattern. One of the busy bus routes of the city is from Tambaram to Parrys (Route 21G), this congested bus corridor is chosen as a case study. The route is 32 km long, has 19 bus stops, and connects Tambaram (a suburb) to Parrys (the CBD). The scheduled headway during the weekday evening peak is 10 minutes.

Data Collection

The spatial data of all signalized intersections and bus stops were collected with a hand-held GPS unit (GARMIN GPS). With the data thus obtained, all the signalized intersections and the bus stops were digitized. GPS data was collected during the evening peak hour along the MTC bus (No.21G) from Parrys to Tambaram. GARMIN GPS receivers with a set reception frequency of 4 seconds were used for this purpose. While, the first two buses act as probe vehicles and give the speeds of the links ahead, the third bus information is used for building the model. Along with the GPS data passenger data was also collected. Passenger data includes: number of people ingress and outgress at both front door and back door. This data was collected by manual counting.

Data Extraction

The GPS data collected cannot be used as such for model building, it is modified for further use. The entire GPS data was plotted on a digitized route map as shown in the Figure 1. Time stamp of data points which fall before and after bus stops and intersections were considered as arrival times and departure times at the bus stops and intersections respectively. Travel time between bus stop i and j is obtained as difference of departure time at bus stop i and arrival time at bus stop j. Dwell time at stop 'i' is obtained as difference of arrival time and departure time at bus stop 'i'.

The passenger ingress and outgress variables are important as they influence the bus stop dwell time. However, these variables cannot be used directly as inputs to the travel time model. So, this passenger data is classified in to the following meaningful variables: total passenger ingress (TPI), total passenger egress (TPE), Total front door passenger activity (TF), Total back door passenger activity (TB), section load (PL), maximum of front and back door passenger activity (MFB). To select the set of variables which will accurately explain the dependence of passenger data on bus stop dwell time, preliminary models are developed. The R^2 values of the various combinations are listed in Table 1.

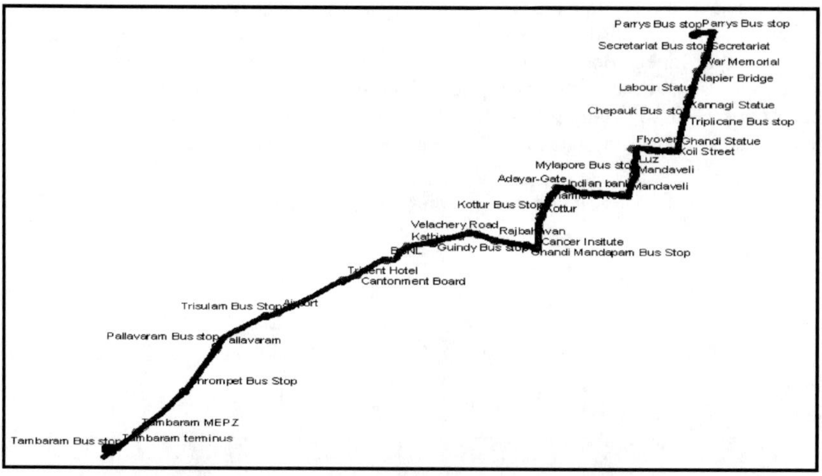

Figure 1. Identification of Bus stops and Intersections

Table 1. R^2 Values of the Different Combinations of the Passenger Data

Model	Independent Variable Combinations	R^2 Value
1	Total Passenger Ingress, outgress	0.635
2	Total Passenger Ingress, outgress , Passenger Load	0.642
3	Maximum Passenger Ingress, outgress	0.602
4	Maximum Passenger Ingress, outgress, Passenger Load	0.613
5	Total Passengers through Front door, Back door	0.582
6	Total Passengers through Front door, Back door, Passenger Load	0.609
7	Max Passengers through (Front door, Back door)	0.687
7	Total Passenger Ingress and Egress	0.571
8	Max Passenger (Ingress or Egress)	0.425

It is found that MFB best explains the Bus Stop Dwell time. Hence, MFB is taken as the input variable in the travel time model. The equation explaining the dwell time is shown:

$$DT_i = 7.46 + 1.977 * MPFB_i$$

Data Characteristics

After the data was extracted in the required format, its characteristics were examined. Specifically, the day to day variability of travel times, bus stop dwell times, intersection delays, and Passenger delays were examined. It is found that the above variables have less day to day variation among the weekdays. Moreover, the variation along the route for all the variables is recurrent in nature. Thus, prediction models like Multiple Linear Regression (MLR) models which do well in recurrent traffic conditions is believed to be appropriate. Figure 2. shows the plot of intersection delays at each of the major intersection along the test route. Although, the variations in the intersections delays are high, the trend seems to be recurrent in nature.

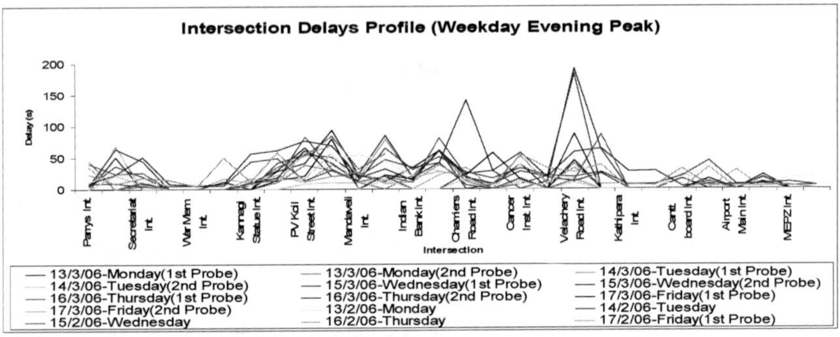

Figure 2. Weekday Intersection delay variations along MTC Bus Route No.21G

Model Development

Two separate Multiple Linear Regression (MLR) models were developed using SPSS (ver.13.0) software (SPSS Inc., 2004), to study the effect of probe vehicle speed data and passenger data on travel time prediction.

Input Variables

The extracted input data from the previous section is used for travel time prediction models. The availability of probe vehicle speed data is limited up to the probe bus. Hence, the target bus stop in all cases is with in the probe bus location. The following variables were chosen as input variables for the model development to study the influence of probe vehicle speed:

- Length of two lane stretch from bus stop 'i' to the target bus stop 'j' in km. (D_{2ij})
- Length of four lane stretch from bus stop 'i' to the target bus stop 'j' in km. (D_{4ij})
- Length of six lane stretch from bus stop 'i' to the target bus stop 'j' in km. (D_{6ij})
- Average probe vehicle speed from the bus stop 'i' to the bus stop 'j' in km/h (S_{ij}).
- Number of bus stops from current bus stop 'i ' to the target bus stop 'j ' (BS_{ij})
- Number of intersections from current bus stop 'i' to the target bus stop 'j' (I_{ij})

- Bus stop dwell times from bus stop 'i' to the target bus stop 'j' in minutes (DW_{ij})
- Intersection delays from bus stop 'i' to the target bus stop 'j' in minutes (ID_{ij})

The variable $AMFB_{ij}$ is chosen as input variable instead of the probe vehicle speed data and bus stop dwell times, to study the influence of passenger data. The output variable is travel time from the current bus stop 'i' to the target bus stop 'j' in minutes (TT_{ij}).

Inter-Correlation & Serial-Correlation

In regression models, the input variables should not be inter-correlated. The correlation coefficient close to one implies that the two independent variables are highly correlated. It is to be seen that all the correlations are logical. Another important assumption of MLR model is that the residuals are statistically independent. A formal statistical test for serial correlation is based on the Durbin-Watson statistic (d), which is given in equation 1.

$$d = \frac{\sum_{t=1}^{t=n-1} (\varepsilon_{t+1} - \varepsilon_t)^2}{\sum_t \varepsilon_t^2} \tag{1}$$

where,

ε = Residual or the error term of the dependent variable, t = Observation number, and N = Total number of observations.

To minimize serial correlation, the probe vehicle data was organized such that: if one data point is from Bus Stop 'i' to Bus Stop 'j', the next data point would be from Bus Stop 'j' to Bus Stop 'k'. Although, this particular selection results in minimum overlap of the dependent variable for probe vehicle data. The same selection would underutilize the very large data set available for passenger data. Hence, the following four data sets were used for studying the effect of passenger data in travel time prediction. The first data set has all combinations of stop 'i' and target stop 'j' (1887 Points). The second data set has every alternate data point from the first set (943 Points), the third data set has every third data point (629 Points), and the fourth set has every fourth data point (471 Points). For these four data sets, Durbin Watson test is used to evaluate the Regression Models.

When there is no serial correlation, the expected value of the d is around 2.0. Values of d less than 1.5 or greater than 2.5 imply positive or negative serial correlation. The summary of the Durbin Watson test for passenger data is given in Table 2.

Table 2. Durbin Watson Coefficient for the four data sets

Data Set	d
1	1.341
2	1.738
3	1.961
4	2.042

The first model is not valid as its'd' is less than 1.5. The other models are valid as their 'd' lies between 1.5 and 2.5. The valid data sets are used to study the influence of passenger data on travel time prediction.

Multiple Linear Regression Models

To study the influence of probe vehicle speed data, a total of 240 data points were used for the model development. In the preliminary model it was observed that the variables Remaining Number of Bus Stops and Intersection Delay were statistically insignificant.. Hence, these variables were removed in the further models.

Model Evaluation

In order to evaluate the performance of the models, the Mean Absolute Percentage Error (MAPE) (given by equation 2) was used as measure of closeness between predicted and observed values.

Table 3. MAPE of the MLR Models

Model Number	Predictors	Mean Absolute Percentage Error (MAPE)
1	I_{ij}, D_{ij}	23.37
2	I_{ij}, D_{6ij}, D_{2ij}, D_{4ij}	19.59
3	I_{ij}, D_{6ij}, D_{2ij}, D_{4ij} , S_{ij}	13.11
4	I_{ij}, D_{6ij}, D_{2ij}, D_{4ij} , S_{ij}, DW_{ij}	9.00

$$\frac{1}{n}\sum_{i=1}^{n}\frac{|y_{ip} - y_{iob}|}{y_{iob}} \times 100\% \qquad (2)$$

where,
y_{ip} = Predicted travel time from current bus stop to target bus stop
y_{iob} = Observed travel time from current bus stop to target bus stop
n = Number of validation data points

A separate validation data set was used to evaluate the models. A total of 200 data points were used for model evaluation.

Model Comparison

The MLR models developed to study the influence of both probe vehicle speed and passenger data (or equivalently the dwell times) are compared with respect to MAPE, as highlighted in Table 3. It is observed that use of probe vehicle speed improves the prediction, which results in MAPE value of 13.11 It is also observed that the dwell times variable improves the prediction, which results in MAPE value of 9.00. The regression equation is given below.

$$TT_{ij} = 3.721 + 0.214^{*} I_{ij} + 1.716^{*} D_{6ij} + 2.47^{*} D_{2ij} + 1.94^{*} D_{4ij} - 0.127^{*} S_{ij} + 1.044^{*} DW_{ij} \quad (3)$$

Variation of the Travel time prediction along the route

Further, the variation of the Travel time prediction error along the test route is studied. If the Travel time prediction error (minutes) improves significantly along the route then real time prediction can be recommended. To study this variation, the initial point is varied from stop to stop as the bus moves along the route and the destination point is fixed at tambaram bus terminus. Figure 3 highlights this variation. It is observed that there is no significant improvement in prediction along the study route. Thus, no real time prediction is required.

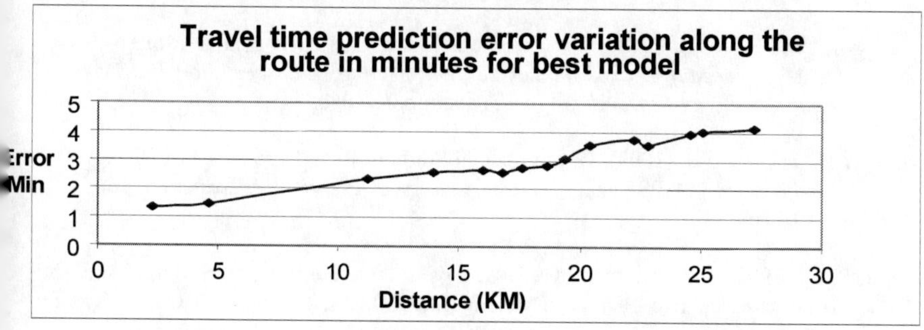

Figure 3. Variation of travel time prediction MAPE along the route

Conclusions

This paper has highlighted the need to consider the probe vehicle data and passenger data for bus travel time prediction in urban arterials in developing countries like India. The data analysis revealed that similar traffic conditions prevail over the route during the peak hours on all weekdays. Thus, Multiple Linear Regression (MLR) models which do well in such recurrent traffic conditions were developed.
The key contributions of this work are:

- Distance remaining from the current bus stop to the target bus stop was classified in terms of, six lane, four lane and two lane,' which led to the MAPE reduction from 23.37 to 19.59 (in model 2)
- The use of probe bus speed data improved the accuracy of the models. The MAPE has reduced from 19.59 to 13.11 (in model 3) by the use of probe vehicle data.
- The use of passenger data improved the accuracy of the model. The MAPE has reduced from 13.11 to 9.00 by the use of passenger data.

One of the likely shortcomings of this model is the transferability of the models on to other urban routes. Although most of the variables used for developing the model are generic in nature, the model should be recalibrated before applying on other routes.

References

Chen, M., and Chien, S. I. J. (2001). "Dynamic Freeway Travel Time Prediction Using Probe Vehicle Data: Link Based vs. Path Based." *Proceedings, 80th annual general meeting, Transportation Research Board, National Research Council*, paper no.01-2887.

"GPS 12 Personal Navigator-Owners Manual and Reference." *GARMIN International, Inc.*, Kansas, USA.1999.

Jeong, R., and Rilett, L. R. (2004). "The Prediction of Bus Arrival Time using AVL data." *Proceedings, 83rd annual general meeting, Transportation Research Board, National Research Council*, paper no.04-4060.

Lin, W.H., and Zeng, J. (1999). "Experimental Study of Real-Time Bus Arrival Time Prediction with GPS Data." *Transportation Research Record*, paper no. 1666, p-101-109.

Ojili, S. R. (1999). "A Prototype of a Bus Arrival Time Notification System Using Automatic Vehicle Location Data." Master's thesis, Texas A&M University, College Station, USA.

Quiroga, C.A., and Bullock, D. (1998). "Travel time studies with global positioning and geographic information systems: an integrated methodology." *Transportation Research Part C: Emerging Technologies*, Volume 6, 101-127.

Schrader, C.C., and Kornhauser, A. L. (2004). „Using Historical Information in Forecasting Travel Times." *Proceedings, 83rd annual general meeting, Transportation Research Board, National Research Council*, Washington D.C.

Shalaby, A., Lyon, C., and Sayed, T. (2001). "Transferability of Travel Time Models and Provision of Real-Time Arrival Time Information." *Proceedings, IEEE Intelligent Transportation Systems*, Oakland (CA).

Shalaby, A., and Farhan, A. (2003). "Bus Travel Time Prediction for Dynamic Operations and Control and Passenger Information Systems." *Proceedings, 82nd annual general meeting, Transportation Research Board, National Research Council*, Washington D.C.

SPSS Inc. (2004). SPSS 13.0, Chicago, Illinois, USA.

Wall, Z., and Daily, D. J. (1999). "An Algorithm for Predicting the Arrival Time of Mass Transit Vehicle Using Automatic Vehicle Location Data." *Proceedings, 78th annual meeting, Transportation Research Board, National Research Council*, paper no. 99-0870.

Modeling Neighborhood Environment, Bus Ridership and Air Quality: A Case Study of Chicago Bus Service

J. Lin,[1] and M. Ruan[2]

[1]Corresponding author. Ph.D., Assistant Professor. Department of Civil and Materials Engineering & Institute for Environmental Science and Policy, University of Illinois at Chicago, 842 W. Taylor Street (M/C 246), Chicago, IL 60607; PH (312) 996-3068; FAX (312) 996-2426; email: janelin@uic.edu
[2]Ph.D. student, Department of Civil and Materials Engineering, University of Illinois at Chicago, 842 W. Taylor Street (M/C 246), Chicago, IL 60607; email: mruan2@uic.edu

Abstract

Public transit accessibility and ridership are affected by neighboring environment along the transit lines. This paper investigates the relations between neighborhood features and bus ridership and emissions of the Chicago Transit Authority (CTA) bus lines. A mixed regression model of bus emissions/ridership model is created to quantify the direct relations with both internal bus operation indicators such as route length and number of stops and external factors including neighborhood land use and population socio-demographics. In Chicago, there are a large number of serviced urban Black neighborhoods with low population in vicinity of bus routes, poor connectivity to transit, and low transit users to work in general, resulting in high per unit ridership bus emissions (grams/ridership) for routes servicing those areas. As the city and the transit agency have been making great effort to increasing bus ridership in those neighborhoods, more collaborative work must be done by the city, the local planning agencies and the transit agency to improve accessibility to bus service.

Key words: bus ridership, neighborhood type, mixed regression model, NOx, PM_{10}

Introduction

Public transport has been recognized as a growing alternative mode to automobile travel. After a decline in the recession years of the early 1990s, public transit use in the United States rose sharply from 1995 to 2000 (Pucher, 2002). Unlinked (transit station to transit station) passenger trips increased by twenty-one percent, the highest level in forty years (Pucher, 2002). Public transportation has been identified as a mean in earlier studies to reduce traffic congestion, parking shortage, traffic accidents (Litman, 2004). Public transit improvements also help in reducing road construction costs, inadequate mobility for non-drivers, and excessive energy consumption and pollutant emission from automobile travel (Litman, 2004).

Transit ridership is one of the most important measures of services provided by a transit agency. Internally, a transit agency measures its transit service performance by accessibility to stops at the origin and destination (via walk, bike, or park and ride, etc.), provision of transit availability information, and temporal availability of transit service (Beimborn et al., 2003). These factors are sometimes referred to as internal factors. External factors that influence transit ridership are population density, private vehicle ownership, topography, freeway network extent, parking availability and cost, transit network extent and service frequency, transit fares, and transit safety (Litman, 2001; Azar and Ferreira, 1994; Kuby and Barranda, 2004). Kain and Liu (1999) studied the growth of San Diego transit service and found factors like household and employment location and densities influenced transit travel characteristics. Similarly, after studying 12 transit agencies in the U.S. Yoh et al. (2003) concluded that the most cited factors in explaining ridership growth were external factors such as population and employment growth.

Modeling transit ridership requires precise understanding of demand market and potential of service provider (Larwin, 1999). The obstacle in estimating accurate ridership is the non-availability of accurate data. Taylor and Fink (2003) discussed two different approaches to ridership modeling: (a) *Descriptive analysis*: descriptive analysis techniques use survey and interview data, mainly focused on internal factors to transit industry to predict ridership; and (b) *Causal analysis technique*: causal analysis technique, also known as direct ridership modeling, uses multivariate regression analysis with both internal and external factors to the transit system. Other approaches include ridership models based on fare structure (Hirsch et al., 2000) and logit ridership approach (Nichesen et al., 1984) etc. These models generally do not contain land use information of areas close to transit lines, which has been addressed in the new model developed as part of this study.

Transit service improvement is one of the transportation control management (TCM) alternatives recommended by Federal Highway Administration (FHWA) for local congestion mitigation and air quality improvement (Federal Highway Administration, 1999). The rationale is reducing vehicle miles of travel (VMT) through transit service improvement to achieve environmentally sustainable transportation system. FHWA recommends several off-model[1] emission reduction estimation approaches for TCM programs (Federal Highway Administration, 2000). Most of those approaches are based on spreadsheet models. Among them, off-Net/PAQONE model developed by Illinois DOT predicts emissions impacts of transit, non-motorized travel, and traffic flow strategies and TCM analysis tool developed by Sierra Research, Inc. is an elasticity-based cost–effectiveness screening model.

---

[1] off-model analysis does not require running a travel demand models or traffic simulation models

This paper presents a mixed regression model coupled with the geographic information software (GIS) to quantify the direct relations between bus emissions (only NOx and PM_{10} are considered) and transit service, surrounding neighborhood land use, and population socio-demographics. The Chicago Transit Authority (CTA) bus service between 2001 and 2003 is chosen for the case study. The mixed regression model structure allows explicit modeling of heterogeneity in route service or emissions over time. GIS provides spatial information of the surrounding land use along the CTA bus routes. The model findings will have important practical and policy implications to urban land use planning and public transit planning and operation.

Study Area

CTA serves City of Chicago and 40 surrounding suburbs with over 150 routes (see Figure 1) and more than 2000 buses and 2,273 route miles[2]. CTA buses provide about 1 million passenger trips a day and serve more than 12,000 posted bus stops.

As shown in the background of Figure 1, CTA service area is classified into ten different neighborhood types by using the Census Transportation Planning Package (CTPP) 2000. Detailed explanation of how the neighborhood types were defined can be found in Lin and Long (2006). A statistical clustering method coupled with geographic information system (GIS) spatial analysis was used to classify 65,315 census tracts nationally into ten classes defining neighborhood types. Hence, the term neighborhood is interchangeable with census tract in this study. The neighborhoods were clustered by sixty-four variables describing socio-demographics, land use, and journey-to-work features of a census tract. They were derived from the CTPP and Tiger line files.

The ten neighborhood types are: (1) Urban/2nd city non-Hispanic Black dominant, (2) Rural , (3) Non-Black Hispanic dominant, (4) Natural scenic, (5) Suburban young, (6) Suburban retired, (7) Suburban middle income working neighborhood, (8) Urban elite, (9) City poor, primarily minority, and (10) Suburban mid-age wealthy.

Each neighborhood type is named after the inclusive census tracts' common features describing the neighborhoods' socio-economics, demographics, and geographic/land use characteristics. For example, *Urban elite* (in dark brown) represents an urban neighborhood with primarily young non-Hispanic White working in professional, managerial or technical fields and earning $45,000 and more. Compared to the suburban households, many of the urban elites (17.7%) use transit to work and have a small household size (an average of 1.91 persons per household). South Chicago has a large number of urban Black dominant neighborhoods (in black). City West has a large Hispanic community (in light brown). The west and north suburbs are concentrations of suburban wealthy neighborhoods.

[2] Statistics are from the CTA official web page: http://www.transitchicago.com

Each of the ten neighborhood types uniquely defines a neighborhood lifestyle and travel behavior of households within the neighborhood (Lin and Long, 2006). Their impact on the CTA bus ridership and emissions is the focus of this study, as to be shown later in the paper.

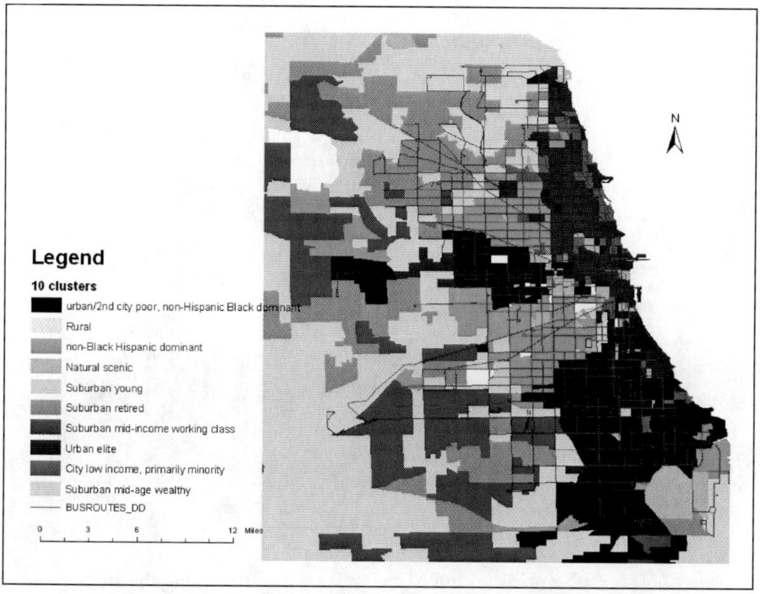

Figure 1-CTA bus routes and serviced neighborhoods

Specifically for this study, fifty-five CTA bus routes are studied. The study period is three years between 2001 and 2003. Data for the fifty-five routes include demographic, land use, bus boarding and transit service data, in spatial or non–spatial form:

- Census tract coverage
- GIS coverage of CTA bus routes and train stations
- The Census Transportation Planning Package (CTPP) 2000 that contains aggregated (zonal) household socio-demographic attributes and journey to work travel characteristics
- Tiger line files
- Route-based CTA bus daily ridership between 2001 and 2003,
- Number of bus stops of each route

Non-spatial data (e.g., population, ridership etc) were mapped to each bus route within a quarter mile distance by using GIS (see Figure 2). A quarter-mile distance is a commonly perceived accessible distance to bus stops. In the rest of the paper all attributes are measured within the half-mile buffer area of a bus route unless otherwise noted.

Figure 2- Bus buffer areas within a quarter mile distance along the studied bus routes

Table 1 shows the average census tract population and transit users to work by neighborhood type. Not surprisingly, relatively high city low income minority population lived close to the bus lines, followed by the suburban mid-income working class and the suburban retired population groups. What was unexpected was the relatively low urban Black population living with the bus buffer area. CTA buses do not serve rural areas. Table 1 also shows less urban elite population living in proximity to the CTA bus routes even compared to the suburban mid-age wealthy group. However, the urban elite neighborhoods had disproportionally high average per-census-tract transit users to work along a bus route, second to the City low income minority group.

Table 1- Average tract population and transit users to work by neighborhood type

Neighborhood type	Population		Transit users to work	
	Mean	Std Dev	Mean	Std Dev
Urban/2nd city non-Hispanic Black dominant	2,183	1,671	199	164
Non-Black Hispanic	4,461	2,668	334	223
Natural Scenic	526	530	43	19
Suburban young	3,546	2,223	261	159
Suburban retired	5,366	2,574	301	217
Suburban mid-income working class	5,716	2,865	283	197
Urban elite	4,261	2,766	543	363
City low income, minority	6,338	2,032	1538	835
Suburban mid-age wealthy	4,600	1,394	385	126

Table 2 summarizes street coverage within a neighborhood type. Street coverage was defined by the ratio between total street length within the same neighborhood type and the bus route length (in kilometers/mile). The higher the metric, the better street coverage in that neighborhood type in vicinity of a bus route. Hence, street coverage to some extent measures the connectivity of streets in a neighborhood to the bus route. Interestingly, the urban elite neighborhoods had orders of magnitude higher street coverage than any other neighborhood types. The difference ranged from over thirty times to hundreds and even thousands of times higher. All suburban groups have poor street coverage in vicinity of a bus route.

Table 2- Street coverage in a neighborhood type per mile bus route (in kilometers/mile)

Neighborhood type	Mean	Std Dev.
Urban/2nd city non-Hispanic Black dominant	2.71	5.20
Non-Black Hispanic dominant	1.76	3.18
Natural scenic	10.78	23.97
Suburban young	.44	1.10
Suburban retired	.47	1.97
Suburban middle income working neighborhood	.08	.41
Urban elite	358.56	723.37
City poor, primarily minority	6.11	11.90
Suburban mid-age wealthy	.043	.135

Table 1 lists the monthly average daily per bus route ridership between 2001 and 2003. Only weekday ridership was considered. As expected, July, August,

December, and January had relatively light bus ridership because of school breaks and holidays. Generally speaking, ridership in the fall was highest in a year.

With respect to daily bus emissions, Figure 4 shows almost an inverse trend of bus ridership in Figure 3 since the vertical axis in Figure 4 is in terms of per unit ridership bus emissions (in grams/ridership). Bus emissions here only included only running exhausts and starts emissions of NOx and PM_{10}, which are determined by the following calculation:

$$\textit{Daily Bus emissions (g/ridership)} = \textit{Daily bus run miles}*(EF_{NOx}+EF_{PM10}) \,/\, \textit{ridership} \quad (1)$$

The emission factors were estimated with MOBILE6.2 by using the Chicago-specific input parameters (e.g., VMT distribution, age distribution, fuel type, I/M program, etc.) for regional emission inventory estimation. Both summer and winter emission factors were estimated. For running exhaust, an average bus running speed of 15 mph was assumed. Finally, these emission rates were used to determine the total NOx and PM_{10} emissions: 13.6 g/mi (winter) and 13.021 g/mi (summer) for NOx, and 0.4012 g/mi (winter) and 0.3872 g/mi (summer) for PM_{10}.

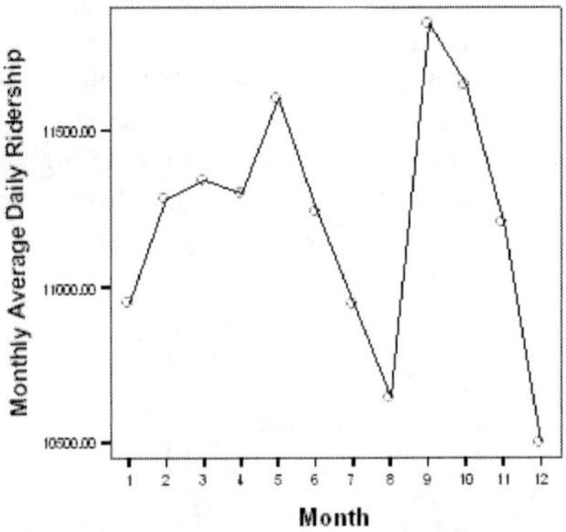

Figure 3. Monthly average daily per route bus ridership

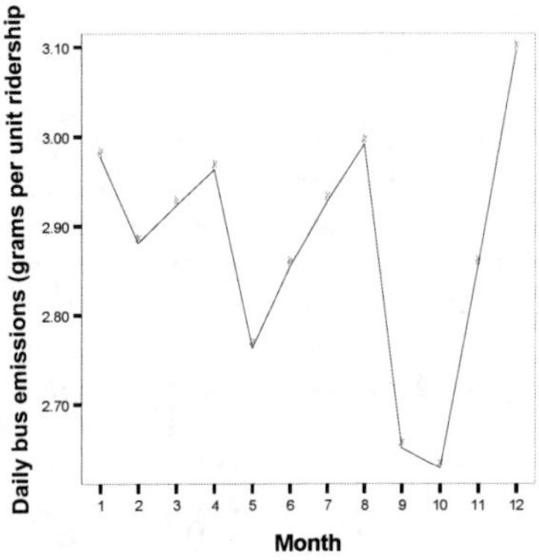

Figure 4. Monthly average daily per route bus emissions (grams/ridership)

Model structure

A mixed regression model with auto correlated errors was found to be an appropriate form for this study. Because the panel data structure is composed of repeated measures on fifty-five CTA bus routes over three years. The mixed regression structure explicitly models heterogeneity among routes, via random effects, and autocorrelation over time, via autoregressive error terms. The model can be written in the matrix form as:

$$\mathbf{y}_i = \mathbf{X}_i \boldsymbol{\beta} + \mathbf{v}_i + \boldsymbol{\varepsilon}_i \tag{2}$$

where, $\mathbf{y}_i$ = 12×1 monthly average daily bus emissions (in grams/ridership) vector for route i (i=1, 2,.....N), a repeated measure over a three-year period of 2001-2003

$\mathbf{X}_i$ = $12 \times p$ covariate matrix for route i,

$\boldsymbol{\beta}$ = $p \times 1$ vector of fixed regression parameters,

$\mathbf{v}_i$ = 12×1 vector of random effects,

$\boldsymbol{\varepsilon}_i$ = 12×1 error vector.

The error terms and the random effects are assumed to have the following properties:

$\boldsymbol{\varepsilon}_i \sim N(\mathbf{0}, \sigma^2 \boldsymbol{\Omega}_i)$

$\mathbf{v}_i \sim N(\mathbf{0}, \sigma_v^2 \mathbf{I})$

where, $\mathbf{I}$ is a unity matrix; $\boldsymbol{\Omega}_i$ is a user-specified auto correlated error structure. Among various types of auto correlated (AC) errors, the first-order autoregressive error structure, AR(1), and Toeplitz (TOEP(h)) error structure are tested. AR(1) assumes the following relation between current and previous time lag errors.

$$\varepsilon_{i,t} = \rho\, \varepsilon_{i,t-1} \tag{3}$$

In contrast, TOEP (h) assumes constant and decreasing covariance between the immediately lagged errors and zero covariance beyond h lags:

$$Cov(\varepsilon_{i,t}, \varepsilon_{i,t-k}) = \begin{cases} \rho_{i,t-k}, & k = 1, ..., h-1 \\ 0, & k \geq h \end{cases} \tag{4}$$

where, $\rho_{i,t-k}$ is decreasing constants for k=1, ..., h-1. The final selection is based on Akaike Information Criterion (AIC) and Bayesian Information Criterion (BIC) values. (Hedeker and Gibbons, 2006)

The variance-covariance matrix of the repeatedly measured dependent variable $\mathbf{y}_i$ is of the form:

$$\mathbf{V}(\mathbf{y}_i) = \sigma_v^2 \mathbf{I} + \sigma^2 \boldsymbol{\Omega}_i \tag{5}$$

Equation (5) illustrates the two components in the variance-covariance of the repeated measures model: random individual effects and the auto correlated errors. The former concerns about the heterogeneity in individual bus route, while the latter displays an autocorrelation structure for the repeated error structure $\boldsymbol{\varepsilon}_i$.

A set of possible covariates were assembled and listed in Table 3. Transit service related variables considered were: *per square mile number of train stations* within the quarter-mile buffer to account for the level of train service in the vicinity of the bus route, *number of bus stops along a route*, *average automobile travel time* (in contrast to bus travel time), and *average bus headway*. Month dummy variables were included to estimate monthly/season effect on bus emissions and ridership. January was taken as the baseline time point, and 11 variables were created representing the other months. The socio-demographic features, including *average household size*, *average household income*, *average vehicles per household*, were average values for each neighborhood type in vicinity of a route. The land use features, *unitpop*, *unitworkers*, *unitintersect*, *unitroad* and *unithouse*, were measures of total population, workers, intersections, road length and households per mile bus route length. Lastly, the percentages of each neighborhood type within the route buffer areas were generated.

Table 3- List of possible covariates

	Independent Variable	**Description**
Transit service	AVG_TRAVEL_TIME	Average auto travel time within ¼ mile buffer
	STOP_CNT	Total number of stops on a bus route
	AVG_HEADWAY	Average headway of all trips on a route in a service day
	TRAIN_STATION_CNT	Number of train stations per square mile within ¼ mile buffer
Month	MON_1	=1, if February
	MON_2	=1, if March
	MON_3	=1, if April
	MON_4	=1, if May
	MON_5	=1, if June
	MON_6	=1, if July
	MON_7	=1, if August
	MON_8	=1, if September
	MON_9	=1, if October
	MON_10	=1, if November
	MON_11	=1, if December
Socio-demographics and land use (by neighborhood type)	AVGSIZE	Average number of household members
	AVGHHINC	Average number of household income
	VEHRATE	Average number of vehicles per household
	UNITPOP	Total population (in thousands) per mile route length
	UNITWORKERS	Total workers (in thousands) per mile route length
	UNITINTERSECT	Total number of intersections per mile route length
	UNITROAD	Total road length (in kilometers) per mile route length
	UNITHOUSE	Total household (in thousands) per mile route length
	PERCENTAREA	Percent of area

Model Results

It was hypothesized in this study that bus ridership and emissions in Chicago were determined by socio-demographics and land uses within a quarter-mile distance to a bus route as well as bus service indicators. Only weekday ridership was considered.

The final model specification is shown in Table 4. Model parameters were estimated using the stepwise linear regression method. Only the covariates with significant coefficients were included in the final model. Logarithmic transformation was found necessary for the dependent variable after the initial model residuals were examined. The model goodness-of-fit null model likelihood ratio Chi-square statistic is 1699.59, significant at the 0.01 level, which means the final model form significant improves the predictive ability from the null model form.

All model coefficients' signs are intuitively correct. Among all the month dummy variables, only month December has a positive sign compared to month January, which means the daily per unit ridership bus emissions in December is higher than that in January. This is consistent with what has been observed in Figure 4.

A negative coefficient for number of bus stops suggests bus stops associated positively with ridership and thus (log) bus emission per unit ridership decreases by 0.00545 if one stop is added.

Total population in the urban non-Hispanic Black neighborhoods is positively correlated with per unit ridership daily bus emissions because these neighborhoods have generally low transit users compared to other neighborhood types. An example is the transit users to work shown in Table 1, where the Black neighborhoods have the second lowest (only to natural scenic) transit users to work due to their low worker rates.

High road length in the urban elite neighborhoods means good street coverage and good connectivity to the bus route. Good connectivity is positively related to bus ridership and thus per unit ridership bus emissions dropped. Again, the model result is consistent with the finding of orders of magnitude higher street coverage in the urban elite neighborhoods in Table 2. Furthermore, the high coefficient value indicates high unit impact of street coverage on per unit ridership bus emissions.

In the final model, the AR(1) autoregressive error structure was chosen by comparing the Akaike Information Criterion (AIC) and Bayesian Information Criterion (BIC), both of which indicated AR(1) was an adequate error term structure. The AR(1) parameter, $\rho=0.4555$, is shown statistically significant, indicating significant autocorrelation among the repeatedly measured daily bus emissions (grams per unit ridership). The significant random effects (indicated by the significant σ_v^2 statistic) indicate the existence of heterogeneity in per unit ridership bus emissions across the fifty-five study bus routes.

Table 4-Final model specification

Variable	Estimate	Standard Error	t value	Pr >|t|
INTERCEPT	1.6543	0.2036	8.12	<.0001
MON_1 (Feb)	-0.03051	0.01051	-2.90	0.0039
MON_4 (May)	-0.06484	0.01151	-5.63	<.0001
MON_5 (Jun)	-0.02819	0.01153	-2.44	0.0148
MON_8 (Sep)	-0.08408	0.01153	-7.29	<.0001
MON_9 (Oct)	-0.09304	0.01151	-8.08	<.0001
MON_11 (Dec)	0.07252	0.01194	6.07	<.0001
STOP_CNT	-0.00545	0.001620	-3.36	0.0008
UNITPOP_Urban Black	0.07702	0.02939	2.62	0.0090
UNITROAD_Urban Elite	-0.2739	0.1094	-2.50	0.0126
Covariance Parameter Estimates:				
σ_v^2	0.1940	0.04079	4.76	<.0001
ρ	0.4555	0.04756	9.58	<.0001
c^2	0.008233	0.000721	11.42	<.0001
Model goodness-of-fit:				
Null Model Likelihood Ratio Test: Chi-Square (2) = 1699.59, Pr <.0001				

Conclusions

This paper has presented a mixed regression model that describes the relations between bus emissions and land use, socio-demographic characteristics as well as bus performance measures. The model was applied to fifty-five CTA bus routes for their service between 2001 and 2003.

Bus ridership and emissions are found to be strongly impacted by both internal bus service measures such as number of bus stops along a bus route and external factors related to neighborhood types and socio-demographics in vicinity of bus routes. Specifically to Chicago, CTA bus service covers a variety of neighborhoods with distinct characteristics as seen in Figures 1 and 2. There are many serviced urban Black neighborhoods with low population in vicinity of bus routes, poor connectivity to transit, and low transit users to work in general. This results in high per unit ridership bus emissions (grams/ridership) for routes servicing those areas. The urban elite neighborhoods, on the other hand, have good connectivity to bus routes and high transit usage, resulting in low per unit ridership bus emissions.

These findings are worrisome as low income neighborhoods are generally believed to rely heavily on public transit for their daily activity. As the city and the transit agency have been making great effort to increasing bus ridership in those neighborhoods, our analysis results simply suggest that the neighborhood's low street coverage and low population in vicinity of bus routes do not support ridership. More collaborative work must be done by the city, the local planning agencies and the transit agency to improve accessibility to bus service.

There are several limitations associated with the method presented in the paper. Ridership data for modeling is available at route level on a daily basis. No ridership information is currently made available at each bus stop. Hence, the model is not capable of providing direct explanation between the type of surrounding neighborhood and ridership at a bus stop, nor is it able to take into account waiting time and transfer connectivity within CTA system.

Acknowledgement

We would like to thank CTA for providing the data. The views expressed in the paper are those of the authors solely.

References

Azar, K., and Ferreira J. (1994). "Using GIS Tools to Improve Transit Ridership on Routes Serving Large Employment Centers: The Boston South End Medical Area Case Study." *Computer, Environment and Urban Systems*, Vol. 18, No. 3, pp. 205-231.

Beimborn, E. A., Greenwald, M. J. and Jin, X. (2003). "Accessibility, connectivity, and captivity – Impacts on transit choice." *Transportation Research Record* (1835), pp.1-9, Transportation Research Board National Research Council, Washington D.C.

Federal Highway Administration. (1999). "Off-Model Air Quality Analysis: A Compendium of Practice." Southern Resource Center, Atlanta, GA 30303.

Federal Highway Administration. (2000). "A Sampling of Emissions Analysis Techniques for Transportation Control Measures, final report." Prepared for Federal Highway Administration, prepared by *Cambridge Systematics, Inc.*, October.

Hedeker & Gibbons. (2006). "Longitudinal Data Analysis", Wiley-Interscience, A JOHN WILEY & SONS, INC., Hoboken, NJ.

Hirsch, L., David. J., Hickey R., and Cravo V. (2000) "Effects of Fare Incentives on New York City Transit Ridership." *Transportation Research Record* No. 1735, pp.147-157, Transportation Research Board National Research Council, Washington D.C.

Kain, J.F. and Z. Liu. (1999). "Secrete of success: assessing the large increases in transit ridership achieved by Houston and San Diego transit providers." *Transportation Research Part A – Policy and Practice*, Vol 33 (7-8): 601-624

Kuby, M., and Barranda, A. (2004). Upchurch, C. "Factors Influencing Light-Rail Station Boarding In the United States." *Transportation Research Part A* 38 No. 3, pp.223–247.

Larwin, Thomas F. (1999) "Urban Transit In: Transportation Planning Handbook." *Institute Of Transportation Engineers*, pp.468.

Lin, J. and Long, L. (2006). "What Neighborhood Are You In? Empirical Findings of Relationships between Residential Location, Lifestyle, and Travel", *Compendium CD ROM of the 85th Annual Meeting of Transportation Research Board*, Washington D.C.

Litman, T. (2001). "Evaluating Transportation Choice." *Transportation Research Record* 1756, pp.32-41, Transportation Research Board National Research Council, Washington D.C.

Litman, T. (2004). "Evaluating Public Transit Benefits And Costs: Best Practices Guidebook." *Victoria Transport Policy Institute*, 1250 Rudlin Street, Victoria, BC V8V 3R7, Canada.

Nickesen A., Meyburg, A., and Turnquist, M. (1984). "Ridership Estimation for Short-Range Transit Planning." *Transportation Research Part B*, Vol 17B, No.3, pp 233-244.

Pucher, J. (2002). "Renaissance of Public Transport in the United States?" *Transportation Quarterly*.

Taylor, B. and Fink, C. (2003). "*The factors influencing transit ridership: a review and analysis of ridership literature*." University of California Transportation Center (UCTC) working paper, UCLA Department of Urban Planning.

Yoh, A. C., Haas, P. J. and Taylor, B. D. (2003). "Understanding transit ridership growth – Case studies of successful transit systems in the 1990s." *Transportation Research Record* 1835, pp 111-120, Transportation Research Board National Research Council, Washington D.C.

Bus Public Transport Energy Consumption and Emissions versus Individual Transportation

C. Silva,[1] J. Bravo,[2] G. Gonçalves,[2] T. Farias,[2] and J. Mendes-Lopes[3]

[1]IDMEC-Institute of Mechanical Engineering, Instituto Superior Técnico, Technical University of Lisbon, Portugal; email: carla.silva@ist.utl.pt
[2]IDMEC-Institute of Mechanical Engineering, Instituto Superior Técnico, Technical University of Lisbon, Portugal
[3]IN+Center for Innovation, Technology and Policy Research, Instituto Superior Técnico, Technical University of Lisbon, Portugal

Abstract

Urban sustainable mobility solutions are being discussed throughout the world to deal with problems of high use of land, congestion situations, energy consumption and emissions derived from road transportation. Part of these solutions concerns the promotion of public transportation as an alternative to individual transportation. Public transport only becomes an attractive option if, for the mobility provided (passenger x km), emits less pollutants and consume less energy.

This paper focuses on comparing the fuel consumption and tailpipe emissions (CO_2, HC, CO, NO_x, PM) between urban Diesel buses (10 litres and 2.7 litres turbocharged Diesel) *and* typical gasoline and diesel European light duty vehicles (1.2 litres gasoline, 1.9 litres turbocharged Diesel). Results are analyzed as a function of transported passengers. For the proposed analysis, typical low traffic and congested traffic driving cycles were selected as an input to the EcoGest model, combined with driver behaviour characteristics and vehicle characteristics. Main results are easy-to-use charts highlighting where each mode is more attractive as a function of the number of passengers inside the bus and the number of people inside the vehicle.

Results were obtained using EcoGest, a numerical model developed in our research institute. This model allows simulating the energetical and environmental performance of a road vehicle in a predefined route, accounting for the driving behaviour, vehicle specifications, powertrain specifications, occupation rate and road grade. The model has been used to simulate light duty and heavy duty vehicles, including gasoline, natural gas and hydrogen cars, and diesel and natural gas buses. Predicted results have been compared with measured fuel consumption and tail-pipe emissions in real-traffic situations, standard cycles (FTP75, NEDC), and predicted results of similar models (e.g. ADVISOR). Predictions of EcoGest are, usually, very good.

KEYWORDS – public transport, individual transport, energy consumption, emissions, urban transportation planning

Introduction

All industrialized countries are facing a high dependency in foreign oil, which is mainly consumed by the transport sector. Besides this energetic problem, anthropogenic green-house gas emissions, and climate change/global warming (derived from the combustion of fossil fuels) have become more and more a cause of concern. In 2005, the transport sector contributed to 25% of all the world anthropogenic CO_2 emissions released into the atmosphere. Approximately 80 % of those emissions are from road transport, of which 60 % is from automobiles, sports utility vehicles, and pick-up trucks used to meet commuter and other household transport needs 0. Also road transport sector contributes significantly to other emissions responsible to harm our health: hydrocarbons (HC), carbon monoxide (CO), nitrogen oxides (NO_x) and particles (PM). For example, in 2003 the European Union (Euro 25) accounted for a contribution from this sector of 44% for total anthropogenic CO; 40% total NO_x, 26% total HC and 16% of total PM 0.

Linked with road vehicle growth of individual transportation, congestion problems, traffic hazards and urban pollution are also increasing.

Solutions for urban sustainable mobility are being discussed throughout the world in order to deal with problems concerning the high use of land, congestion situations, energy consumption and emissions derived from road transportation. These solutions cover:

- promotion of public transportation as an alternative to individual transportation;
- tax incentives to drive more environmentally friendly vehicles;
- introduction of Intelligent Transportation Systems (ITS) such as Tolling (based on using Satellite Positioning on Highways and Urban Areas), congestion pricing, traffic lights, etc;
- urban planning;
- technological developments, such as improving vehicle energetic efficiency, going for more efficient alternative propulsion systems (hybrids, fuel cells), increasing availability of possible feeding fuels (bio fuels, natural gas, hydrogen).

The first solution is seen as part of a decisive factor to reduce congestion and emissions in urban areas. However, if the number of passengers is low, public transportation is not always the best solution. This paper focus on that measure by analysing the energy and environmental performance of public and individual transportation on a basis of the number of people transported.

Table 1 Main features of the LDV and buses used for the simulations

Vehicle	Car	Car	Mini bus	Regular bus
# of seats	5	5	32	80
Fuel	gasoline	diesel	diesel	diesel
Engine displacement (dm^3)	1.2	1.9	2.7	10
Engine maximum Power/Torque	54kW@ 5600rpm 110Nm@ 4000rpm	74kW@4000rpm 240Nm@1800rpm	115kW @ 3800rpm 330Nm @1400-2400rpm	180kW@2200rpm 1100Nm@2300rpm
Air induction	atmospheric	turbo	turbo	turbo
Catalytic converter	three-way	oxidation	oxidation	oxidation
Emission standard	Euro III	Euro III	Euro III	Euro III
Transmission (5-speed)	manual	manual	auto	auto
Weight (kg)	1010	1400	2360	10300
Frontal area (m^2)	2.0	2.2	3.8	6.5
Drag coefficient	0.32	0.32	0.36	0.6
Tires	185/55 R15	195/65 R15	195/70 R15	275/70 R22.5
Coefficient of rolling resistance	0.010*	0.010*	0.007*	0.007*
Coefficient of rotational inertia	1.05*	1.05*	1.15*	1.15*

* Data assumed by the authors, based on reference values from the literature 0

Methodology

Four vehicles, typical of European cities, were chosen for this study: two cars (light-duty vehicle – LDV) and two buses. Their characteristics are summarized in Table 1.

Two traffic situations were considered, both for a distance of 1.0 km:

- Low traffic (Figure 1) – the cars start from 0 km/h, accelerate to cruising speed (50 km/h), and brake to 0 km/h at a traffic light at 1.0 km, where they stay for 30 s. The buses also start from 0 m/s and stop at the same traffic light where they stay for 30 s too, but their cruising speed is lower (40 km/h); furthermore, they stop twice for 20 s at bus stops;
- Congested traffic (Figure 2) – .the cars stop for 30 s at two traffic lights, and stop three times for 10 s in stop-and-go situations. The buses stop similarly to the cars, but again they stop twice for 20 s at bus stops.

The acceleration and deceleration rates for the cars and buses were obtained from real-traffic measurements.

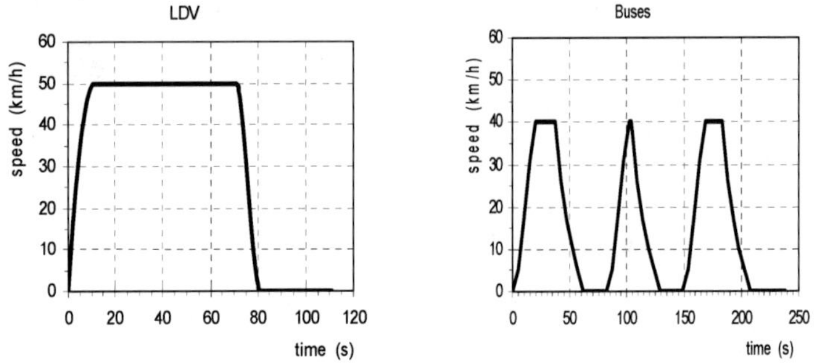

Figure 1. Driving cycles under low traffic for the cars (left) and buses (right). Distance: 1.0 km. Average speeds of 33 km/h and 15 km/h, respectively

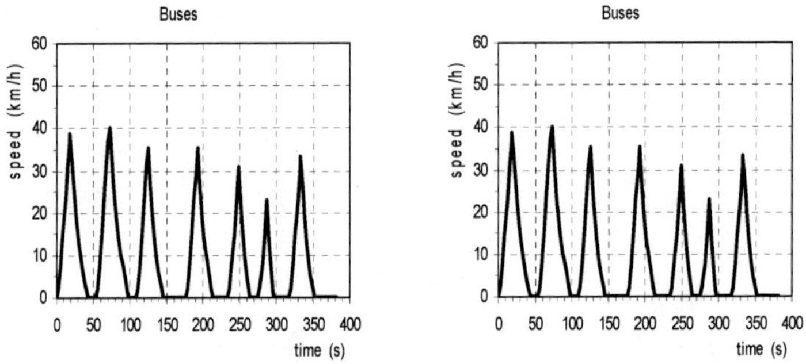

Figure 2. Driving cycles under congested traffic for the cars (left) and buses (right). Distance: 1.0 km. Average speeds of 17 km/h and 9 km/h, respectively.

The simulation tool used for the calculations is EcoGest code 0, based on equations of longitudinal dynamics behaviour of vehicles combined with engine and catalytic converter numerical simulation (see Figure 3). This numerical code, developed in our research institute, allows simulating the energetical and environmental performance of a road vehicle in a predefined route, accounting for driving behaviour, vehicle specifications, powertrain specifications, fuel specifications, occupation rate and road grade. The model has been used to simulate light duty and heavy duty vehicles, including gasoline, natural gas and hydrogen cars, and diesel and natural gas buses. Predicted results have been compared with measured fuel consumption and tailpipe emissions in real-traffic situations, standard cycles (FTP75, NEDC), and predicted results of similar models (e.g. ADVISOR) 0. Predictions of EcoGest are, usually, very good for fuel consumption and CO_2 emissions. In terms of average trip-based values, fuel consumption and CO_2 emission predictions are typically within 10% of experimental data, and in many cases within 5%. HC, CO and NO_x emission predictions are poorer, in many cases within 20% of the data, with up to 50% deviations from experimental data, but the model captures the qualitative trends and is similar to current worldwide used models such as ADVISOR. It should be noted that the present study compares results for the LDV and buses, all obtained by simulation on EcoGest. Therefore, the relative errors on those comparisons are much lower than the values mentioned above.

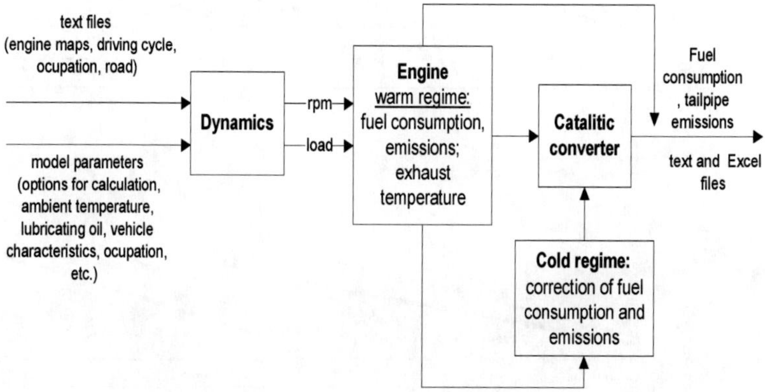

Figure 3. Schematics of EcoGest model 0 and 0.

The output data of EcoGest for all the vehicles and driving cycles was treated and final results were presented per *passenger*km*. Easy-to-use charts were drawn, highlighting where each mode (public or private) is more attractive as a function of the number of passengers inside the bus and number of people inside the LDV. A future scenario is drawn considering more advanced aftertreatment technologies such as particle filters and selective catalytic reduction of NO_x emissions.

Results

Each chart stands for fuel consumption and for emissions (CO_2, HC, CO, NO_x, PM), under low congested and highly congested traffic. Each line draws the border between the regions where fuel consumption and emissions are lower in each vehicle.

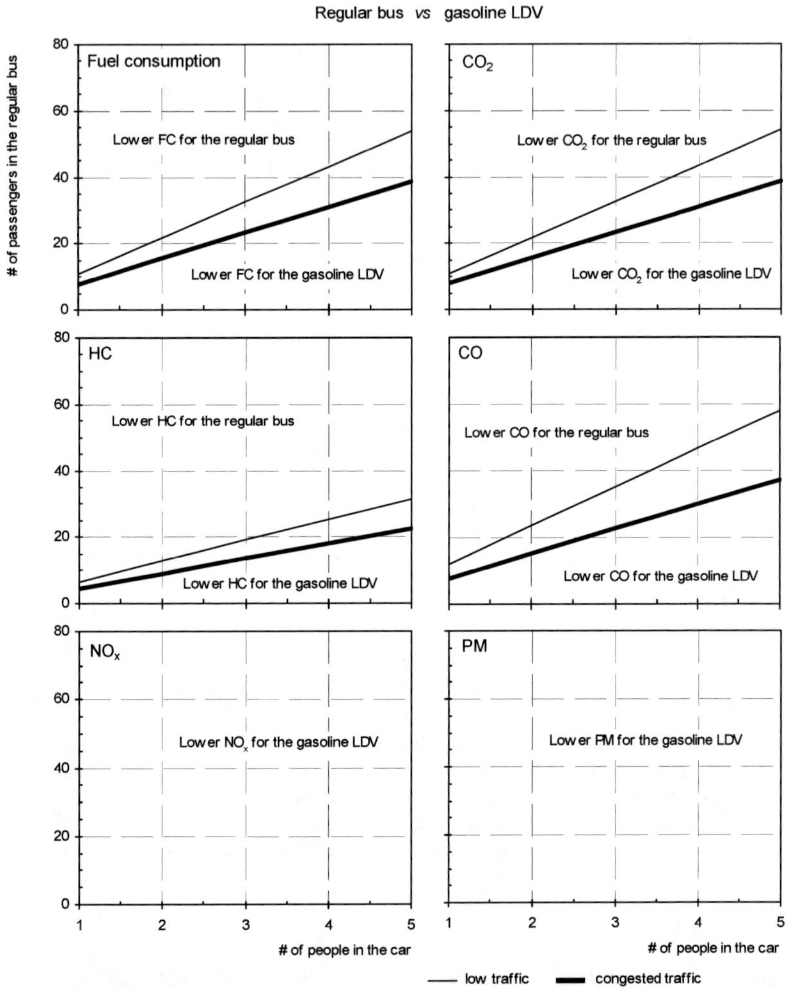

Figure 4. Equivalence between the regular bus and the gasoline LDV for fuel consumption (FC) and for emissions, computed per ***km*passenger***

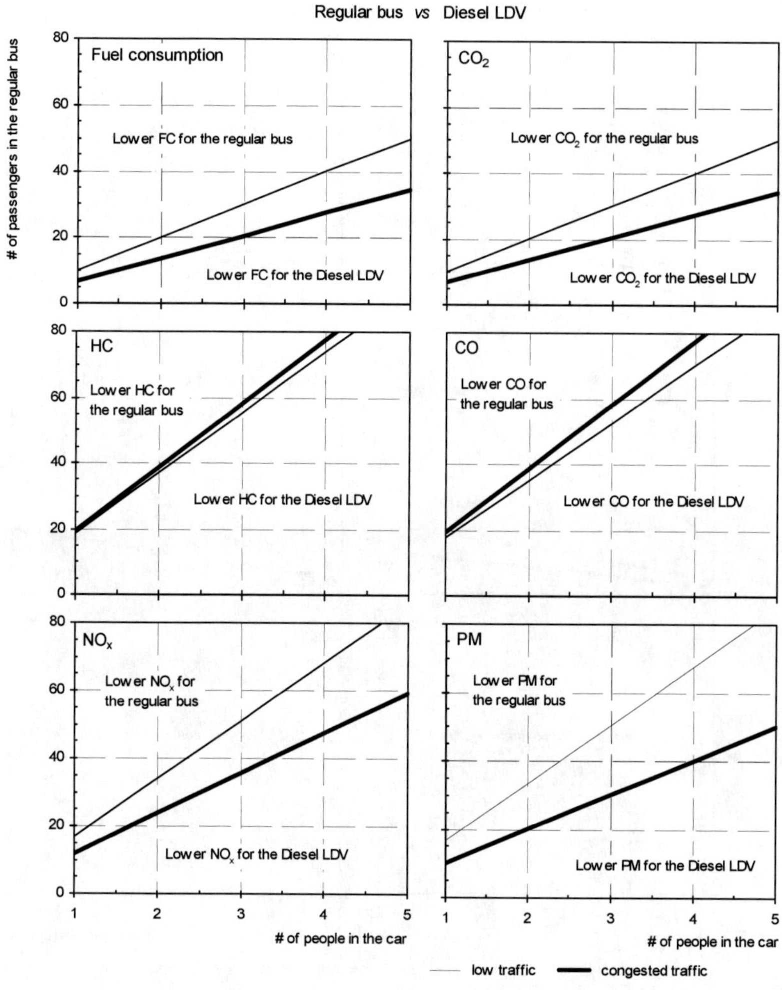

Figure 5. Equivalence between the regular bus and the Diesel LDV for fuel consumption (FC) and for emissions, computed per ***km*passenger***

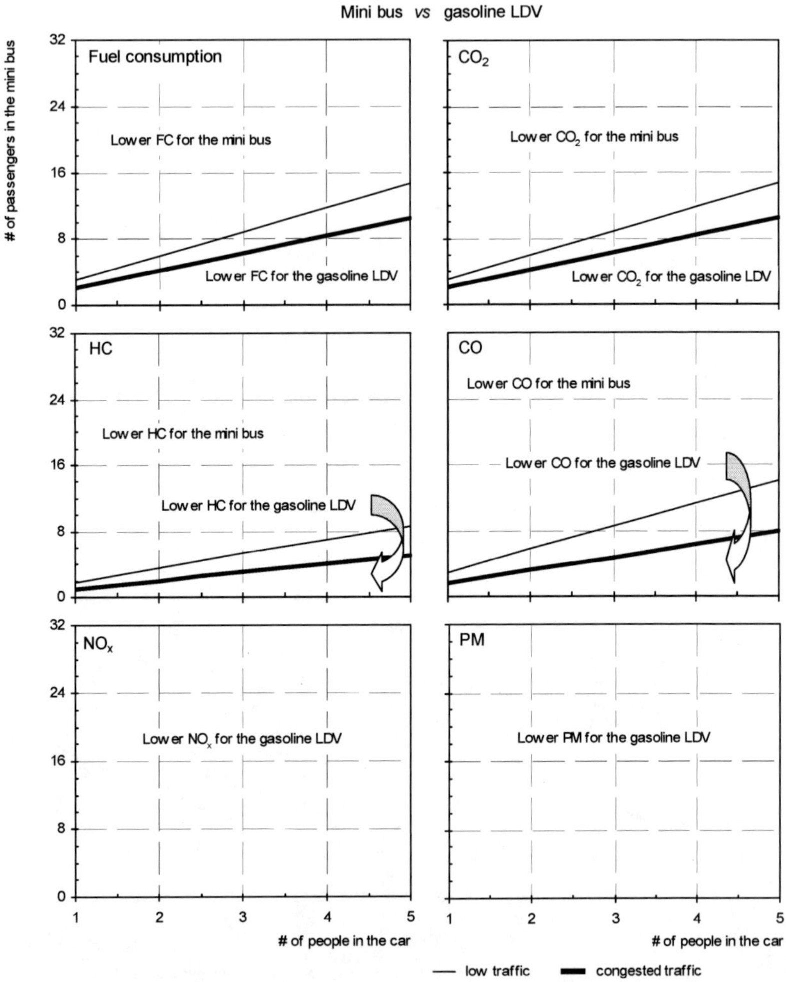

Figure 6. Equivalence between the mini bus and the gasoline LDV for fuel consumption (FC) and for emissions, computed per ***km*passenger***

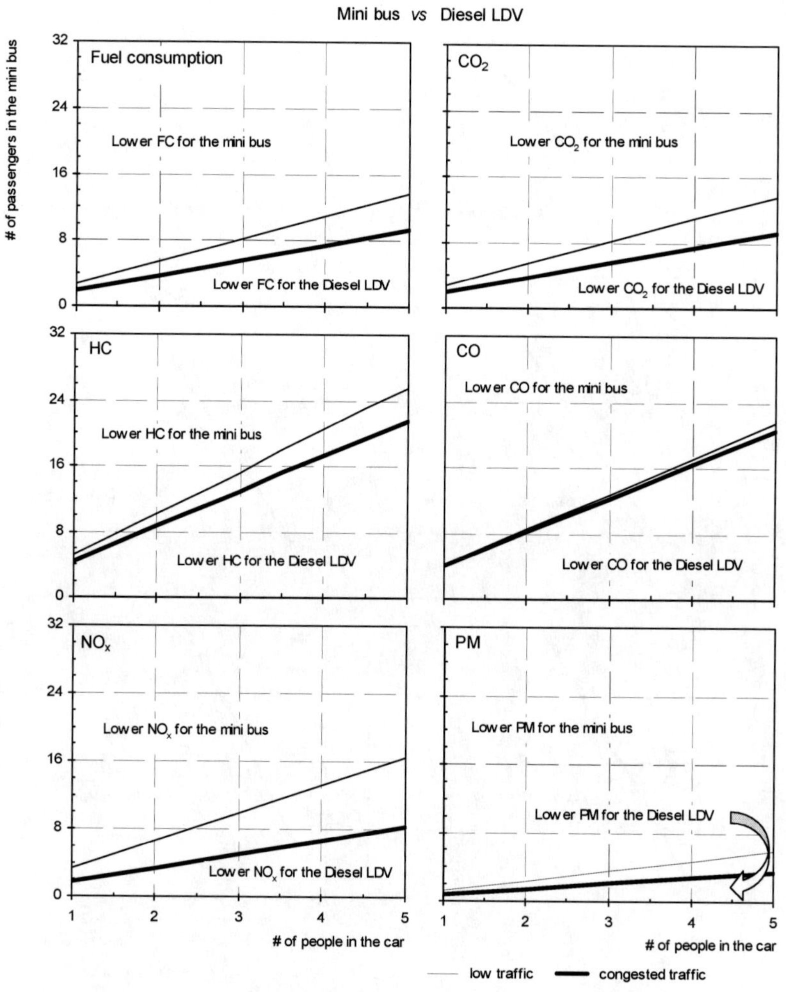

Figure 7. Equivalence between the mini bus and the Diesel LDV for fuel consumption (FC) and for emissions, computed per ***km*passenger***

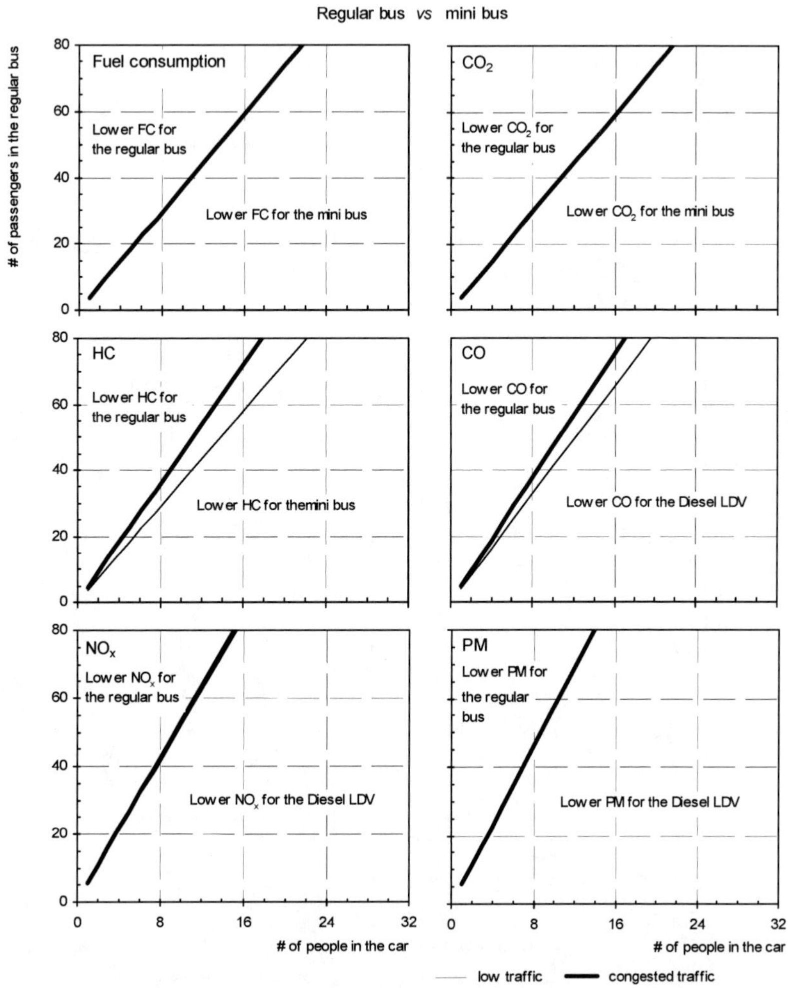

Figure 8. Equivalence between the regular and the mini buses for fuel consumption (FC) and for emissions, computed per ***km*passenger***

Figures 4 to 8 show that the equivalences are virtually linear and with intercept very near zero. Therefore, the results can be summarized as shown in Tables 2 and Table 3.

Table 2 - Number of passengers in the buses equivalent to one person in the cars for the same energetical and environmental performance per *km*passenger*

Low congested traffic

Number of passengers in the regular bus						
	FC	**CO2**	**HC**	**CO**	**NOx**	**PM**
Gasoline LDV	10.9	10.9	6.4	11.8	> 80	> 80
Diesel LDV	10.1	10.1	18.5	17.6	17.0	16.8
Number of passengers in the mini bus						
	FC	**CO2**	**HC**	**CO**	**NOx**	**PM**
Gasoline LDV	3.0	3.0	1.8	2.9	> 32	> 32
Diesel LDV	2.8	2.8	5.1	4.3	3.3	3.0

High congested traffic

Number of passengers in the regular bus						
	FC	**CO2**	**HC**	**CO**	**NOx**	**PM**
Gasoline LDV	7.8	7.8	4.5	7.6	> 80	> 80
Diesel LDV	6.9	6.9	19.4	19.5	12.0	10.0
Number of passengers in the mini bus						
	FC	**CO2**	**HC**	**CO**	**NOx**	**PM**
Gasoline LDV	2.1	2.1	1.0	1.6	> 32	> 32
Diesel LDV	1.9	1.9	4.3	4.1	1.7	1.8

Table 3 Number of passengers in the regular bus equivalent to one person in the mini bus for the same energetical and environmental performance per *km*passenger*

Number of passengers in the regular bus						
	FC	**CO_2**	**HC**	**CO**	**NO_x**	**PM**
Mini bus	3.7	3.7	3.6	4.1	5.1	5.7

A Glimpse Into The Future

A future scenario was also drawn considering more advanced aftertreatment technologies such as selective catalytic reduction of NO_x emissions and particle filters. According to catalyst companies, typical conversion efficiencies for NO_x emissions and particles using a Selective Catalytic Reduction + Continuously Regenerating Trap are, respectively, 70% and 90% 0. Typically there is an increase in fuel consumption due to increase backpressure of 5% 0.

Considering such kind of devices installed in the diesel vehicles, their comparative results remain similar.

For the gasoline LDV, the values relative to fuel consumption and CO_2 emissions in Table 2 should be increased approximately by 5%. As far as NO_x is concerned, the number of passengers equivalent to one person in the gasoline LDV, in the buses would still be above their maximum capacity. However, if an increase of conversion efficiency of the SCR is considered, the results would be as shown in Table 4.

Table 4 Number of passengers in the buses equivalent to one person in the gasoline LDV for the same NOx emissions per *km*passenger*

Number of passengers in the buses			
Conversion efficiency of the SCR	**80%**	**90%**	**99%**
Regular bus	> 80	70	7.8
Mini bus	27	14	2.1

PM emissions should also be more comparable to gasoline LDV PM emissions. However, the model does not calculate PM for gasoline vehicles.

Conclusion

The main results obtained for two buses and two LDV, typical of European cities, have shown that the more efficient mode of transport (energetically and emissions wise) can be either the LDV, a mini bus, or a regular bus, depending on the number of people transported. The results expressed per *passenger*km* for low traffic situations show that

- each person in the gasoline LDV is equivalent to approximately 11 to 12 passengers in a regular bus for fuel consumption, CO_2 and CO emissions;
- that number falls to 6 to 7 for HC emissions, but the gasoline LDV is always preferable for NO_x and PM emissions if the bus is not equipped with Selective Catalytic Reduction and Continuously Regenerating Trap;
- each person in the Diesel LDV is equivalent to approximately 10 passengers in a regular bus for fuel consumption and CO_2;
- that number goes up to 17 to 18 for the remaining emissions;
- in the comparison between the mini bus and the LDV, each person in the gasoline LDV is equivalent to approximately 3 in the bus for fuel consumption, CO_2 and CO emissions;
- again, that number falls to approximately 2 for HC emissions, but the gasoline LDV is always preferable for NO_x and PM emissions if the bus is not equipped with Selective Catalytic Reduction and Continuously Regenerating Trap;
- each person in the Diesel LDV is equivalent to approximately 3 passengers in a mini bus for fuel consumption and for CO_2, NO_x and PM emissions;
- that number goes up to 4 to 5 for HC and CO emissions.

If traffic is congested, the equivalent number of passengers in the buses drops to approximately 60 to 70% of the values above.

In the comparison between the mini and the regular buses, each person in the mini bus is equivalent to approximately 4 in the regular bus for fuel consumption, CO_2, HC and CO emissions, and to 5 to 6 for NO_x and PM emissions. These results do not depend significantly on the traffic situation, although differences in the response of the automatic transmission of the buses may account for some small differences in those equivalences between the numbers of passengers.

The results presented illustrate one aspect to be considered (energetical and environmental performance) when deciding whether buses or private cars are preferable for each specific situation. Other criteria (technical and political) have to be considered to take into account a more global view of the problem.

Nomenclature

HC	hydrocarbons
CO	carbon monoxide
CO_2	carbon dioxide
FC	fuel consumption
NO_x	nitrogen oxides
PM	particles (Particle Matter)
RPM	engine speed
SCR	Selective Catalytic Reduction
CRT	Continuously Regenerative Trap

Acknowledgments

The authors would like to acknowledge FCT (National Science and Technology Foundation) through the national project POCI/ENR/57450/2004. The second author is supported by a postdoc grant of the FCT and European Social Fund (FSE), within the Third Framework Program.

References

World Resources Institute 2006.

EUROSTAT 2006.

Bosch. Automotive Handbook. 6ª Edition, Bosch, 2004.

Silva, C. M., G. A. Gonçalo, T. L. Farias, and J. M. Mendes-Lopes (2006), "A Tank-to-Wheel Analysis Tool for Energy and Emissions Studies in Road Vehicles. *Journal of Science for the Total Environment, Vol.367, Issue 1, Pages 441-447, 2006.*

C. M. Silva, T. L. Farias, H. Christopher Frey and N. M. Rouphail (2006) "Evaluation of numerical models for simulation of real-world hot-stabilized fuel consumption and emissions of gasoline light-duty vehicles." *Transportation Research part D: Transport and Environment, Volume 11, Issue 5, Pages 377-385, 2006.*

http://www.eminox.com/products/emissions-control.shtml

Pischinger, S., M. Lamping, and A. Sehr (2006), "The effect of pollutant emission reduction." Research and Development, Auto Technology.

CUTE Hydrogen Bus Project: Findings and Extension to Hyfleet

G. A. Gonçalves,[1] J. O. Portugal,[2] and T. L. Farias[3]

[1]Instituto Superior Técnico, Universidade Técnica de Lisboa, Departamento de Engenharia Mecânica, Av. Rovisco Pais, 1049-001 Lisboa, Portugal; email: gonçalo.goncalves@ist.utl.pt
[2]Instituto Superior Técnico, Universidade Técnica de Lisboa, Departamento de Engenharia Mecânica, Av. Rovisco Pais, 1049-001 Lisboa, Portugal; email: joana.portugal@ist.utl.pt
[3]Instituto Superior Técnico, Universidade Técnica de Lisboa, Departamento de Engenharia Mecânica, Av. Rovisco Pais, 1049-001 Lisboa, Portugal; email: tiago.farias@ist.utl.pt

Abstract

The use of hydrogen as an alternative fuel has been suggested as a key element in promoting the improvement of urban air quality and to reduce GHG emissions, while diversifying and guaranteeing the security of energy sources supply. Several projects have been undertaken to develop a strategy towards a European hydrogen-based transport system. The present paper describes the part of two of those projects: CUTE and HYFLEET:CUTE, sponsored by European Commission's 6th Framework Research Programme, involving several partners, mainly from industry, governments, academic and transport operators.

The CUTE project was officially closed on March 2006, while HyFLEET:CUTE is an ongoing project. Both projects have the objective of developing and demonstrating a hydrogen based public transport system. This paper will look at the main conclusions of CUTE project and at the life-cycle analysis of the hydrogen used in the filling station and the fuel cell buses that ran in the city of Porto. The life-cycle analysis covers al phases from production, transport, and storage of hydrogen, up to the operation in the municipal fleet. This included the study of the energy efficiencies and environmental impact (the so called well-to-tank and tank-to-wheel studies) of different hydrogen production pathways. Regarding the bus operation, the energy and environmental vehicle performance is assessed and compared with current technologies.

Furthermore, in the framework of the Hyfleet:CUTE project, a new filling station was built by the French oil company Total in Berlin that uses on-site hydrogen production with a LPG reforming system, combined with liquid hydrogen trucked from a Linde factory. The municipal fleet vehicles used are built by MAN and use an internal combustion engine and compressed hydrogen storage.

Introduction

The European Union is facing a serious energy crisis, struggling with important problems regarding oil shortage, increasing of energy demand and environmental concerns. The transport sector is one of the main energy consumers (and unlike many other activities it is almost entirely dependent on oil fuel sources) and simultaneously it is one of the main sectors responsible for green house gases (GHG's) emissions. Accordingly, in 2004 the transport sector accounts for 21% of total EU-15 emissions, but it is projected that the current EU-15 emissions continue increasing to reach 35% above 1990 levels (EEA, 2006).

Driven by security of supply and environment issues, European Union State Members have committed themselves to the ambitious target of introducing 20% of alternative fuels, including hydrogen, by 2010 (CE, 2000). Therefore, alternative fuels, such as hydrogen, have come to the forefront of research in recent years as it is considered the key element in promoting the improvement of urban air quality and to reduce GHG emissions, while diversifying and guaranteeing the security of energy sources supply.

Hydrogen is a clean energy carrier (in terms of emissions resulting from its use) that is generated from a wide range of energy sources, both fossil and renewable. Thereby European Union has been conducting several projects to develop a strategy towards European Hydrogen-based Transport System. CUTE and HyFLEET:CUTE are two of those projects, sponsored by European Commission, involving several partners, mainly from industry, governments, academic and transport operators. The present paper describes the Life Cycle Assessments (LCA's) of both hydrogen filling infrastructure and the hydrogen powered vehicles that were/are carried out in Porto (Portugal) and Berlin (Germany). This included the study of the energy efficiencies and environmental impact (the so called well-to-wheel analysis) of different hydrogen production pathways and vehicle technology options.

Hydrogen Supply Infrastructure

One of the most important factors when developing an alternative fuel transportation system is the capacity to supply the fuel. Hydrogen is the most abundant element on Earth, though it is not considered a primary energy source, as it can rarely be found in its molecular state (H_2). The CUTE and HyFLEET:CUTE projects explore wide range pathways to produce hydrogen as a transport fuel, including steam reforming and water electrolysis, both on-site and centrally hydrogen production. At the present stage, steam reforming from natural gas is the most feasible and cost-effective technology to produce hydrogen on a large scale. Nevertheless, regarding a GHG's emissions free system and the fossil resources dependency, alternative pathways were analyzed using renewable energy sources (solar, hydro and geothermal power) for hydrogen production.

The volume energy density of hydrogen gas under ambient conditions is much lower than that of conventional fuels. Hydrogen must therefore be compressed at a high pressure (350 bars or higher) in order to ensure enough range for daily bus operation. The hydrogen may be either stored at a high pressure before the refuelling (overflow filling) or compressed to a suitable pressure during the filling (booster filling system). Both systems were used and in both cases, the filling time is expected to be no longer than 15 minutes. During the CUTE and HyFLEET:CUTE projects it was decided to cover a broad range of alternatives, regarding hydrogen production routes and filling station technologies, as it allowed concluding what the best route to provide hydrogen, regarding energy consumption, hydrogen losses and costs. The characteristics of CUTE and HYFLEET:CUTE filling stations are summarised at Table 1.

Table 1 Hydrogen filling stations characteristics.

	Primary energy source	H_2 production path	Max. filling time (min.)
Amsterdam	Renewable (green electricity)	On site (electrolysis)	15
Barcelona	Renewable (solar)	On site (electrolysis)	20
Berlin	Crude Oil + GPL	On site (steam reforming) and centralised (transport in gaseous form)	–
Hamburg	Renewable (wind)	On site (electrolysis)	<10
London	Crude Oil	Centralised (transport in liquid form)	30
Luxembourg	Natural Gas	Centralised (transport in gaseous form)	10
Madrid	Crude Oil + Natural Gas	On site (steam reforming) and centralised (transport in gaseous form)	10-15
Porto	Variable (grid mix)	Centralised (transport in gaseous form)	12-15
Stockholm	Renewable (green electricity)	On-site (electrolysis)	20-35
Stuttgart	Natural Gas	On-site (steam reforming)	<15

The Cute and HyFLEET:CUTE refuelling facilities supplied so far the fuel cell buses with over 192,000 kg hydrogen in more than 8,900 refuellings. The following figure shows the amount hydrogen dispensed at each site. Amsterdam, Barcelona, Hamburg, Stockholm, Madrid and Stuttgart use on-site hydrogen production, accounting for more than 120,000 kg of hydrogen. The other sites use exclusively external supply.

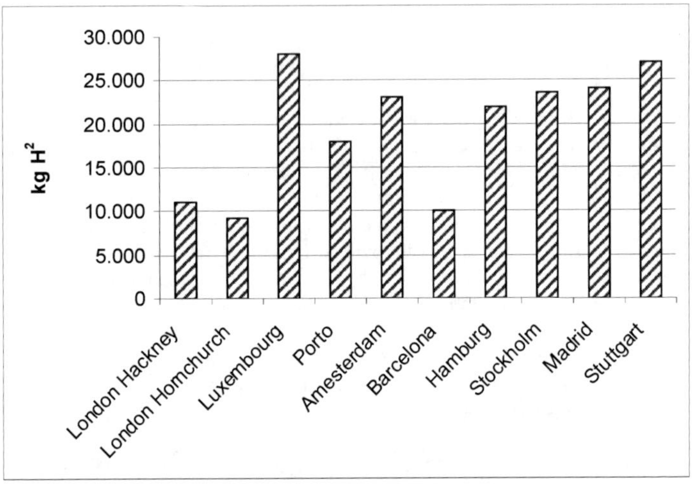

Figure 1. Amounts of Hydrogen dispensed at each site (CUTE, 2006)

Figure 2 shows the specific hydrogen losses relative to the sum of external supply and on-site generation. The sites with only a few problems during the operating phase had a standard value for hydrogen losses in the range of 5% to 10%, due to purging of system components and background leakage. Though sites with significant components failures, such as London and Stuttgart, display a higher level of loss.

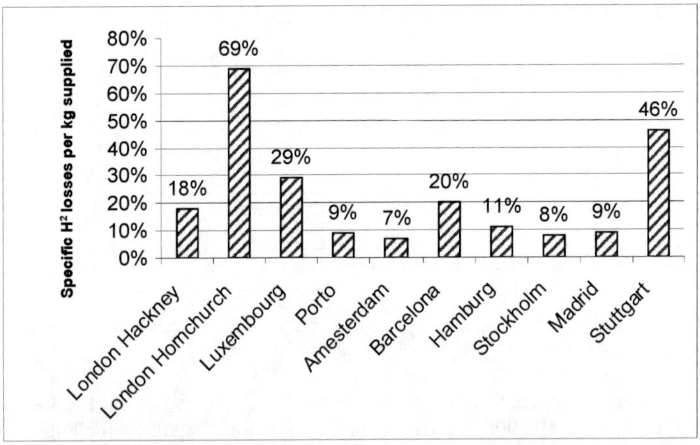

Figure 2. Specific hydrogen losses (CUTE, 2006)

Hydrogen Infrastructure in Porto

The hydrogen for the Porto filling station was produced in the Linde plant in Alenquer (300 km south of Porto) through an electrolysis process using grid electricity. It was then transported by road in tankers (in gaseous form at 200 bar) to the filling station. After that the trailers were connected to the station, serving as stationary storage. As the storage pressure in the buses has a maximum of 350 bar, in order to transfer the hydrogen to the fuel cell buses an additional compression stage was required, raising the filling pressure to 400 bar (Booster filling system).

Hydrogen Infrastructure in Berlin

Berlin hosts the most ambitions HyFLEET:CUTE demonstration project. In fact, its refilling infrastructure is capable of refuelling a fleet of up to 20 buses, which aims to achieve higher energy efficiency than the current status CNG technology. The hydrogen is produced on-site or it is externally supplied by Linde. The on-site production is carried out with LPG (Liquefied Petroleum Gas) from oil refineries, by a steam reforming process. It is also been considered the possibility of reforming Bio-DME (Bio Dimethyl Ether) in substitution of LPG, as it is a renewable resource, therefore it environmental impact is much lower.

Furthermore two stationary fuel cell CHP (combined heat and power) systems were installed. These use surplus hydrogen produced in the station as well as 'boil off' from stored liquid hydrogen to generate heat and electricity to supply the filling station. Any surplus electricity generated is fed into the electricity grid. This arrangement will greatly increase the total efficiency of the filling station. Figure 3 presents the diagram of the Berlin hydrogen infrastructure:

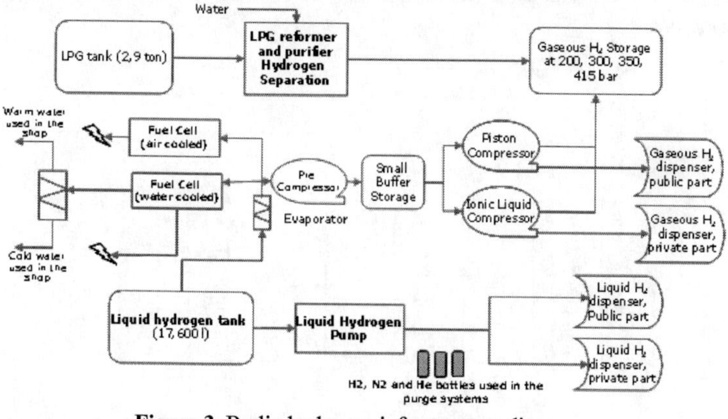

Figure 3. Berlin hydrogen infrastructure diagram

Bus Technology

Fuel Cell Buses

On of the main goals of the CUTE program was to demonstrate the viability of fuel cell technologies in transports. The vehicle used in the project was based on the standard Diesel powered Citaro bus commercialized by EvoBus (a subsidiary of the DaimlerChrysler group). The engine and transmission are located in the same position and the vehicle offers the same passenger space. In order to increase the reliability and serviceability of the vehicle, the changes have been reduced to a minimum, keeping as much of the conventional bus parts as possible.

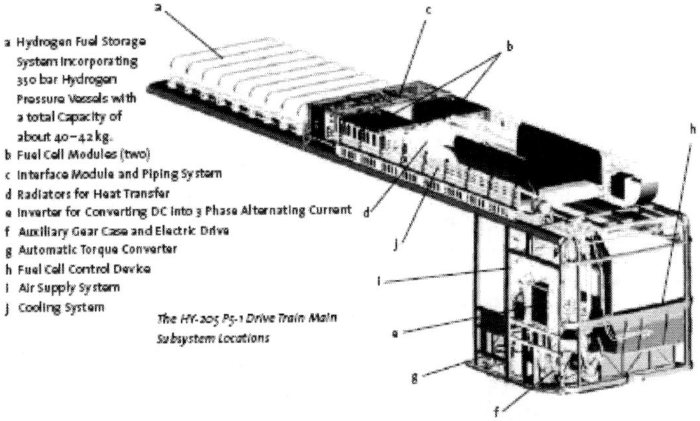

Figure 4. Location of the main subsystems (CUTE, 2006)

After the two years project the fuel cell buses completed an average of 94,000 km and operated for 2,330 hours. It average speed is showed in the Figure 5 where it can be observed a large difference between the nine cities due to it different traffic conditions.

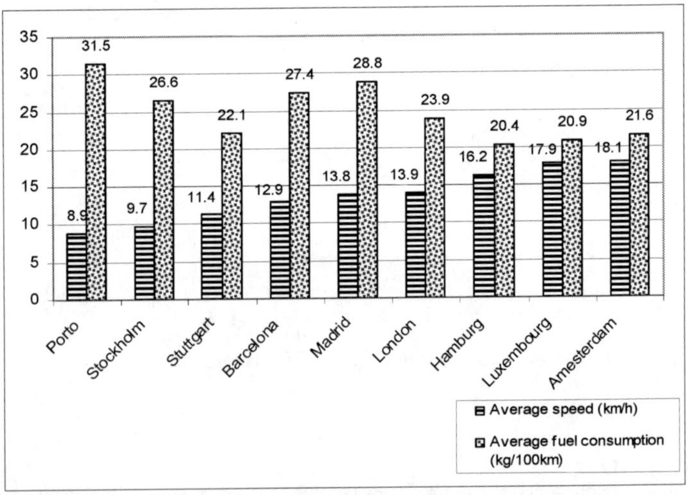

Figure 5. The average speed and fuel consumption for the nine CUTE cities

The buses availability was extremely high (Figure 6) varying from 60% (Barcelona) to 99.6% (Stuttgart). The overall average in all the nine CUTE cities was 81.6%.

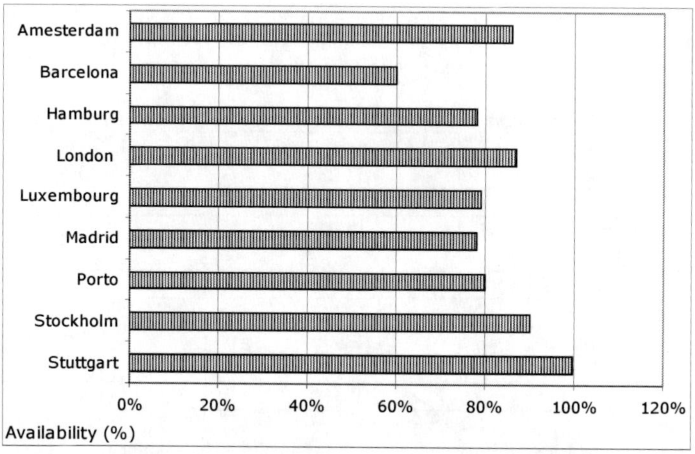

Figure 6. Vehicle availability in the nine CUTE cities

ICE Buses

The ICE buses that operate in Berlin were developed by MAN and are loosely based on the company CNG buses. The engine is a four stroke six cylinder inline, horizontally installed. Empty weight is 12,600kg. Two batches of vehicles are manufactured: the first uses an indirect injection normally aspirated engine developing 150kW. The second generation uses a direct injection, lean burn, turbocharged engine developing 200kW. Hydrogen is stored compressed at 350bar in rooftop tanks (10 cylinders, ~50kg H2 stored).

Energy and Environmental Analysis of Hydrogen Supply Infrastructure and Fuel Cell Buses (Well-To-Wheel Analysis) in Porto

Well-To-Tank Analysis

Regarding the Well-to-Tank analysis life cycle analysis tool was used, GABI 4. (GABI, 2003), using information available at it database and country profile specific information. The analysis considers the energy consumption and GHG's emissions (CO_2, CH_4, N_2O). The following results are related with the life cycle analysis that was conducted in Porto. As it was stated before, in Porto station the hydrogen was produced off-site through a electrolysis process. Two options where considered, in the first the electricity used is obtained directly from the Electricity grid, and represents an average of the total electricity production plants in the country. In the second option, the electricity for the electrolysis process is of renewable source; all other processes (compression, transport) are similar.

The processes for electricity production where modelled according to the database of the GABI software. The efficiencies considered reflect the technology used. Renewable energy sources (wind, hydro, solar) are considered to have an efficiency of 100%, as they convert renewable energy directly into electricity, without a thermal cycle; solid waste use is also considered to have an efficiency of

100% due to the fact that it uses waste and not primary resources. The sources of energy for electricity production where obtained from ERSE (ERSE, 2001). Table 2 provides an overview of the electricity production in Portugal.

Table 2 Sources of Energy for electricity Production in Portugal

Power Plant	Contribution	Efficiency
Coal	34.28%	30%
Hydro	25.55%	100%
Natural gas (simple and combined cycle)	14.07%	43%
Fuel oil + Diesel oil	11.71%	29%
Co-generation	10.90%	58%
Imports	2.00%	33%
Solid Waste	1.17%	100%
Wind	0.31%	100%

According to ERSE (2001) losses in the transport grid amount to 1.7% of total electricity production, the losses in the distribution grid are 8.4%. Global efficiency of production, transport and distribution is 42,7%, though when only renewable electricity is used, the overall efficiency raises up to 90%. According to GM (2002) efficiency of electrolysis plant is 65%, with an energy consumption of 0,43 kWh/MJ of hydrogen. In the case of the equipment used in the Linde Production Plant, the final pressure of the hydrogen is 10 bar. This hydrogen suffers a primary compression up to 250 bar for transport (the tanker pressure is 200 bar). It is assumed that the efficiency of the compressor is the same than the one used for natural gas: 68%. The compressed hydrogen is transported by road using diesel tankers, the total distance covered is 300 km, the return trip also considered. The Secondary compression raises the pressure from 200 to 400 bar to fill the vehicles.

The results regarding the energy consumption and GHG's emissions of the hydrogen production and storage are presented in Figure 7 and Figure 8.

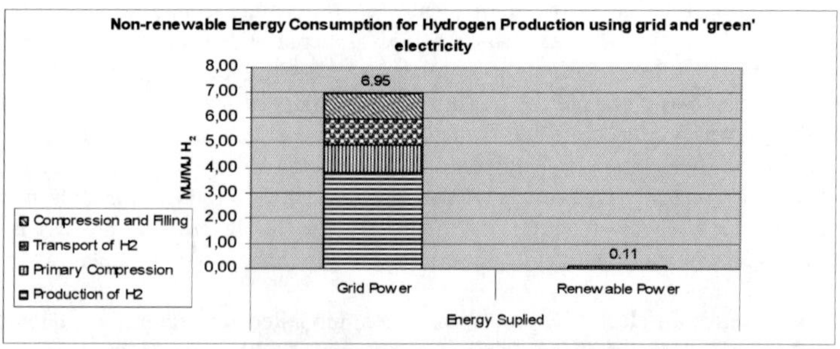

Figure 7. Energy Consumption for hydrogen production and storage applying grid and 'green' electricity.

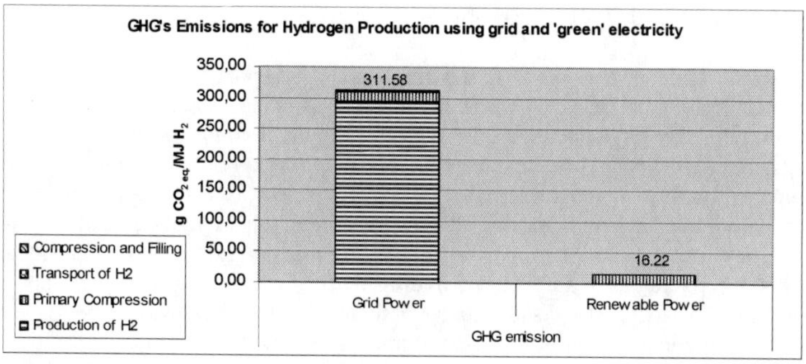

Figure 8. Energy Consumption for hydrogen production and storage applying grid and 'green' electricity.

Tank-To-Wheel Analysis – CUTE-Porto

Emissions and the energy consumption of the different vehicles (diesel, natural gas and hydrogen) where calculated using two different software: Copert (2000) for Diesel vehicles and Mobile 6 (2002) for CNG vehicles. Consumption, mileage and vehicle characteristics where obtained directly from STCP. Regarding hydrogen, it was admitted that it emits no harmful emissions and an energy consumption of 23.9 MJ/km (for vehicle with a 200km autonomy it was accepted a 40 kg of hydrogen).

Table 3 Energy Consumption and TTW GHG's emissions for a regular Euro III Diesel Bus, a CNG bus and the CUTE Hydrogen bus.

	Energy Consumption MJ/Km	TT WGHG's Emissions Kg $CO_{2\ eq.}$/Km
Diesel (Euro III)	18.4	1.370
CNG	22.9	1.443
Hydrogen	23.9	0

Tank-To-Wheel Analysis - Berlin

The emissions resulting are mostly water with traces of pollutants. At the time of writing the vehicles were being subjected to approval testing by TUV and the results are not yet available. Preliminary results (at the date of writing four buses were already running) indicate emissions much lower than those of conventionally powered Diesel or CNG vehicles. Fuel consumption is also lower than the fuel cell bus for the CUTE project.

Conclusion

Hydrogen is seen as an alternative fuel capable of improving the energetic crisis and environmental and economic consequences that European Union is presently facing. The CUTE and the HyFLEET:CUTE projects have demonstrated that a hydrogen and fuel cell based transport system offers a realistic promise of achieving the fuel oil independence. Though the energy input into the hydrogen production system is crucial in determining the efficiency and the extent to which climate change and fossil fuels resources dependency are improved.

The main result of this work shows that public transport by buses consuming hydrogen produced by conventional technologies does not as yet present a solution for the reduction of GHG emissions. Still, hydrogen from renewable energy resources shows a clear benefit, but it must be noted that this solution seems to stumble on economic barriers, as renewable power generation does not stand alone, at least without subsidies. The production of hydrogen from grid electricity is clearly an intermediate solution, as it is currently the most practical way of producing low amounts of hydrogen, but in the long term it is not a sustainable strategy. The disadvantages associated with hydrogen can be greatly reduced if the local effect of the less polluting vehicle is taken into consideration. Whether this is enough to offset the disadvantages is a matter for further studies.

References

C.E. (2000), "Green Paper - Towards a European strategy for the security of energy supply."

COPERT III (2000), European Environment Agency.

CUTE (2006), "Detailed Summary of Achievements: A Hydrogen Fuels Cell Bus project in Europe 2001-2006.

EEA (2006), "Green House Gas emissions Trends and Projection in Europe 2006." *EEA report no 9/2006.*

ERSE (Entidade Reguladora dos Serviços Energéticos) (2001), "Caracterização do Sector Eléctrico – Portugal Continental 2001."

GABI 4 (2003), Institute for Polymer Testing and polymer Sciences.

MOBILE 6. (2002), US Environment Protection Agency.

Calhau, K.R., Gonçalves, G.A., and T. L. Farias (2004), "Environmental Impact Of Hydrogen In Urban Transports, Renewables 2004." New and Renewable Energy Technologies for Sustainable Development, Évora, Portugal.

Gonçalves, G. A., Farias, T. L., Teixeira, R. and A. Silva (2004), "Operation of Hydrogen Fuel Cell Urban Buses in the City of Porto." *12th PanAmerican Conference Traffic & Transportation Engineering, Albany, New York.*

Estimating Mobile Source Emissions Benefits from Land Use Change – A Review of the State of the Practice

J. S. Lane[1]

[1]AICP, GISP, Director Of Planning, The Louis Berger Group Inc., 1513 Walnut Street, Suite 250, Cary, North Carolina-27511, Ph (919) 467-3885; FAX (919) 467-9458; email: slane@louisberger.com

Abstract

During 2004-2005, the Federal Highway Administration (FHWA) contracted with a private consulting firm to conduct a review of the state of the current practice of estimating the changes in mobile source emissions from land use changes. The study systematically reviewed over 50 independent research items; conducted numerous interviews with various public and private agencies including US EPA and FHWA Division Offices; and formulated a taxonomy of the existing methodologies available to metropolitan planning organizations and other agencies to address this issue.

The study concluded with a description of the obstacles in the areas of travel demand modeling, emissions modeling, research deficits, and organizational barriers that would need to be breached in order to substantially improve the state of the practice. Simple elasticity-based models are the near-universal form of estimating change at this point, but the various context-related issues of the environment make these models very difficult to calibrate with any degree of surety.

This paper reviews the state of the practice discovered during this exercise and highlights some of the methodological barriers and recommendations to overcoming these obstacles. An additional review of the literature produced since the study was completed at the beginning of 2005 will also be conducted. The paper intends to provide the audience with a better grasp of the current methodological issues, applied research, and the general approach to estimating mobile source emissions from land use change.

Context and Purpose

Research and applied practice have been used to attempt to define the nature of land use and travel behavior for several decades. Seminal figures in planning, architecture, human health, engineering, and environmental fields have each placed their imprint on the understanding of how human living environments shape the actions of people within those environments, as well as measuring the consequences of human-environment interactions on both parties. This paper provides an overview of how the research has responded to increasing the understanding of this field, and how the current state-of-the-practice operates and, furthermore, can be improved.

Since the 1970 and 1990 amendments to the federal Clean Air Act of 1963, an increasing number of studies have suggested that land use can also indirectly influence emissions of airborne pollutants. These pollutants are largely produced through the use of internal combustion engines operated in private automobiles and trucks, among other point and area sources. The theories have a similar origin, suggesting that land use patterns influence trip making frequency, trip lengths, choice of what mode of transportation to take, and so forth. A wide variety of approaches and technical strengths are exhibited in this body of literature, but the overwhelming majority of conclusions cite that land use patterns do (1) influence the trip making behavior of individuals; and (2) when measured, these travel changes in turn exert some positive influence in the aggregate level of emissions from private automobiles and trucks when measured over a broad area. There have been a minority of studies that found these changes to be very small, and, on occasion, some studies were sufficiently detailed as to suggest that these changes actually have a negative impact (increased VMT and emissions) within a small area where the proposed land use change is the most concentrated.

To understand the past research and current practice, the Federal Highway Administration (FHWA) sponsored a research project that addressed the following objectives:

- Critically review existing literature and research on the topic of land use, travel behavior, and emission interactions;
- Conduct a series of interviews and case studies with acknowledged experts and practitioners that are currently conducting or supporting work in the area of quantifying the effects of land use on emissions; and
- Outline the state-of-the-practice and recommend improvements and research that can aid practitioners in the future.

The report findings were oriented towards practitioners that are or might be conducting land use and emissions studies.

Key Findings: Review of Past Research

Forty-seven literature items were critically reviewed as part of this project, although more items were rejected after an initial review indicated that they did not present quantifiable results. The project workplan expressly called for a database management system to store all of literature items, based on the recognition that sorting and assimilating information from many studies gathered over a period of months would make the task of distilling their contents problematic. Another reason for creating a database as opposed to a more traditional "flat-file" format in a word processor was that each literature item was to be rated on various factors that described the item's usefulness to the project goals. The Research Team described each literature item based upon rating factors using a one-to-five scale with one being the lowest utility and five being the highest (the score was left blank where the variable did not apply to the particular literature item). These variables described the research's handling of internal/external factors influencing outcomes; the validity of the research approach; how transferable the research is to other areas; if and how the research handled air quality/emissions; and the data requirements for the methodology. By scoring each paper, it was possible to prioritize each research effort by various areas of emphasis. Although no attempt was made to assess inter-rater reliability (the degree to which two or more reviewers may interpret the same information differently), each paper was reviewed twice by different people to add validity to the reviews. The scores were averaged and commentary merged to form a single record of the review.

During the course of the literature review, there were four distinct research tracks identified; an emergent fifth track is that of synthesis research. Each track is characterized by different hypotheses, and, to some degree, by the technical capacity of the agencies conducting the study. These are explained, briefly, below:

- *Health and Human Environment.* The general premise of this research consists of attempts to relate the built environment to human health. A typical hypothesis of this strain of research examined the effects that higher density, mixed-use developments had on the propensity of people to walk more than would be the case in a low-density, homogenous suburban environment. Lower body weight, decreased incidence of cardiopulmonary disorders, and other health benefits were typically cited as the dependent variables in these studies.
- *Transit-Oriented Development:* As the name suggests, transit-oriented development is focused on encouraging people to make more use of transit through compatible neighborhood designs, typically centered on commuter or light rail terminals. A typical area of exploration for this type of research is to determine a minimum threshold of housing units/acre above which mode share significantly tilts toward the use of transportation other than single occupant, private automobiles.
- *New Urbanist Design:* The New Urbanism is a term often used by those in the architecture and urban land planning fields to describe ways of accommodating mixed-use and pedestrian-friendly activities in close proximity. Descriptions include common building designs, relaxing restrictive zoning standards, changing street setback standards, requiring rear lot vehicle access, and adapting street cross-sections to slow vehicular traffic while accommodating pedestrian and bicycle travel. Often, these community designs call to older, pre-World War II neighborhoods that were situated on rectilinear street systems and serviced (initially) by a trolley or other transit system. These studies typically attempt to compare two or more neighborhoods with different design features.
- *Travel Demand Modeling:* The fourth research track is that of achieving improvements to travel demand modeling methods. Most often, these studies focus on the generation of trips and mode choice, and are enhanced by the presence of statistically robust travel behavior surveys, or diaries. Using these surveys allows considerably more control over some of the variables that cross-correlate with other common independent variables that plague the other three tracks of research, particularly household income and auto ownership. On very rare occasions a metropolitan planning organization responsible for conducting travel behavior surveys develops a regular schedule for surveying. The resulting historical profile of travel behavior would be one of the single largest improvements to the land use-travel behavior linkage, since a historical comparison would be able to make a much stronger statement

about cause-and-effect relationships than the cross-sectional studies that have dominated the field on this subject.

- *Synthesis Studies:* Although not described as a separate category in the original research, a fifth literature area has emerged which seeks to synthesize and explain a broad cross-section of past research. These works provide good starting points for initiates into the field of study. Particularly notable are the following three works:
  - Apogee/Haigler Bailly, 1998, *The Effects of Urban Form on Travel and Emissions: A Review and Synthesis of the Literature.* This paper was an unpublished draft completed for the Environmental Protection Agency. The authors concluded that while there had been sufficiently robust research to determine and quantify effects on travel behavior from land use strategies, elasticities should be avoided in favor of setting thresholds where these changes are likely to occur. Many of the highest-quality studies occurred after its release.
  - Reid Ewing and Robert Cervero, 2001, *Travel and the Built Environment – Synthesis.* This report picks up where the Apogee/Haigler-Bailly work leaves off by proposing elasticities for the effects of land use strategies on travel behavior. To do this, the authors compiled past research that had significantly (probability less than or equal to 0.05 of systematic error) proven relationships between certain land use strategies and trip frequencies, trip lengths, mode split and VMT. These elasticities were additive, and hence could be combined to form a composite measure from multiple strategies.
  - Kuzmyak, J. Richard; Pratt, Richard H.; Douglas, Bruce G.; Spielberg, Frank, 2003, *Land Use and Site Design – Chapter 15 of TCRP Report 95, Traveler Response to Transportation System Changes.* This work supports the land use strategy/travel behavior models noted earlier, particularly the effects of density on lowered vehicle miles of travel. The report does note that these effects do not occur in a vacuum and that other, supporting characteristics are important to the overall results.

The literature generally recognized the importance of density, diversity, and design elements (the "three D's" referenced by several authors) on trip-making behavior and therefore emissions. Researchers were able to develop elasticities to describe these effects, which ranged from near-zero to 0.35. Typical values ranged from 0.03 to 0.10. Although all of these figures imply that the relationship between land use characteristics and travel are relatively inelastic, they nevertheless indicate some responsiveness to change, particularly when there are multiple causes for change. Many researchers acknowledged problems with data collection/availability, cross-correlation of key variables, objectivity in measurement, and spatial boundary effects.

Key Findings: Interviews and Case Studies

In addition to the critical review of the literature, the Research Team interviewed a number of researchers and individuals from private agencies, the USEPA, and metropolitan planning organizations. A number of FHWA staff reviewed the draft report and offered suggestions as well.

The case studies provide a window on the state-of-the-practice with respect to modeling the effect of land use changes on travel and emissions. The case studies show some patterns in the driving forces that motivated the studies and analytical methods used to carry them out. The cases are drawn from medium and large urban areas and are meaningful as guideposts on how to estimate the types of travel and emission changes that might occur if certain land use policies are implemented within a metropolitan area. Interestingly, it is not only governments which have embarked upon these studies; the private sector is a lead partner in two of the 11 cases.

Originally, the Research Team hypothesized that the motivator for all of these studies would be an attempt to comply with the regulatory framework imposed by the Clean Air Act Amendments of 1990. As information was collected, it became clear that there were at least two motivators for undertaking this type of work. First, the regional planning organizations generally embarked upon comparative studies linking land use and transportation in the quest of more livable communities. The regional scale case studies are essentially visionary attempts at quantifying propositions to improve livability or quality of life. When livability is the motivator, emissions, transit use, and congested travel serve as measures of the livability of a community. Second, the private sector has been motivated by the desire to use community livability as a product differentiator to attract live-work-play residents and businesses while minimizing the environmental footprint of proposed private developments. These studies occur at a sub-regional or corridor level of scale, and are more likely to be undertaken in order to work within the regulatory framework of the Clean Air Act.

Each case study was evaluated based on several factors to allow more rapid reviews of the cases; the factors are briefly discussed, below:

- *Scale* is the scale for which the analysis was performed. This factor is intended to give end-users some indication of whether or not this particular case study matches the scale of analysis in which they are interested. There are case studies at three scales: site-specific, corridor, and regional level. Note that there is some upward overlap between the scales so that a site scale analysis may include elements of corridor or regional analyses.

- *Level of Practice* is an indicator of how the case study compares to the others in its application of tools. The level of practice is characterized as fair, good, and best. It should be noted here that no individual method is automatically acceptable in any future application, and that the goals of these case studies may influence the choice of approach in other, future applications.

- *Credit Received* indicates whether or not the project sponsor received formal emissions credit in either a state implementation plan or from the congestion mitigation air quality program.

- *Key Player* shows the driving party of the case study. While the key player may not have done the work, they are the entity that is creating the desire to move forward with the study.

The results of these interviews and the 11 case studies were synthesized to outline a state-of-the-practice methodology (see Figure 1, following page) for conducting quantifiable analyses of land use changes and their impacts on mobile source emissions.

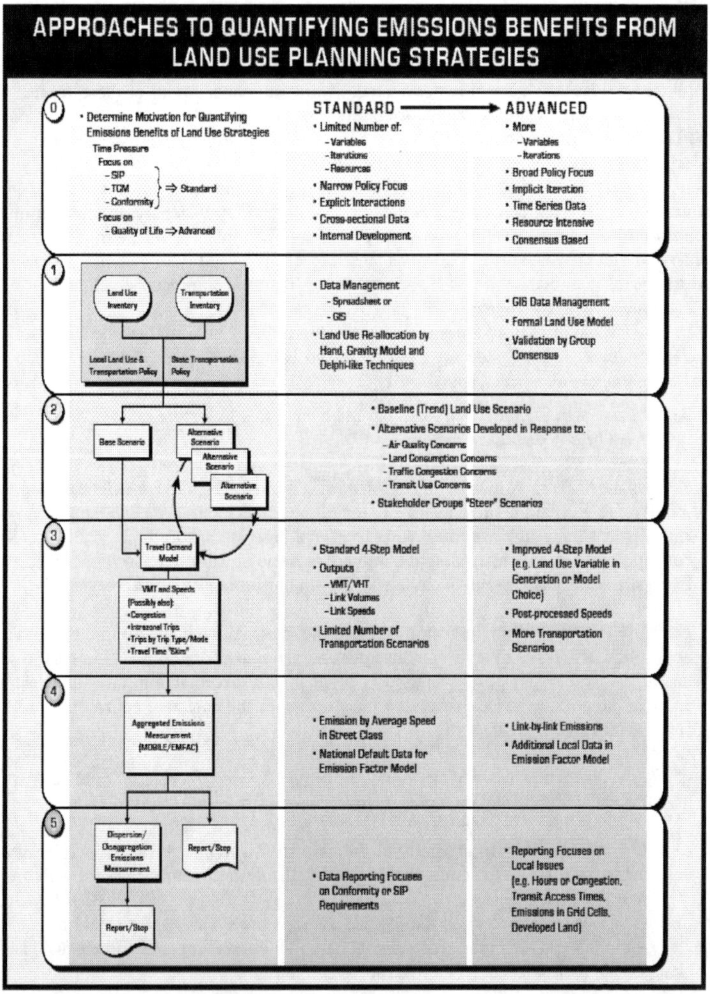

Figure 1. Existing / Advanced State of Quantifying the Emissions Benefits from Land

This land use/emissions analysis process generally follows a four- or five-step progression, the number of steps depending on the need/desire to disaggregate emissions into small subareas:

- Develop inventories of land use and transportation infrastructure according to modeling needs;
- Create a baseline (or "trend") scenario describing how future land uses might look if existing policies remain unchanged, and develop one or more alternative scenarios;
- Input land use and transportation information for all alternatives into a travel demand model or other gravity-based tool;
- Extract vehicle miles of travel by transportation facility, vehicular speeds, and other information required to estimate emissions into an emissions model (e.g., MOBILE or EMFAC); and
- If the objective of the study includes examining emissions benefits conferred to subareas, disaggregate emissions into individual grid cells or other small units of geography. This can be done by some travel demand models and emissions packages such as MODELS3 and CALINE.3

More advanced applications make use of sophisticated land use models, integrated land use-transportation models, modified travel demand modeling techniques, or dispersion modeling to refine impacts on small areas and populations. Only two of the 10 case studies applied the results of their testing to attempt to receive an emissions reduction credit in a conformity or state implementation plan (SIP). In the Atlantic Steel case, while the project was included as a transportation control measure (TCM) in the Georgia SIP, no emissions credits were taken for the project. All cases were either "visionary" exercises to study quality of life issues undertaken by MPOs and local governments, or were generated by a proposed large development that might affect the air quality standing in a transportation conformity maintenance area.

Findings and Recommendations

The following discussion deals specifically with the most common limitations to the existing state-of-the-practice and research methods reviewed for this study.

- *Cross-Correlation:* Numerous studies have cited the tendency (usually in other studies) to omit uncontrolled variables that may exhibit strong cross-correlation to the independent variables being examined. This is particularly true for socioeconomic variables like income, household composition, head-of-household characteristics, and car ownership. Although not as often recognized as a flaw in the research, uncontrolled operational characteristics such as parking pricing or other impediments to choosing a certain mode of travel appear to have a similar confounding role. The implication for the results of these studies is profound, and bring into question the validity of the results unless external or cross-correlated variables are dealt with explicitly in the study methodology.

- *Synergistic Effects:* Although not as frequently mentioned in the literature as the previous issues, there is an inherent synergy between some independent variables. In essence, this is the opposing face of the cross-correlation problem. Individually, a change in land use density or mixing may be insufficient to trigger a change in travel behavior, but collectively the results may be quite different. Hence, the attempt to isolate individual elements of land use strategies may be serving as an accomplice in masking these synergistic effects.
- *Objectivity in Measurement:* Studies have cited the difficulty of devising a consistent, objective metric for assessing urban design features. While the primary goal of these metrics is to capture the sense that walking and bicycling are safe and productive means of travel, the variety of design features that can impart this sense of acceptability may be too great to adequately capture over a large geographic area. This has also hampered efforts to operationalize the effects of urban design in trip generation or mode choice modules of travel demand models. Most studies have instead focused on comparing two or more communities that have different urban forms. However, following this strategy limits the statistical robustness of the effort, and introduces greater potential for cross-correlation to other, uncontrolled variables.
- *Data Availability/Reliability:* The paucity of data, especially at large scales (small units of geography), has often hampered the ability of researchers to adequately populate independent variables, particularly those relating to urban design features. The lack of time series data for a specific study area has implied that almost all of the empirical studies described are limited to a "snapshot" of how one or more study areas are behaving. One of the few exceptions is a Swedish study (Vilhelmson, 1999) comparing the travel behaviors of differently-sized towns and cities. Local scale studies utilizing temporally-stratified data could not be located to be a part of the literature review. Another, related problem is that of travel surveys under-counting secondary trips and walk trips. The implications of this problem are two-fold: first, some types of studies that ideally involve large data sets, data that does not come "pre-packaged" in the correct format or scale, or longitudinal data sets are seldom undertaken. The lack of longitudinal data particularly implies that many studies cannot confidently determine causality. The second result of this issue is apparent as researchers attempt to use surrogate variables in place of hard-to-get data items, usually without much discussion of the appropriateness of using the surrogate variables.
- *Boundary and Edge Effects:* Some studies of urban design acknowledge that major freeways or other physical barriers may exert an edge or halo effect on the rest of the area being studied. These portions of a community may have a very different set of urban design and travel characteristics that distort the results for the study area. Edge effects may also be present in communities with different (for example, more car-oriented) characteristics are dampening the effects of urban design for the sub-area under study. The most straightforward solution for either issue is to utilize smaller study areas, but this approach dramatically increases the overhead required to collect and manipulate data.

A second concern relates to edge and boundary effects, or how one cell (or other analysis unit of geography) affects other cells. An example of this is when a pedestrian-friendly older neighborhood is bounded by one or more adjacent, newer neighborhoods that are automobile dependent. The boundaries between the different land use types are not impervious, linear, or sharp. The effects of different land uses tend to 'bleed' across the boundaries. Few studies attempted to measure the size and strength of the cross boundary effects.

Studies that accounted for edge and boundary effects often used rectilinear overlay grids to define consistent manageable cells and to evaluate the interaction between adjacent cells. The strength of the interaction was often assumed to be proportional to the shared length at the cell boundary. Under this paradigm, adjacent cells with similar characteristics tend to reinforce one another while adjacent cells with dissimilar characteristics tend to have weaker effects. One difficulty with the 'gridiron' model is how to estimate the effect where four cells come together at their vertices. Because the shared boundary length is zero this model would assume that the interaction between any two cells that touch only at their corners is minimal, or even zero. The traditional way of overcoming edge effects is to create and populate very small units of geography. The difficulty of collecting data increases as the size of the individual study unit (e.g., traffic analysis zone) shrinks, making this solution difficult to apply for a large study area. An interesting side note to this discussion is that at least one researcher has pointed out that hexagonal input units of geography would provide equal edges all the way around a cell's perimeter thus eliminating many of the difficulties associated with grid systems.

- *Scalar Effects:* Land use strategies that have been considered as having an effect on travel behavior and environmental quality generally assume one of two forms: a regional scale, policy-oriented directive on controlling growth at the margins of existing, developed areas; or a microscopic focus on individual communities stressing features of the built environment including buildings, streets, sidewalks, and streetscaping. The implications of both scales are discussed.

  At a regional level, additional mixing of land uses, densification (more dwelling units or square feet of employment space per unit of land area), or more closely aligning employment opportunities with residential developments are typical categories of strategies. The general hypothesis is that by bringing common origins and destinations closer together, trip lengths are reduced and the opportunities for alternative (to single occupant, privately-owned vehicles) modes of transportation are enhanced. In practice, this has been difficult to prove or disprove since a number of other factors often play a significant role in trip-making decisions. These include the availability of various modes of travel, climate, and perceptions of personal safety. Many of these confounding factors take place at a much smaller scale than a regional initiative is likely to capture.

  At the smaller, community-oriented scale, researchers have supported the concept that specific design elements in the built environment affect the type and quantity of travel undertaken by residents, and in some cases, workers. Features that are commonly supposed to alter travel behavior include the presence or absence of sidewalks; spacing and setbacks of buildings; proximity of complimentary uses such as shopping opportunities near homes; streetscaping and traffic delineation; and the street layout. The metrics of street layouts include connectivity, number of curved segments, number of cul-de-sacs, and block lengths. There are reliability concerns at this level of input as well, and various researchers have expressed concerns about consistency and objectivity in measuring accessibility and urban design features.

- *Data Inconsistency and Availability:* Both regional- and local-scale analyses have faced significant challenges overcoming data shortfalls and inconsistent or subjective measurements of input data. The research did not reveal any attempts to establish forward or backward links between land use strategies and changes in vehicle fleet composition, percent grade, percent cold starts, ambient air temperature, or patterns of acceleration and deceleration. Most of these omissions are understandable since addressing them would have required making assumptions that are both not intuitive and difficult to test about long-term vehicle ownership trends or changes in the physical space of a study area. However, a significant research topic worth exploring deals with the number of cold starts that occur in various local scale studies of built environments. One of the few papers to deal with this specific research variant (Frank, et al, 2000) found that employment densities are inversely related to trip generation (cold and hot soak), travel time, and distance traveled. One hypothesis from such research would be that closer proximity to compatible uses might cause more trip chaining and hence fewer cold starts, leading to lower overall emissions.

To address these issues, and to improve the current state-of-the-practice, a number of recommendations were put forth for researchers, practitioners, and support agencies:

- Develop tools for managing large data sets and forecasting growth and development scenarios;
- Creation and maintenance of appropriate data inventories;
- Travel model changes (manipulation of intrazonal trips; zone composition and size; sensitivity to design features; and technical improvements);
- Changes to mobile emissions estimation (link-level inputs/outputs; acceleration event handling); and
- Improving accessibility (technical tools, training, data resources and analysis research).

Acknowledgements

The Research Team wishes to acknowledge the support of the Federal Highway Administration (especially Cecilia Ho), Environmental Protection Agency Regional Office personnel, the Association of Metropolitan Planning Organizations, and the staffs of the metropolitan planning organizations cited in this study. Ken Hess, AICP and David Hyder, PE of The Louis Berger Group, Inc. and High Point, NC Metropolitan Planning Organization, respectively, provided critical assistance refining the study framework and providing content in research and analysis. The complete report provided to FHWA has a more thorough treatment of the issues, delineation of the state-of-the-practice, and recommendations for improving the current practice. The full paper is available with the permission of FHWA.

A Simulation of the Effects of Transportation Demand Management Policies on Motor Vehicle Emissions

W. Harrington,[1] S. Houde and E. Safirova

[1]Resources for the Future 1616 P St. NW, Washington, DC 20036; PH (202) 328-5112; e-mail: Harrington@rff.org

Introduction

Increasingly, planners and policy analysts are paying significant attention to the interactions connecting transportation, land use and air quality in a metropolitan setting. These efforts are driven largely by transportation conformity, a Clean Air Act provision requiring both total emission budgets and mobile-source emission sub-budgets in nonattainment areas. All transportation plans in these areas must pass a conformity review to ensure that plan implementation over a 20-year planning horizon will not violate any emission budget in any year. Many nonattainment areas have adopted or have under consideration a set of transportation demand management (TDM) tools that can be adopted to earn emission "credits" toward meeting the transportation emission budget. These measures may include smart growth policies, which are aimed primarily at the supposedly inefficient use of land in suburban development and which are believed by many observers to reduce VMT and emissions as well. They may also include a variety of transportation policies designed to reduce if not completely internalize the congestion externality.

In this paper we focus on the latter type of policies. We compare the emission-reducing properties of a number of transportation policies designed primarily to address congestion. The modeling platform we use is LUSTRE, a spatially detailed and behaviorally complex simulation of transportation, land use and economic activity in the Washington, DC region. LUSTRE also calculates an overall welfare measure that is consistent with utility theory, and we add that measure to the comparison.

Introduction to LUSTRE

This section presents the salient features of LUSTRE, the modeling platform used in the simulations. For a more detailed discussion the reader is referred to Safirova et al. (2006a). LUSTRE integrates RELU, a spatially disaggregated general equilibrium model of a regional economy and land use that incorporates decisions of residents, firms, landlords and developers, with START, a spatially disaggregated strategic transportation planning model that features mode, time period and route choice. The integrated model has a high degree of realism, including agents of different skill levels, real estate and other taxes, and a detailed transportation network combining roads and a congestible transit system.

In RELU, individuals maximize their utility based on a series of discrete and continuous choices. After deciding whether to work or to be unemployed, individuals choose a triple consisting of their zones of residence and employment and their type of housing (single vs. multifamily). Conditional on these discrete choices, individual agents decide how much housing to rent and the quantity of retail goods and services to purchase at each available retail location.

START computes the generalized cost of travel taking into account the time and monetary elements of traveling. Time elements range from the time spent traveling and transit waiting time to parking search time and transit crowding penalties. Monetary elements include car operating costs (fuel tax), car depreciation costs, tolls, parking and transit fares. The value of time is a function of the travelers' wage rate, and varies by trip purpose. The transportation network is disaggregated in 40 travel zones (START's travel zones correspond to RELU's economic zones). Each zone has three stylized transportation links (inbound, outbound, and circumferential) and a number of other "special" links that represent the principal highway segments and bridges of the region. The traffic quality on each link is determined by a speed/flow•distance curve. The rail network of the region, which combines the Washington Metrorail system and suburban heavy rail systems [Maryland Rail Commuter (MARC) and Virginia Railway Express (VRE)], and the bus network are represented. Agents choose, mode, time of day, and route. Trip generation, origin and destination are delegated to RELU; they are then fully endogenous.

Equilibrium

Unlike many other land use-transportation models, the integration in LUSTRE is achieved at the behavioral level of individual agents, not at the aggregate level. Thus, the trip demands of each RELU agent to meet journey-to-work and shopping needs are summed and loaded onto the START transport network, which computes equilibrium generalized costs for each pair of zones, by mode and time of day, and passes those costs to RELU. Using the updated travel costs, RELU finds a new equilibrium prices (rents, wages and goods prices) and quantities (employment, travel demands, goods purchases, land allocations and housing types) by zone. The updated travel demands are passed back to RELU, and the system iterates until a fixed point is reached. The model is calibrated for the greater Washington DC metropolitan area for the year 2000. More details can be found in the mathematical appendix in Safirova et al. (2006a).

It should also be noted that while LUSTRE is an equilibrium model, so that all markets clear, it is not a fully dynamic model. In the real world, the various markets represented in LUSTRE clear at vastly different rates, from the daily clearance of the transportation route and mode choice "market" to the decades-long clearance of the land use market. Further model development now underway will allow partial adjustment in some markets, especially land use, in future versions of the model. In the current version, the evaluation of various policies is achieved by comparison of a long-run equilibrium to the baseline.

Welfare

One of LUSTRE's strengths resides in its ability to compute welfare measures that account for the changes in transportation as economic variables. LUSTRE's welfare measures are provided by RELU based on utility function for individuals. In RELU, utility is agent-based and therefore valuation of travel purposes is internalized. Since the discrete consumer choices in RELU are multinomial logit, the welfare measure for workers of skill class f in the model can be expressed as a logsum:

$$W_f = \frac{1}{\lambda_f} \ln \sum_{ijk} e^{\lambda_f \bar{U}_{ijk|f}}$$

where, λ_f is a dispersion parameter for the distribution of unobserved characteristics of workers of skill class f and $U_{ijk|f}$ the indirect utility function for such workers conditional on residential location i, employment location j, and housing type k. Again, see the mathematical appendix in Safirova et al. (2006a) for more details.

Mobile Emissions

Mobile emission scenarios are produced in a three step process using the output of START coupled with emission rates obtained from the EPA's model MOBILE6.2. In the first step, emission rates specific to the region are produced with MOBILE6.2. These rates take into account the vehicle fleet, the mandatory inspection and maintenance programs and the physical characteristics of the region (altitude and average temperature). In a second step emission rates are post-processed to be consistent with the level of aggregation of Washington START. Finally, the volume of emissions emitted over the study area is computed by coupling START's transportation data and the composite emission rates. Three pollutants are considered: volatile organic compounds (VOC), carbon monoxide (CO) and nitrogen oxide (NOx). The details of this process can be found in Houde et al. (2007).

Comparison of the LUSTRE base case estimates with the 2002 emission inventory developed by the Metropolitan Washington Transportation Planning Board (TPB) (MWCOG 2002) shows that the estimates are reasonably close in total.[1] LUSTRE estimates exceed the COG estimates by 5.5% for VOC and by 1.6% for NOx, and are 7% lower for CO. However, some counties are entirely included in LUSTRE are not in the Washington nonattainment area, so that they are not in the COG emission estimates. If we exclude these counties, we find that LUSTRE estimates are lower than the COG estimates by 5.9% for VOC, 15.3% for CO and 7.4% for NOx. The differences are attributable to the failure of our estimates to include some emissions categories included by the COG (such as diurnal and resting losses for evaporative VOC and emissions of all pollutants on access ramps). In addition, minor differences in link speeds can have major differences in emissions, especially for CO, as shown in Figure 1. For CO the lower estimates in our model could be attributable to the difference in zone structure. Whereas LUSTRE has 40 zones, the MWCOG model has 800 or 2000 zones, depending on task. Our zones are therefore much larger and may be averaging out the differences in speed in the smaller zones. Because

[1] The TPB totals have been adjusted to account for the slight difference between the area modeled in LUSTRE and the actual nonattainment area.

emissions, especially of CO, are highly nonlinear functions of the speed, it is possible that we are unable to capture the diversity in speeds across zones, and that may result in underestimation of CO emission levels in the base case, as well as underestimation of changes in emissions in the policy cases.

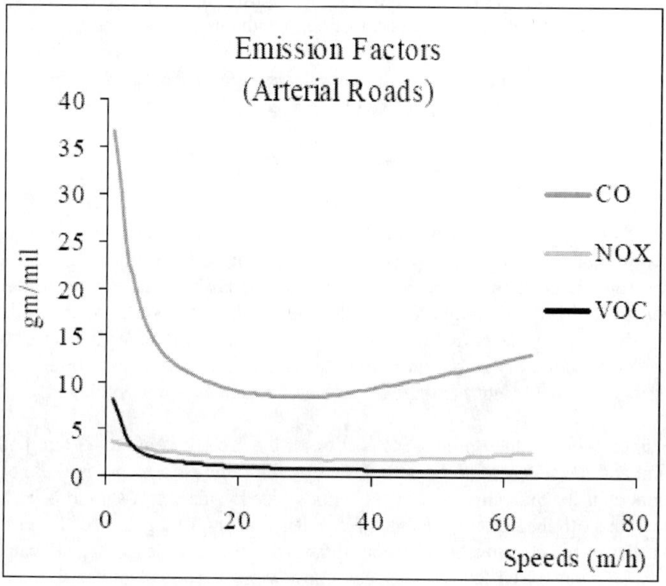

Figure 1. Composite Emission Factors for Travel on Arterial Roads in Washington DC

Policies Description

We analyze six second-best transportation pricing policies for their emission-reduction potential. The pricing policies, it should be noted, are more vigorous than any such policies currently in place anywhere in the US, and thus provide a fair indicator of the potential for such policies to achieve both emission reductions and welfare gains. For all six policies, the revenue collected from tolls and increases in the gas tax are redistributed lump sum to all individuals. The policies are:

- A small cordon, levying a toll on all vehicles entering the downtown core during morning rush hour (Downtown cordon).
- A big cordon, levying a toll on all vehicles entering the area surrounded by the Washington Beltway during morning rush hour (Beltway cordon).
- A double cordon, consisting of the two cordons above (Double cordon).
- A road toll charging users by distance and time of day for travel on all freeways and Potomac bridges in the metropolitan area (Freeway toll).
- A road toll charging users by distance and time of day for travel on all links in the metropolitan area (Comprehensive toll).

- A gasoline tax.

For the cordon tolls, by second best, we mean that each of these pricing policies is optimized within the constraints imposed by the instrument to achieve the highest gain in welfare as computed by LUSTRE. For each cordon, the toll is the same at all entry points. Cordon locations are not necessarily optimal but are chosen for convenience of administration and compatibility with our modeling structure.

For the road tolls, the optimal charge rate is equated to the marginal cost of congestion (MCC) on each applicable link in each of the three time periods:

$$MCC_k = \left(\frac{1}{S_{k1}} - \frac{1}{S_{k0}} \right) \times FD_k \times VOT_k \, .$$

where S_{k0} and S_{k1} are correspondingly the initial and resulting speed levels on the link k after adding one VMT to the link, FD_k the number of link users, and VOT_k the average value of time. For the freeway toll, the charge only applies to freeway travel, while for the comprehensive toll all links of the network are charged.[2] The comprehensive toll is probably close to a first-best optimum with respect to the congestion externality only, but it fails to internalize other externalities (such as emissions, accidents, etc.) and the recycling of revenues is not optimized.[3]

The sixth policy is the optimal gas tax for all the area. It is assumed that both the federal and the state gas taxes are adjusted by the same percentage until the overall welfare gain for the entire study area is maximized. This tax, it should be noted, is an increment to the current gasoline tax, which is about 40 ¢/gal. We also note that this gasoline tax operates in the model rather more like a mileage tax.[4] Because the current version of LUSTRE does not allow vehicle choice, motorists can only respond to the tax by reducing VMT; they cannot gain by choosing a vehicle with better fuel economy. However, they can reduce VMT in numerous ways: by switching to transit or carpool modes, by choosing less distant shopping locations, by switching employment or residential locations, or by leaving the labor force altogether.

Results

Tables 1 and 2 display the modeling results for the six transport pricing policies. Table 1 contains some information on the scale and scope of each policy and its anticipated effects on total travel. The second and third columns give the percent of VMT affected by the policy after implementation and the rates charged on each of those VMT, and the last two columns give the effect on VMT. The ranking of

---

[2] Note that the $MCCs$ for all tolled links are computed simultaneously and account for the network effects (see Safirova et al. 2006 for a discussion of the network effects caused by road pricing). To do so, LUSTRE is run iteratively, where drivers are charged with the $MCCs$ as computed at the precedent iteration, until the $MCCs$ converge.

[3] In addition, LUSTRE features two other types of frictions -- taxes and alternative congestible transit modes – that make finding the global optimum even more complex.

[4] It is not a pure distance tax either, because fuel economy depends on link speed, but it is probably closer to a VMT tax than to a fuel tax in its effects.

policies in the table is the same regardless of which variable is sorted on (except for the "toll rates were charged," which cannot be ranked). By far the most intrusive policy on an average basis is the gasoline tax; at 7.9 ¢/mile it is over twice as costly on average than the comprehensive toll, which is the second most costly. These two policies reduce VMT by 14 and 7 percent, respectively.

As shown, the freeway toll and cordon policies have much lower average costs and affect ever smaller percentages of trips. Only 7 percent of trips cross the beltway during the morning rush-hour and only 1 percent the downtown cordon. Yet those policies, especially the cordons, are the most variable in their impacts and for certain types of trips—trips into downtown DC in particular—the charges can be substantial. For example, a trip from Fairfax, VA to the core, a distance of about 20 miles, would cost $5.61 (28.1 ¢/mile) for the double cordon, $2.77 (13.9 ¢/mile) for the beltway cordon, and $4.70 (23.5 ¢/mile) for the downtown cordon, higher than the average rates for any policy considered (column 4). As the example suggests, cordon policies can be very inequitable in that most of the vehicles on the road during morning rush hour contribute to congestion, but only a small minority have to pay.

Table 1. Six Second-best Transportation Policies: Optimum fees and effects on VMT

Policy	Percent of VMT affected	Toll rates, where charged	Average cost per VMT (¢/mi)	Total estimated VMT (millions per day)	% change
Base Case				172.7	
Gas Tax	100%	$2.02/gal	7.9	147.4	-14.6
Comprehensive toll:	100%	Variable	3.3	160.5	-7.1
Freeway toll	26%	Variable	0.7	169.0	-2.1
Double Cordon	7%[a]	D: $2.18 B: $3.43	0.4	170.5	-1.3
Beltway Cordon	7%[a]	Beltway $2.77	0.3	171.1	-0.9
Downtown Cordon	1.1%[a]	Downtown $4.70	0.2	171.3	-0.8

[a]Percent of trips, not VMT.

The fifth column of Table 1 provides the reduction in VMT for each of the policies examined. Four of the policies—again, the three cordons and the freeway toll—show very small reductions in total VMT. It is of interest that the double cordon has a stronger effect on VMT than the Beltway cordon, not only because it is more costly, but because transit is a more attractive substitute for the double cordon trips, which tend to be the radial trips that the Metrorail system was designed to serve (See Safirova et al. 2006b for further discussion).

The Freeway toll also exerts a small mileage reduction of 3.7 million total daily VMT, or 2.1 percent. Freeway VMT decline by 2.5 million, or 5.6 percent, so that mileage on the untolled zonal links declines by 1.2 million VMT. Thus the switch from tolled to untolled road links is outweighed by the shifts out of vehicular travels altogether. Part of the reduction in total VMT is a mode shift to transit (increase of 15.6% in number of trips) or carpooling (increase of 11.3%). The remainder is attributable to the general equilibrium effects: workers move closer to their jobs and some leave the labor force altogether.

Further information on these six policies is given in Table 2. The emission reductions (columns 4-6) track closely with VMT changes, so once again the gas tax produces the greatest emission reductions and the cordon policies the least. VOC reductions are slightly greater than VMT reductions, except for the gas tax, while NOx and especially CO reductions a bit less. These slight departures from proportionality are a consequence of speed changes, which affect emissions in a manner illustrated in Figure 1. Emission rates are U-shaped as a function of speed, particularly for CO. Thus, increases in speeds will bring about lower emissions at low speeds, but higher emissions at high speeds. It is not surprising, therefore, that for all six policies the percentage changes in emissions are very close the percentage changes in VMT.

Table 2 Welfare Gains and Emission Reductions for a Set of Second-Best Transportation Policies (LUSTRE)

Policy	Welfare gains (millions of 2000$)	Total Congestion Externality Cost (millions of 2000$)	Average MCC per Link (¢/mile)	Emission Reduction Relative to Base Case		
				VOC	CO	NO_x
Base Case	-	3182.2	7.4	-	-	-
Gas Tax (2.02 $/gallon)	304.9	2496.6	6.86	13.74%	-13.8%	-14.51%
Comprehensive toll	659.6	1361.1	3.3	-7.98%	-5.14%	-5.90%
Freeway toll	385.0	2422.9	5.8	-2.34%	-1.18%	-1.43%
Double Cordon (D: $2.18; B: $3.43)	150.9	2996.4	7.1	-1.50%	-1.05%	-1.13%
Beltway Cordon ($2.77)	104.3	3032.5	7.2	-1.05%	-0.70%	-0.77%
Downtown Cordon ($4.70)	91.2	3085	7.3	-0.93%	-0.64%	-0.69%

Table 2 also presents the welfare improvement associated with each policy. Note that the ranking for this outcome differs from the rankings with respect to VMT and emissions, for which the gasoline tax is at the top. In terms of welfare change the gasoline tax ranks behind the road tolls and ahead of the cordon tolls. If we disregard the gas tax, however, the order is the same as before. One thing to keep in mind is that at present our welfare calculator does not include the benefits from emission reductions. That omission will be rectified in the future, when we replace the current welfare calculator with a more comprehensive version that will enable us to examine the implications of full social cost pricing in the manner of Proost and van Dender (2001), but in a spatially disaggregated network model.

Table 2 also presents the average marginal congestion cost per link and the total congestion cost after implementation of each policy, the latter being sum of the products of marginal congestion costs and flow-distances on each link.[5] By far the smallest value of each belongs to the comprehensive toll. The main reason the freeway toll and the cordons fail to achieve the same amount of congestion reduction is that they only apply to a fraction of trips (see Table 1). That explanation does not apply to the gasoline tax, however. Even though it is levied on all VMT, its average MCC is much worse than the comprehensive toll and the freeway toll, and not much better than the cordon policies, which are levied on only a small minority of trips.

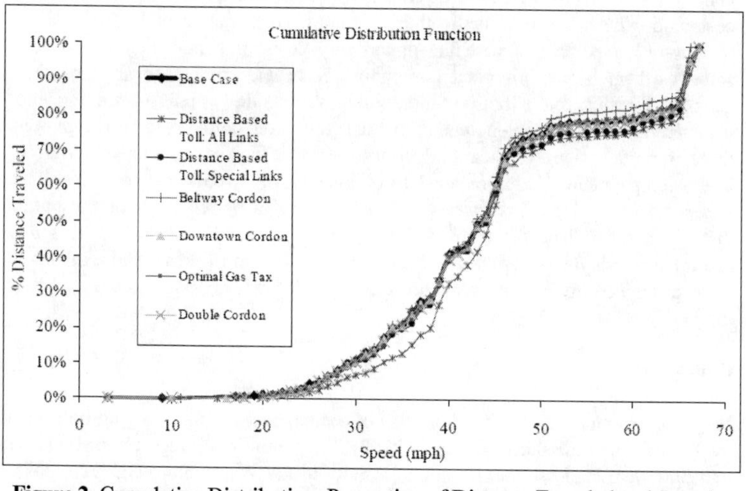

Figure 2. Cumulative Distribution: Proportion of Distance Traveled and Speeds

The optimal gas tax is less effective at internalizing the congestion externality than the comprehensive toll because the gas tax is levied uniformly, without regard to congestion. In contrast the comprehensive toll rate is specific to each link and equals the MCC. Thus, under a gas tax some links end up being overpriced and others, with

5 The size of the congestion externality is computed post-simulation, i.e. once the policy has been implemented. This measure corresponds to the marginal congestion costs computed on all links at given time. It does not take into account the network effects that result from implementing congestion pricing. However as discussed in Safirova et al. (2007), the present measure provides an accurate estimate of region-wide average levels of marginal congestion costs.

high congestion, are underpriced.

It remains to be explained why the gas tax scores so well on welfare improvement even though it is not very effective at reducing congestion. One reason is that it generates a vast amount of revenue—nearly $1.18 billion per year. The assumption of lump sum distributions is highly unrealistic. It is more likely that the funds will be given over to reductions of existing taxes, new public works spending, or some combination of the two.

For example, some of those who now support transportation pricing, especially in the environmental community, condition their support on the use of a substantial portion of the revenues for investment in transit. Others look to funding for new road construction. The welfare effects of these various dispositions of revenues depend crucially on the details. This is true of every one of the revenue instruments analyzed in this paper, but the size of the gasoline tax revenues makes that instrument particularly sensitive to the disposition of the revenue.

The revenue redistribution also has a not insignificant effect on employment, and hence on VMT through the reduced demand for commuting. In LUSTRE, commuting costs and even more importantly the lump-sum redistribution of the revenue of a transportation policy decrease the opportunity costs of being unemployed. Thus, some workers facing important commuting costs and simultaneously receiving a generous transfer from a transportation policy would simply prefer unemployment. It is particularly true for low-income individuals for whom commuting costs represent a more important share of their budget and derive a higher marginal utility from a lump-sum payment. Therefore, the 1.18 billion of revenue collected and redistributed under the optimal gas tax causes higher unemployment relatively of the optimal distance based toll on the entire network, which collects 700 millions. This drive toward unemployment results in an important reduction of VMT and explains the large difference in emission reductions between the gas tax and the distance based toll.

Conclusion

We have examined the effect of a range of second-best transport pricing policies on motor vehicle emissions, using LUSTRE, a spatially disaggregated model of transportation, land use and economic activity in the Washington, DC metropolitan area. Such policies are thought to affect both, vehicle use and pollutant emission rates, the former by discouraging vehicular travel and the latter by improving vehicle flow and increasing average vehicle speed. We find that these policies consistently have benign effects on emissions, but only mildly so. We also find that the reductions are driven almost entirely by reductions in VMT.

With the exception of the gasoline tax, the ranking of policies vis a vis welfare was the same as the ranking with respect to VMT.

Among these transport policies, the effect on VMT, and hence emissions, depends primarily on the coverage of the policy. Thus the cordon policies, which affected only 1 to 7% of total regional VMT, had a very small effect on all outcomes, although they strongly affect trips into the core. The two policies affecting all VMT had the largest effect. Both these policies were pretty drastic, though: a large tax on fuel and a congestion toll levied on every trip. No one, as far as we are aware, has seriously proposed a congestion management policy as intrusive as either of these. In this respect, perhaps the most interesting policy is the freeway toll, which resembles the "HOT networks" proposal put forward by the Reason Foundation's Robert Poole (2003). The freeway toll only affects 26% of VMT but achieves over half the welfare benefits of the comprehensive toll. However, it provides emissions reductions of only around 2%.

References

Houde S., E. Safirova and W. Harrington (2007), "A Simulation of the Effects of Transportation Demand Management Policies on Motor Vehicle Emissions." *RFF Discussion Paper, forthcoming.*

Metropolitan Washington Transportation Planning Board (TPB) (2002), "Periodic Inventory Emissions." http://www.mwcog.org/uploads/committe-documents/uF5XVIo2004524133912.pdf

Proost, S. and K. V. Dender (2001), "The Welfare Impacts of Alternative Policies to Address Atmospheric Pollution in Urban Road Transport." *Regional Science and Urban Economics 31(4): 383-411.*

Poole, R. (2003), "HOT Networks: A New Plan for Congestion Relief and Better Transit." Testimony before the Joint Economic Committee, US Senate (May, 2006).

Safirova, E., S. Houde, D. Lipman, W. Harrington, and A. Baglino (2006), "Congestion Pricing: Long-Term Economic and Land use Effects." *Resources for the Future, Discussion Paper 06-37, 2006.*

Safirova, E., S. Houde, C. Coleman, W. Harrington, and D. Lipman (2006), "Long-Term Consequences of Congestion Pricing: A Small Cordon in the Hand is Worth Two in the Bush." *Resources for the Future, Discussion Paper 06-42, 2006.*

Safirova, E., K. Gillingham and S. Houde (2007), "Measuring Marginal Congestion Costs of Urban Transportation: Do Networks Matter?" *Transportation Research Part A: Policy and Practice, forthcoming.*

Integrated Impacts of Regional Development, Land Use Strategies and Transportation Planning on Future Air Pollution Emissions

S. Bai,[1] D. A. Niemeier,[2] S. L. Handy,[3] S. Gao,[4] J. R. Lund,[5] and D. C. Sullivan[6]

[1]Ph.D., Department of Civil and Environmental Engineering, University of California, One Shields Ave., Davis, CA 95616; PH (530) 752-8460; FAX (530) 752-7872; email: sbai@ucdavis.edu
[2]Ph.D., P.E.,Department of Civil and Environmental Engineering, University of California, Davis, CA. email: dniemeier@ucdavis.edu
[3]Ph.D., Department of Environmental Science and Policy, University of California, Davis, CA. email: slhandy@ucdavis.edu
[4]Ph.D., Department of Environmental Science and Policy, University of California, Davis, CA. email : sgao@ucdavis.edu
[5]Ph.D., Department of Environmental Science and Policy, University of California, Davis, CA. email: jrlund@ucdavis.edu
[6]Project Manager, Sonoma Technology, Inc., Sonoma, CA. email: Dana@sonomatech.com

Abstract

Urban and regional air pollution is an increasingly critical problem in areas where rapid growth s occurring and future development is expected. Within the typical planning horizon, many modeling tools (e.g., land use, travel demand, and emissions models) are used to evaluate the impact of regional development, population trends, and infrastructure improvements on air quality. However, these models were not designed, nor are they generally used as a cohesive system with feedbacks between them. Consequently, there is a gap in our understanding of how different models interact and combine to influence emissions inventories associated with future policy scenarios. In this study, funded by the Environmental Protection Agency, an emissions modeling framework was developed using the UPlan, TP+/Viper, and UCDrive emission models for the San Joaquin Valley (SJV) in central California. Specifically, future land use and subsequent travel patterns and emissions inventories were examined based on four policy scenarios for the year 2030: baseline (following general plans, with no roadway expansion), controlled growth (compact urban boundary, higher residential densities, no roadway expansion), uncontrolled growth (low residential densities, roadway expansion), and as-planned (following general plans, with roadway expansion). The integrated modeling framework links emissions to levels of travel pattern variations under different land use policy and development strategies. For example, compared to a scenario with highly controlled urban growth strategies, less restraint on urban growth results in more than a 20% increase in vehicle miles traveled and about 18% higher emissions of primary pollutants at the regional scale in year 2030. By means of comparing and connecting land use, travel patterns and emissions, the sensitivity of SJV mobile emission inventories to different possible growth scenarios can be assessed.

1. Introduction

Urban and regional air pollutants such as ozone and airborne particles are persistent public health problems with serious economic consequences. Although technology is continuously developed to reduce pollutant emissions per unit activity, population expansion, urban sprawl, and shifts in societal values and lifestyles tend to increase emissions-producing activities, leaving as uncertain long-term progress towards the abatement of air quality problems. A wide range of factors affect pollutant emissions, both in sign and magnitude, and there is a gap in our understanding of how various factors along with potential development policies combine to influence air quality at the urban and regional scale. Without additional knowledge, it will be difficult to develop informed decisions to safeguard public health and economic well-being.

This paper presents an integrated modeling framework that allows us to examine future mobile source air pollutant emissions associated with different regional level land use and transportation development scenarios. Land use policies are seen as an effective tool for improving air quality because these policies have the potential to decrease vehicle travel through trip reduction and alternative mode use (U.S. Environmental Protection Agency 2001). However, land use strategies interact with decisions about transportation investments in terms of influencing air quality. While some land use characteristics (e.g., high density and mix of land use) can directly shape specific travel patterns through changing travel behaviors (Litman 2005, Handy 2005), others (e.g., large scale new development) require new and expanded transportation systems that in turn affect future air quality by, for example, increasing induced traffic (Cervero 2002).

The San Joaquin Valley (SJV) in central California, the focus of this study, currently experiences one of the most severe air pollution problems in the United States (U.S. Environmental Protection Agency 2000, Bedsworth 2004). The SJV region is expected to undergo rapid growth and development over the next 40 years (Teitz et al. 2005) with all the concomitant emissions control challenges associated with population expansion, transportation, industry, agriculture, and power generation. In this paper, we begin by describing four regional policy scenarios based on a range of land use and transportation variables expected to affect San Joaquin Valley mobile source emissions. A cohesive modeling framework, which includes land use modeling, travel demand modeling and emissions modeling, is then used to examine the mobile source emissions impacts. We then provide a discussion of the modeling results and comparative analysis based on the predicted regional mobile source inventories, followed by conclusions in the final section.

2. Policy Scenarios

The San Joaquin Valley, which includes eight counties (Fresno, Kern, Kings, Madera, Merced, San Joaquin, Stanislaus and Tulare), is facing tremendous change in the next few decades. The region's population is projected to grow by 50% from 2000 to 2020 and by 139% from 2000 to 2050 (California Department of Finance 2004), contributing to the conversion of agricultural land to residential uses and to an anticipated increase in traffic. Future emissions are dependent on how expected population growth is spatially accommodated and how traffic patterns change; a process that itself depends on several key factors.

Important variables associated with SJV local and regional land use policies include the extent and type of agricultural and habitat land preservation, the range of local development strategies being considered, the urban service boundaries and local zoning policies. Significant population growth will also trigger increased demand for transportation infrastructure. The transportation network is projected to take shape in two different ways: an expansion to the existing roadway network and the addition of a high-speed rail link. According to the California High Speed Rail Authority (CHSRA), the 700-mile long electric-powered high-speed train system has been proposed to stretch from the Bay Area, through the Central Valley, ending in southern California, with an expected completion of 16 years in terms of engineering, environmental clearance, right of way purchase, and construction (California High-Speed Rail Authority 2000).

Four specific 2030 scenarios were developed in consultation with a project advisory board, comprising a panel of regional air quality and economic experts from the California Air Resources Board, the San Joaquin Valley Air Pollution Control District, the California Department of Finance, the California Department of Transportation and other experts on regional growth in the San Joaquin Valley. As summarized in Table 1, each scenario represents a specific strategy to accommodate future population growth. The *Baseline* scenario assumes that projected population growth is accommodated according to the adopted general plans of cities and counties in the region and no expansion of transportation facilities. The *Controlled Growth* and the *Uncontrolled Growth* scenarios reflect two ends of the development spectrum: briefly, the former scenario assumes all population growth is accommodated at high densities within existing urban boundaries with high-speed rail but no roadway expansion, while the latter assumes population growth is accommodated at low and very low densities and that all planned roadway expansions occur but that high-speed rail does not. The *As-Planned* scenario represents what is currently planned (i.e., from county general plans and regional transportation plans) and includes a mixture of land use and transportation improvement measures identified in both the *Controlled Growth* and *Uncontrolled Growth* scenarios.

3. Modeling Approach

A systematic modeling system was developed and used to test each scenario in terms of future mobile source emissions inventories for the SJV (Figure 1). The modeling system consists of a GIS-based land use model (UPLAN), a four-step travel demand model (specified in TP+/Viper) and a new mobile emission inventory model (UCDrive-MOBILE). The database embedded in the modeling system includes a wide range of socio and economic forecasts (e.g., population and employment), the current and future roadway networks, existing and future land use patterns and a group of parameters reflecting different policy scenarios. The final products of the modeling procedure are projected scenario-specific mobile source inventories.

Table 1. Specification of scenarios with land use and transportation factor levels

Land Use/Variable	Scenario 1 Baseline	Scenario 2 Controlled Growth	Scenario 3 Uncontrolled Growth	Scenario 4 As-Planned
Residential Density	Continuation of 2000 densities for new growth	All new growth at high-density	All new growth at low and very low density	Continuation of 2000 densities for new growth
Development Strategy	No Change	Transit-oriented development, infill, and redevelopment	No Change	Following county general plans
Land Preservation	Residential allowed in agricultural areas	Agricultural areas protected according to county general plans; 50% increase in protected wetlands areas	Residential allowed in agricultural areas	Agricultural areas protected according to county general plans; 25% increase in protected wetlands areas
Transportation				
Roadway network	1998 network	1998 network	2030 network based on RTPs (Regional Transportation Plans)	2030 network based on RTPs
High-speed rail	No high-speed rail	High-speed rail (SJV corridor alignment)	No high-speed rail	High-speed rail (SJV corridor alignment)

Land Use Modeling

We used UPLAN, an urban growth rule-based grid model, to allocate future land use. UPLAN allocates the area of each land use type through a set of rules based on projected population increases, local land use plans, existing cities, and existing and planned transportation routes (Johnston et al. 2003). Using data obtained from the California Department of Finance, regional planning agencies, local governments, and the expert advisory panel, land use scenarios were constructed based on assumptions about the factors listed in Table 1. In UPLAN, the land uses are distinguished by categories such as residential, industrial, commercial and agricultural, and are characterized by the density of development (e.g. population per grid cell or employment per grid cell).

To build the future scenarios, projected population and employment growth are allocated to grid cells based on a series of decision rules that reflect specific assumptions about land preservations and development strategies. For example, in the *Controlled Growth* scenario, population growth is allocated to grid cells within existing areas of development up to density limits appropriate for a smart growth approach; population growth beyond what can be accommodated within areas of existing development is allocated to grid cells contiguous with existing development and based on expected growth rates for different parts of the region.

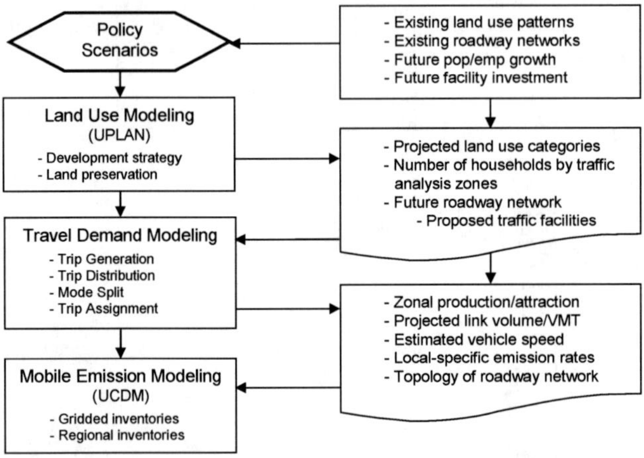

Figure 1. Integrated modeling system for SJV mobile source emissions inventories

Travel Demand Modeling

County-specific four-step travel demand models are currently used in most of the SJV counties. We utilized these same models, with some enhancements, to identify how changes in land use and transportation policies affect future traffic patterns. The four-step travel demand models use data on land use patterns and a description of the transportation network as inputs, along with additional data such as numbers of households, average income and automobiles per household, and employments by type (e.g. agriculture, manufacturing, etc.). Specifically, land use patterns obtained from UPLAN are represented by population and numbers of jobs for predefined traffic analysis zones (TAZ) for each of the eight SJV counties. The transportation network is described by a system of nodes and links that represent the major transportation facilities in the region (e.g. freeways, highways, major arterials) and reflect operating characteristics of the facilities (e.g. capacity, speed limits, directional restrictions).

The two transportation-related factors (Table 1) are directly considered in the travel demand modeling process. The impacts of high-speed rail are modeled based on three trip categories. Typically, at the county level, the availability of high-speed rail tends to decrease through traffic (accounting for between-county long-distance trips), reduce county network gateway to downtown traffic (accounting for within-county long-distance trips), but increase trips between downtown and traffic analysis zones within counties (accounting for within-county park and ride trips). The magnitude of these impacts on the travel demand is specified for the associated traffic analysis zones (TAZs) and reflected in the trip distribution and mode split steps of the modeling process.

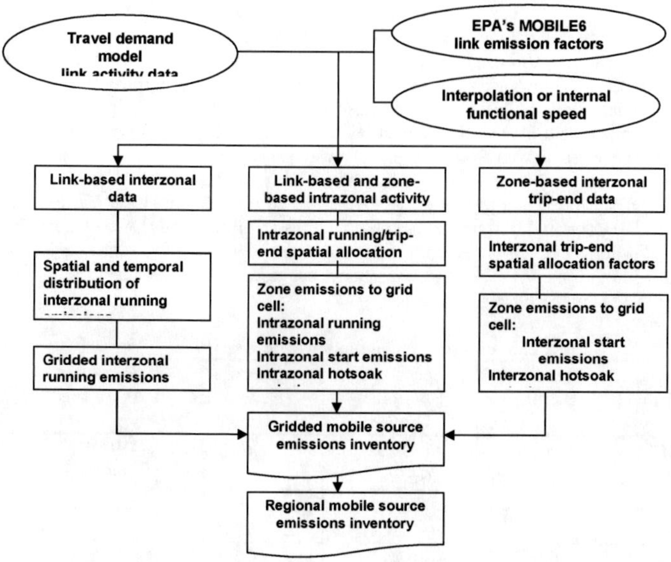

Figure 2. Calculation modules and data flow of UCDM modeling process

Mobile Emissions Modeling

We used UCDrive-Mobile (UCDM), a new mobile source inventory grid-based model, to prepare the SJV regional mobile source inventories (Figure 2). The UCDM model uses facility-specific emission rates from MOBILE6 and combines them with link-based traffic activities obtained from travel demand models. The advantage of UCDM is that using three modules (interzonal running, intrazonal running/trip-end and interzonal trip-end) emissions from mobile sources may be modeled such that the temporal and spatial variabilities are represented simultaneously (Niemeier et al. 2003). Consequently, vehicle emissions inventories for the entire SJV can be produced at the grid level, in addition to the regional level. This allows us to capture microscale changes resulting from variations in travel patterns among different policy scenarios and reliably estimate the emissions impact of highly localized changes in land use patterns as well.

4. Modeling results

To understand how future traffic activities and mobile emissions vary under different land use and transportation policy assumptions, a comparison analysis was performed at both regional and local levels.

Regional Level Comparison

Table 2 presents the projected daily vehicle miles traveled (VMT), vehicle hours traveled (VHT), number of trips and average trip distance for the entire SJV region. Compared to the *Baseline* scenario, the *Controlled Growth* scenario with a high-density land use pattern results in lower VMT (6.5% reduction), lower VHT (12.5% reduction), fewer total trips (3.7% reduction) and shorter travel distance (about 3%) at the regional level. As expected, this suggests that, given a condensed land use pattern as well as controls on roadway expansion, people tend to travel less (i.e., make fewer trips or shorter trips). Conversely, significant increases in VMT (12.3% growth), VHT (7.6% growth) and trip distance (12.2% increase) are found in the *Uncontrolled Growth* scenario, in which highly sprawled land use patterns dominate future development. The changes in travel patterns in the *As-Planned* scenario are basically between that of the *Controlled Growth* scenario and the *Uncontrolled Growth* scenario, suggesting moderate increases in travel activities over that seen in the *Baseline* scenario.

Table 2. Summary of SJV daily traffic activities

Regional Traffic Activity	Scenario 1 Baseline	Scenario 2 Controlled Growth	Scenario 3 Uncontrolled Growth	Scenario 4 As-Planned
VMT (1000 miles)	184,164	172,252 (-6.5%)	206,789 (+12.3%)	193,915 (+5.3%)
VHT (hour)	6,142,470	5,372,364 (-12.5%)	6,609,367 (+7.6%)	6,035,425 (-1.7%)
Trips (1000 trips)	16,653	16,044 (-3.7%)	16,668 (+0.1%)	16,675 (+0.1%)
Trip Distance (mile)	11.06	10.74 (-2.9%)	12.41 (+12.2%)	11.63 (+5.2%)

The regional mobile source inventory results for different policy scenarios suggest a consistent pattern across primary pollutants. That is, the *Controlled Growth* scenario produces lower emissions, while the *Uncontrolled Growth* scenario tends to result in higher emissions. The percent changes between the *Baseline* scenario and other scenarios are calculated for total organic gas (TOG), carbon monoxide (CO), nitrogen oxides (NOx) and particulate matter (PM10) and are presented in Table 3 and graphically shown in Figure 3. The comparison indicates that highly sprawled development, which produces more vehicle traffic and greater levels of roadway congestion, is positively correlated with higher mobile emissions (e.g., in the *Uncontrolled Growth* scenario). Looking at the SJV regional totals, future development with uncontrolled growth leads to increased emissions of about 18% to 19% for criteria pollutants over that produced with the controlled growth policy assumptions. The *As-Planned* scenario, compared to the baseline (do-nothing), tends to result in a moderate reduction of TOG (1.4%) but a 3% to 5% increase in CO, NOx and PM.

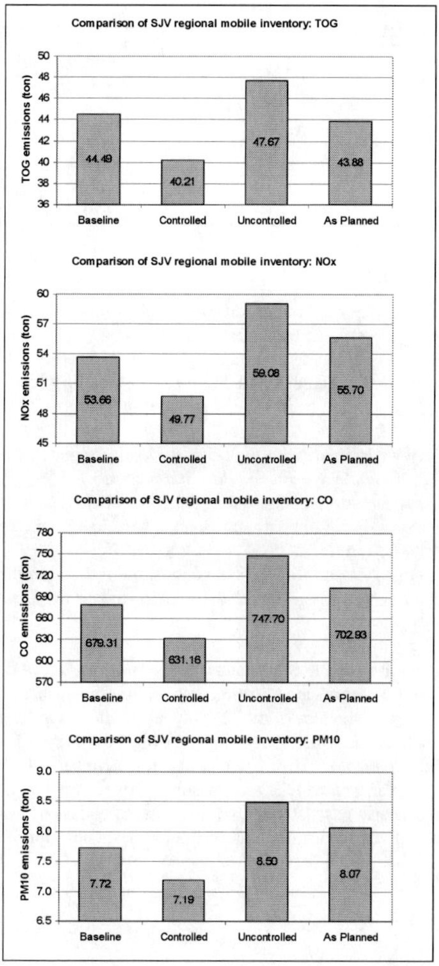

Figure 3. Comparison of SJV regional mobile inventories

Table 3. Summary of SJV regional mobile source inventories (unit: ton)

Regional Totals	Scenario 1 Baseline	Scenario 2 Controlled Growth	Scenario 3 Uncontrolled Growth	Scenario 4 As-Planned
TOG	44.49	40.21 (-9.6%)	47.67 (+7.2%)	43.88 (-1.4%)
CO	679.31	631.16 (-7.1%)	747.70 (+10.1%)	702.93 (+3.5%)
NOx	53.66	49.77 (-7.2%)	59.08 (+10.1%)	55.70 (+3.8%)
PM	7.72	7.19 (-6.9%)	8.50 (+10.0%)	8.07 (+4.6%)

Local Level Comparison

Another way to examine the impacts of different policy assumptions is to zoom in for a comparison of changes in local level travel patterns (e.g., congested speed and volume-capacity ratios) and the gridded mobile source emission outcomes. Using Stanislaus County in the San Joaquin Valley as an example, Table 4 and Figure 4 look at link level comparisons between various policy scenarios. Important observations include:

- The *Controlled Growth* scenario with high-density land use patterns results in nearly a 20% reduction in average travel delay over the *Baseline* scenario.
- The road network appears more congested in the *Uncontrolled Growth* scenario with sprawl development patterns, indicated by expanded areas with lower congested speeds and higher link volume-capacity (v/c) ratios.
- With increased trip distance and travel time, congestion in the *Uncontrolled Growth* scenario results in more than 50% average travel delay than that in the *Controlled Growth* or *As-Planned* scenarios.

The gridded TOG inventories of the Stanislaus County for the *Controlled Growth* scenario and the *Uncontrolled Growth* scenario are illustrated in Figure 5 and based on 4×4 km emission maps. As expected, the spatial pattern of mobile source inventories indicates a strong correlation between expanded roadway network congestion and high pollutant emissions at the grid cell scale. Compared to the *Controlled Growth* scenario, the *Uncontrolled Growth* scenario appears to have a substantially larger area with TOG emissions between 50 and 150 kg per grid cell. The *Uncontrolled Growth* scenario has also resulted in increased emissions (>150 kg per grid cell) in the central urban area (City of Modesto) along the major urban corridor of Highway 99.

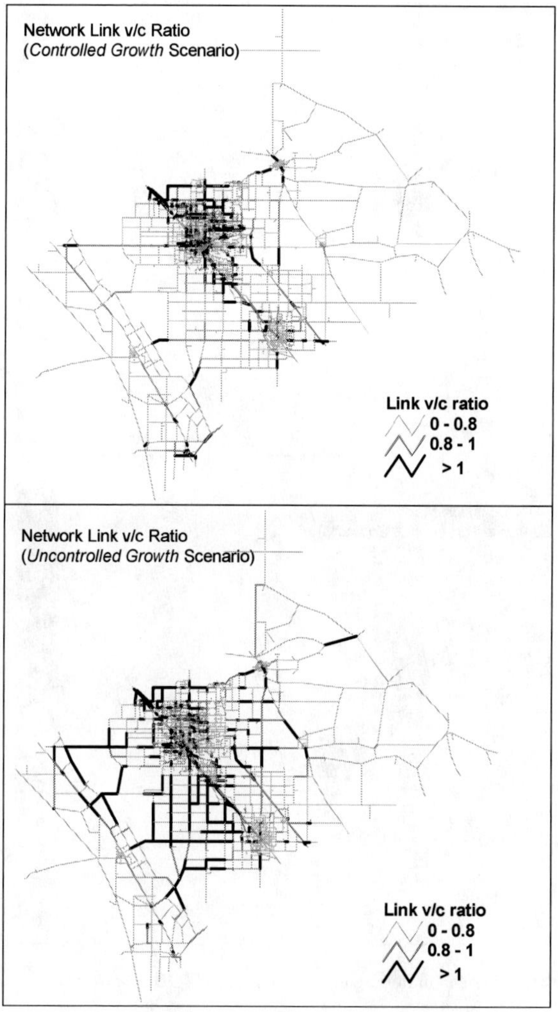

Figure 4. Comparisons of Stanislaus County roadway congestion levels (volume/capacity ratios)

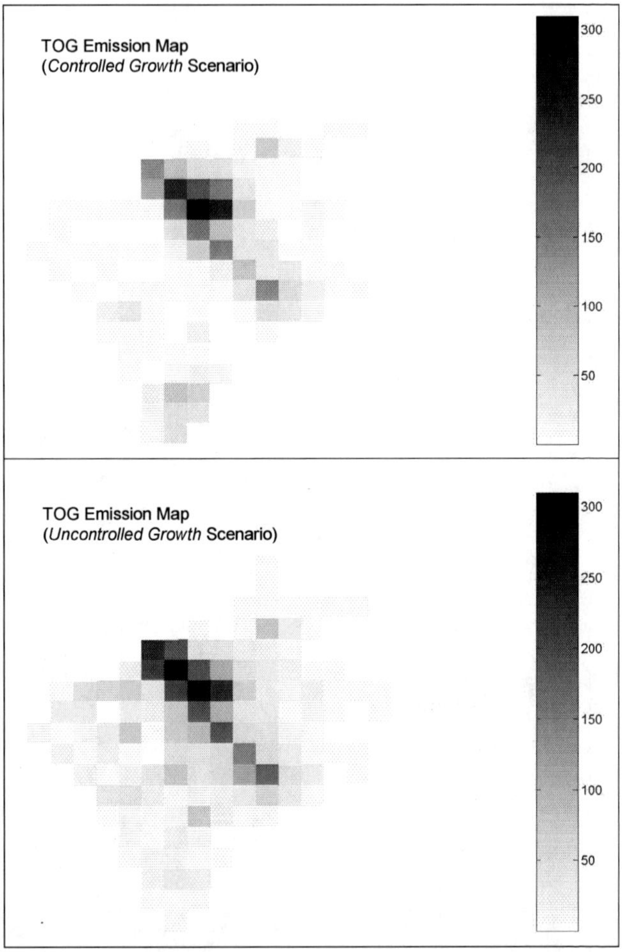

Figure 5. Comparison of Stanislaus County gridded TOG inventories (unit: kg/cell)

Table 4. Summary of Stanislaus County link level traffic statistics

Traffic Statistics	Unit	Scenario 1 Baseline	Scenario 2 Controlled Growth	Scenario 3 Uncontrolled Growth	Scenario 4 As-Planned
Total VMT	mile	16,411,281	15,427,411 (-6%)	22,089,022 (+35%)	18,860,451 (+15%)
Total VHT	hour	712,671	617,969 (-13%)	1,022,107 (+43%)	744,941 (+5%)
Number of vehicle trips	one-trip	2,233,862	2,122,622 (-5%)	2,308,304 (+3%)	2,274,656 (+2%)
Average trip distance	mile	7.35	7.27 (-1%)	9.57 (+30%)	8.29 (+13%)
Average trip length	minute	19.14	17.47 (-9%)	26.57 (+39%)	19.65 (+3%)
Average trip delay	minute	8.18	6.66 (-19%)	12.72 (+56%)	7.62 (-7%)
Freeway average congested speed	miles/hour	37.7	43.9	38.8	42.8
Freeway average v/c ratio	v/c	0.898	0.829	0.915	0.865
Arterial average congested speed	miles/hour	36.1	36.5	35.3	37.3
Arterial average v/c ratio	v/c	0.691	0.654	0.746	0.666

5. Conclusion

Using a systematic land use, transport, emissions modeling system for the San Joaquin Valley, we found, as expected, that long-term land use and transportation development strategies can have significant impact on mobile source pollutant emissions (Table 5). Compared to an uncontrolled growth pattern, effectively applying comprehensive controlled strategies such as smart growth, land preservation and constraints on roadway expansion may contribute to more than a 15% reduction in future regional vehicle traffic and associated mobile emissions. In terms of effect magnitude, the estimated percent reductions in this study tend to represent a maximum for the SJV region, given that both the *Controlled Growth* and *Uncontrolled Growth* scenarios reflect somewhat extreme factor levels (e.g., either very high or very low residential densities). Practical development plans and strategies will more likely result in emissions impact within the estimated range provided in the comparison study.

Modeling alternative emissions scenarios that employ different policies for land use and transportation provides a basis for informed decisions to be made about future growth policies, especially for an area like the San Joaquin Valley with expected dramatic population growth. Further research, such as sensitivity analysis and modeling pollutant concentrations, would be beneficial to further study linkages between policy decisions and future air quality and identify the important factors leading to improved air quality.

Table 5. Impact of different long-term development strategies

Scenario	Baseline	Controlled Growth	Uncontrolled Growth	As-Planned
Growth pattern	Do-nothing with existing conditions	Compact growth, infill, and mixed use	Urban sprawl	Following current plans
Residential density	—	↑↑	↓↓	↓
Travel mode options	—	↑	↓	—
Vehicle miles traveled	↑	↓	↑↑	↑
Vehicle hour traveled	↑	↓	↑↑	↑
Traffic congestion	↑	↓	↑↑	↑
Pollutant emissions	↑	↓↓	↑↑	↑

Acknowledgements

This research is partially funded by the United States Environmental Protection Agency under Grant No. RD-83184201 and NSF grant BES-0302538. The authors would like to thank the expert panel members (Mike Brady, Adrian Fleissig, Dan Leavitt, Carolyn Lott, Karen Magliano, Frank Mitlohner and Marla Mueller) for their valuable comments on policy scenario constructions. The authors also would like to thank Sangho Choo and Julie Ogilvie (UC Davis) and Bryan Penfold (Sonoma Technology, Inc) for their technical assistance at the early stage of this research.

References

Bedsworth, L. W. (2004). "Clearing the Air in the San Joaquin Valley: Developing and Action Plan for Regulators, Legislators, and the Public." Union of Concerned Scientists. <http://www.ucsusa.org/clean_vehicles/california_driving/clearing-the-air-in-the-san-joaquin-valley.html> (Feb. 2, 2007).

California Department of Finance. (2004). "Population Projections by Race/Ethnicity for California and Its Counties, 2000-2050." <http://www.dof.ca.gov/Research/Research.asp> (Feb. 5, 2007).

California High-Speed Rail Authority (CHSRA). (2000). "High-Speed Rail System: Business Plan." <http://www.cahighspeedrail.ca.gov/plan> (Feb. 5, 2007).

Cervero, R. (2002). "Induced Travel Demand: Research Design, Empirical Evidence, and Normative Policies." *Journal of Planning Literature*, 17, 3-20.

Handy, S. (2005). "Smart Growth and the Transportation-Land Use Connection: What Does the Research Tell Us?" *International Regional Science Review*, 28, 146-167.

Johnston, R. A., Shabazian, D. R., Gao, S. (2003). "UPlan: A Versatile Urban Growth Model for Transportation Planning." *Transportation Research Record*, 1831, 202-209.

Litman, T. (2005). "Land Use Impacts on Transport: How Land Use Factors Affect Travel Behavior." Victoria Transport Policy Institute, <http://www.dot.wisconsin.gov/localgov/docs/victoria-transport.pdf> (Feb. 2, 2007).

Niemeier, D. A., Zheng, Y., Kear, T. (2003). "UCDrive: A New Gridded Mobile Source Emission Inventory Model." *Atmospheric Environment*, 38(2), 305-319.

Teitz, M. B., Dietzel, C., Fulton, W. (2005). *Urban Development Futures in the San Joaquin Valley*. Public Policy Institute of California, 2005. San Francisco, CA.

U.S. Environmental Protection Agency. (2000). "USEPA Fact Sheet: San Joaquin Valley Ozone Reclassification." <http://www.epa.gov/region09/air/sjvalley/fact.html> (Feb. 2, 2007).

U.S. Environmental Protection Agency. (2001). *EPA Guidance: Improving Air Quality through Land Use Activities*. EPA420-R-01-001, Air and Radiation, Office of Transportation and Air Quality.

Using TRANUS to Construct a Land Use-Transportation-Emissions Model of Charlotte, North Carolina

B. J. Morton,[1] D. A. Rodríguez,[2] Y. Song,[3] and E. J. Cho[4]

[1]Senior Research Associate, Center for Urban and Regional Studies, Campus Box 3410, University Of North Carolina, Chapel Hill, NC 27599-3410, PH (919) 962-8847; FAX (919) 962-2518; email: bjmorton@unc.edu
[2]Associate Professor, Dept. of City and Regional Studies, Campus Box 3410, University Of North Carolina, Chapel Hill, NC 27599-3140; email: danord@unc.edu
[3]Assistant Professor, Dept. of City and Regional Studies, Campus Box 3410, University Of North Carolina, Chapel Hill, NC 27599-3140; email: ys@email.unc.edu
[4]Graduate Research Assistant, Center for Urban and Regional Studies, Campus Box 3410, University Of North Carolina, Chapel Hill, NC 27599-3410; email: eunjoo@email.unc.edu

Abstract

Integrated land use-transportation-emissions models are necessary to rigorously assess the potential of land use and transportation policies to reduce the vehicular emissions contributing to tropospheric ozone and to fine particulate matter. A theoretically- and empirically-grounded model contains these major components: data on economic sectors, population sectors, and intersectoral flows of commodities and labor; a transportation network; sectoral demands for land, predicting both the quantity and location demanded; elastic trip generation; transportation mode choice including non-motorized modes as a function of built-environment characteristics; a traffic assignment algorithm; and a MOVES-like module that estimates emission factors. That technical approach is incorporated in a TRANUS-based model that is being used to assess long-term development scenarios that could be implemented in Charlotte and Mecklenburg County, North Carolina's largest metropolitan area and part of an ozone nonattainment area. A unique feature of the Mecklenburg County model is use of random-utility theory and a typology of the built environment in the estimation of key parameters describing residential and enterprise locational choices and transportation mode choice.

CE Database subject headings: air quality, emissions, integrated systems, land development, public policy, simulation models, statistical models, transportation models, vehicles.

Introduction

Public policy to address pollution and climate-change challenges remain a vital topic of consideration. This paper summarizes a project, a work in-process, that is examining whether smart growth policies can influence travel patterns and, through them, vehicular emissions so as to have a measurable impact on regional air quality. We describe a simulation model at the traffic analysis zone (TAZ) level to assess the vehicular emissions associated with future development scenarios for the City of Charlotte-Mecklenburg County area, North Carolina's largest metropolitan area and part of an ozone nonattainment area. The primary modeling platform is TRANUS: software that facilitates construction of integrated land-use/transportation models grounded in spatial input-output analysis. We have augmented TRANUS by using random-utility theory and a typology of the built environment in the calculation of key inputs.

We profile the major components of our model: data on economic sectors, intersectoral flows, households, and transportation networks; sectoral demands for land, which are separately specified for quantity and location; elastic trip generation; mode choice including non-motorized modes as a function of built environment characteristics; and traffic assignment. The emission factor model is also briefly described.

In view of our interest in the relationship between land use and air quality, we summarize in more depth the residential and enterprise locational choice models because, given the transportation networks and available transportation modes, they determine the spatial pattern of emissions and the volume of vehicular activity on which the quantity of emissions depend. A unique feature of the locational choice modules is inclusion of explanatory variables representing the built environment. We describe the process of calibrating the model's behavioral parameters. We characterize the model's outputs and explain the coupling between the travel outputs and a bottom-up emissions estimation module relying on data from portable emissions measurement systems.

Project Background and Structure

The project is motivated by the U.S. Environmental Protection Agency Office of Research and Development's request for applications for funding to develop innovative methods to project the air pollutant emissions responsible for fine particulate matter and tropospheric ozone, using tools capable of capturing the influence, over a rather long planning horizon (50 years), of technological change, population trends, and regional development patterns. An interdisciplinary team formed at the University of North Carolina at Chapel Hill and North Carolina State

University: the members represent the disciplines of economics, civil engineering, land use planning, and urban transportation planning. We developed a plan to rigorously test the hypothesis that regional implementation of certain development patterns can significantly influence the spatial characteristics and quantity of direct and indirect emissions from mobile sources, and hence reduce the levels of tropospheric ozone and fine particulate matter. The development patterns of interest may be labeled smart growth and New Urbanist, i.e., compact, mixed-use development providing pedestrian- and transit-friendly built environments. Given the study area's actual land uses, the widespread diffusion of alternatives to sprawl could realistically take decades.

The research team proposed to test the hypothesis with a state-of-the art simulation model comprising the following major modules.

- Cross-sectional land-market equilibrium model.
- Multimodal behavioral travel forecasting model, including nonmotorized modes and incorporating built-environment attributes.
- Modal approach to estimating emissions based upon the conceptual underpinnings of EPA's Multi-Scale Motor Vehicle and Equipment Emissions Estimation System (MOVES).

Mundane considerations strongly influenced our choice of software for the land use-transportation model and of study area—although other considerations also were influential. To stay within the solicitation's budget and time constraints ($750K, three years), the research team needed to develop a strategy of forming additional research partnerships in which project visibility and model results alone would provide sufficient motivation for collaboration. In other words, we could not afford to pay a software license fee, hire modeling consultants, or spend sizable amounts of money on data.

Our modeling platform, TRANUS, is free, effective, accompanied by a tutorial, and documented in a user's manual and a separate mathematical description (www.modelistica.com). Tomás de la Barra, its developer, has been generous with his time, responding to the project team's numerous email requests for technical assistance. Our study area, the city of Charlotte and Mecklenburg County, has also lived up to our expectations. It is data rich, and local government staff in the GIS and transportation departments have readily provided scores of shapefiles and databases (see below), and they have been unfailingly helpful, prompt, and even willing to prepare specialized data extractions that they would not otherwise need to prepare.

At the beginning of the project, members of the research team made two trips to Charlotte to introduce the project to staff with state and local governments, local elected officials, and researchers at the University of North Carolina-Charlotte. Research team members provided reassurance that the simulation model would not in any sense compete with the local Metropolitan Planning Organization's travel forecasting model, and that it would be highly improbable that the research project would thus interfere with the MPO's efforts to predict the mobile source emissions associated with the current long-range regional transportation plan and hence potentially derail demonstration of transportation conformity. The research team has sustained outreach efforts during the ongoing model building phase by, for example, inviting local government staff and UNC-Charlotte researchers to review important parameter estimates.

Research Design

Figure 1 presents a conceptual overview of our project's analytical processes. Our research design entails developing a typology of land-use patterns at the scale of Census block groups; estimating models that explain the demand for land on which to build residences and establishments; estimating mode choice models; constructing baseline (control) and future-year (treatment) scenarios, the latter containing hypothesized innovations in land use and transportation policy and provision; using our TRANUS model to predict each scenario's equilibrium land use and travel patterns; estimating emissions; and running an air quality model to predict concentrations of fine particulate matter and tropospheric ozone.

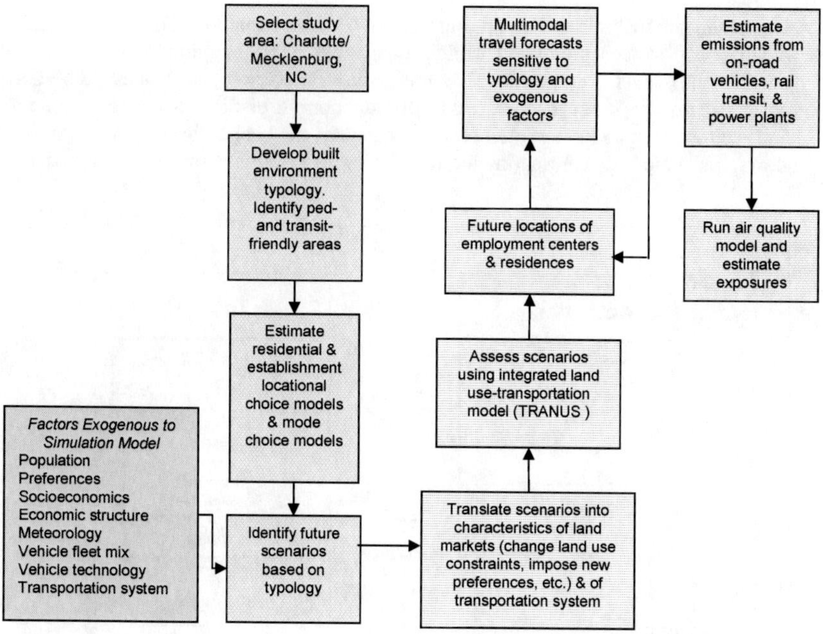

Figure 1. Conceptual Overview of Research Project's Technical Approach

Before describing in some detail the ways in which we are incorporating the built environment into our behavioral models—the focus of this paper—it is helpful to anticipate the final outputs of the entire land use-transportation model because they drive the mobile source emissions endogenous to the scenario assessments: for each included vehicle category, the number of interzonal trips, link-level traffic volume, and link-level average speed. A major objective of the project is to estimate mobile source emission factors using a new tool, the VSPmicro Emission Factor Model. VSPmicro is founded on the use of real-world activity data collected with portable emissions monitoring systems. Thus its emission estimates emerge from a microscale, operating-mode based approach.

The factors that affect real-world microscale emission rates include vehicle speed, acceleration, road grade, vehicle load, ambient temperature and humidity, and vehicle/fuel technology. The first three factors can be used to estimate a surrogate for vehicle load, vehicle specific power (VSP). VSP is a key factor for creating average emission rates for specific operating modes, and it can be linked to the transportation network and vehicle activity information that TRANUS provides. The emissions model can easily accommodate cold starts. VSPmicro will generate emission factors for gasoline, diesel, compressed natural gas, ethanol, biodiesel, and electric highway vehicles and for rail transit vehicles.

Upstream of VSPmicro lies in our TRANUS-based model. To drive emission estimation, for each motorized vehicle type included in the simulation, the model provides link-level average speed. Figure 2 shows in very broad strokes TRANUS's basic structure: a module representing the functioning of the study area's economy and real estate markets, an interface that translates the economic flows into network flows, and a module representing the functioning of the area's transportation system.

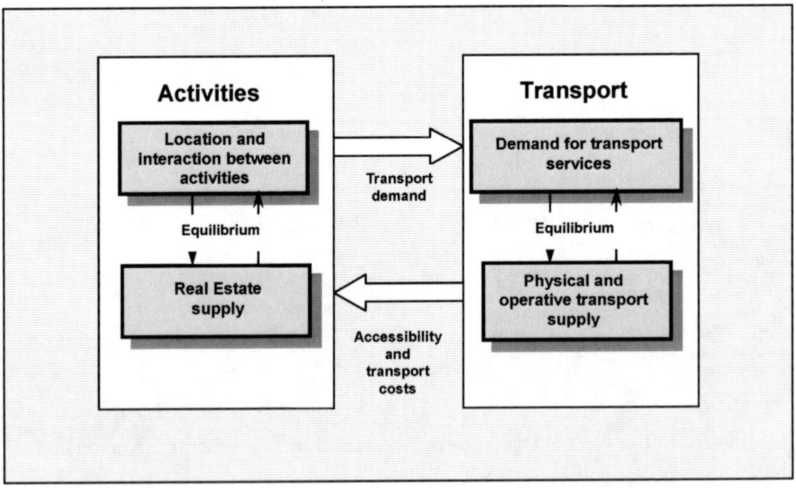

Figure 2. TRANUS's Basic Structure

The role of the built environment

We are incorporating the built-environment determinants of travel patterns into the simulation model's activities and transportation modules. Because land development patterns appear to shape travel behavior (e.g., Ewing and Cervero 2001), such patterns may also affect emissions and air quality at local, regional, and subcontinental scales. The impact on air quality of changing land development patterns may occur through several routes. Changes in personal travel may change the total quantity of emissions, chemical speciation of emissions, time-of-day of emissions, and location of emissions. Theoretically, however, the impact of development pattern on emissions is difficult to ascertain because of the multiplicity of intervening factors and their sometimes opposing directions. For example, the increased proximity afforded by mixing residential, retail, and office land uses appears to support walking trips. However, it is less clear whether such trips complement or substitute for existing trips that rely on motorized transportation (Ewing and Cervero 2001; Handy 1996). Likewise, a higher development density is likely to shorten travel distances, but its impact on travel speeds is less certain; in addition, trip generation may increase (Boarnet and Crane 2001). It follows, at least from theory, that the net emissions and air quality impacts of alternative development forms, expressed through travel behavior, are ambiguous. Therefore, long-range, theoretically-based simulation modeling of land use and travel behavior is necessary to quantitatively forecast the influence of regional development patterns on emissions.

Two equations capture the approach taken by TRANUS to impart the spatial dimension to the representation of productive activities, i.e., the locations of business establishments and the producers of labor power, households. For convenience, the equations are interpreted in terms of business establishments. The first equation expresses the probability that establishment *n* locates in zone *j*, and the probability depends on both the (dis)utility, such as the cost of land, of a location in each zone (U_j^n) and the other attributes that affect a zone's attractiveness (A_j^n). The latter, termed the attractor function, comprises one or multiple factors (X_j^k) lagged one period, each multiplied by a weight (b_k^n) capturing a factor's relative influence. The attractor function is literally and exclusively an attractor—no factors can repel. Further, the weights must be greater than or equal to one, and hence only their relative magnitudes matter.

$$\Pr_j^n = \frac{(A_j^n)\exp(U_j^n)}{\sum_j (A_j^n)\exp(U_j^n)} \qquad (1)$$

$$A_j^n = \sum_k b_k^n (X_j^k) \qquad (2)$$

The attractor function therefore provides the means for representing such nonmonetary determinants of agglomeration as a nearby (in the same zone) supply of labor, nearby establishments in the same sector, and nearby establishments in other sectors. As explained more fully in the next section, the attractor function also provides the means for making the built environment a determinant of locational choice.

Starting with a set of urban form attributes measured at the Census Block Group level, we used factor analysis to identify a small set of dimensions that capture essential differences in the built environment: walkability, accessibility, agglomeration, industry, and property values. Then, in collaboration with stakeholders familiar with the study area, we used the factor scores as inputs to a cluster analysis to identify eight distinct neighborhood types—producing the Neighborhood Transect. For example, Type 1 consists of a unique central business district block group representing the heart of the downtown district, where local and regional access is very high. At the study area's periphery, Type 8 is dominated by industrial and forested uses; residences typically are isolated, large-lot single-family dwellings. Table 1 describes and Figure 3 maps all neighborhood types.

Table 1. Neighborhood Transect

Neighborhood Type	Description
1	Core central business district (CBD)
2	CBD with residences
3	First ring of suburbs
4	Second suburban ring, which is farther from CBD, transit, and retail than first-ring suburbs
5	Urban single family residences, with some commercial and transit
6	Suburbs with low access to green space & low regional access
7	Exurbs with access to green space & low regional access
8	Isolated residences in forested, industrial, & commercial areas

The neighborhood types play an unconventional but crucial role. They are unconventional because they are neither purchased, consumed in any literal sense, nor consciously supplied—they are a kind of public good created by the unplanned co-occurrence (within block groups) of intersections, bus stops, structures, employment, and almost two dozen additional specific attributes. Neighborhood types are crucial because we have found through statistical analysis that they significantly and understandably influence mode choice and location choice.

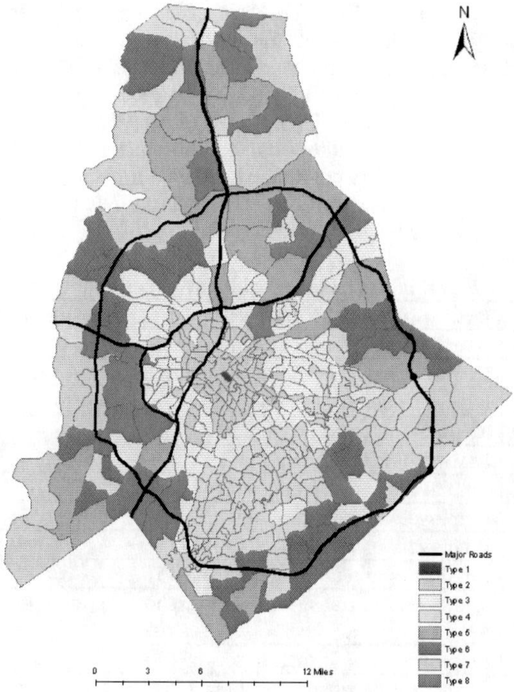

Figure 3. Mecklenburg County's Neighborhood Types

Firm location choice modeling

For purposes of brevity, we focus the description of our locational choice models on business establishments. Similar models were estimated for residential locational choice and are being summarized elsewhere.

As of late February 2007, the activities/land use section of our Mecklenburg County model comprises eight types of micro-environment or neighborhood, three household sectors (differentiated by income), useable land (developed and developable land), and twelve economic sectors (differentiated by two-digit NAICS code). The latter do not include agriculture and resource extraction because employment in those sectors is very small (in the hundreds), and therefore, relative to the entire study area, they presumably generate few trips and mobile source emissions.

We constructed the estimation database from the Neighborhood Transect, the local MPO's Transportation Analysis Zones (TAZs), employment data in the 2000 Census Transportation Planning Package, and Mecklenburg County's tax parcel data and privilege/business license data. The shapefile representing the Neighborhood Transect maps the boundaries of the study area's Census Block Groups and neighborhood types. The Neighborhood Transect and TAZ shapefiles were joined to allow identification of the neighborhood types in each TAZ. We used the tax parcel data to calculate, for each TAZ, the value of nonresidential land. The privilege/business license data allowed us to determine each establishment's location and NAICS code. Thus we were able to generate a database supporting modeling efforts for which the decision-making unit is an individual establishment and the decision is the TAZ in which to locate.

Table 2. Variables in Establishment Locational Choice Models

Variable	Description—All Variables Calculated for Each TAZ
land_price	Average price of nonresidential land
distance	Distance from centroid to nearest freeway on-ramp
inc_high	High-income households
inc_med*	Medium -income households
inc_low	Low-income households
workers_1	Employees in construction (NAICS code 23)
workers_2	Employees in manufacturing (31–33)
workers_3	Employees in wholesale trade (42)
workers_4	Employees in retail trade (44 and 45)
workers_5	Employees in transportation, warehousing, and utilities (22, 48, and 49)
workers_6	Employees in information sector (51)
workers_7	Employees in finance, insurance, real estate, rental, and leasing (52 and 53)
workers_8	Employees in professional, scientific, and technical services; management of companies and enterprises; administrative and support and waste management services (54–56)
workers_9*	Employees in educational services, health care, and social assistance (61 and 62)
workers_10	Employees in arts, entertainment, recreation, accommodation, and food services (71 and 72)
workers_11	Employees in other services, except public administration (81)
workers_12	Employees in public administration (92)
ntype_1	Dummy variable for neighborhood type 1
ntype_2	Dummy variable for neighborhood type 2
ntype_3	Dummy variable for neighborhood type 3
ntype_4	Dummy variable for neighborhood type 4
ntype_5	Dummy variable for neighborhood type 5
ntype_6	Dummy variable for neighborhood type 6
ntype_8	Dummy variable for neighborhood type 8

For each of the 12 economic sectors in our simulation model, we estimated a conditional logit model of locational choice. At the outset of the estimation process, the models included these explanatory variables: average cost of nonresidential land in a TAZ; distance from a TAZ's centroid to the nearest freeway on-ramp; numbers of low-income, medium-income, and high-income households; numbers of employees in each economic sector; and dummy variables for neighborhood type, the reference being the exurbs. Table 2 provides the variable names and a brief description.

The sample sizes range from 79 (Public Administration) to 3,188 (Professional, Scientific, and Technical Services, Management of Companies and Enterprises, and Administrative and Support and Waste Management and Remediation Services). The Public Administration sector is atypically small: the next smallest, Information, has 343 establishments. The largest sector is not atypically large because it is one of five sectors of which the number of establishments exceeds 1,000.

Results

Space permits a detailed discussion of a regression's results for only one sector, manufacturing. Table 3 presents the results for the manufacturing establishments. (We hasten to add that the results are preliminary because, at the time of writing, we had not yet looked for violations of the independence of irrelevant alternatives assumption.) Those results are typical of all models in the sense that at least one neighborhood type has a statistically significant coefficient. In the manufacturing establishments model, neighborhood types 2 and 4 have statistically significant, positive coefficients. Neighborhood type 2 is adjacent to the core central business district (CBD); type 4 is farther from the CBD and corresponds to the second suburban ring. The evidence is strong for concluding that the manufacturing establishments are more likely to locate in either a transitional area between the core CBD and the oldest suburbs or in the second suburban ring.

Two models required special handling because their coefficients for the land price variable have a statistically significant coefficient with an unexpected positive sign. A positive coefficient implies, contrary to our intuition about profit maximization, that, everything else equal, TAZs with more expensive land are more attractive than the zones with less expensive land. The sectors for which we obtained that unexpected result are 1) retail trade and 2) arts, entertainment, recreation, accommodations, and food service. For those sectors, we removed the land price variable and reestimated the coefficients; some models had positive but insignificant coefficients for the land price variable, but we did not alter those models.

Table 3. Conditional (Fixed Effects) Logit Model of Manufacturing Establishment's Locational Choice

Variable	Coef.	Std. Error	z	P>|z|
land_price*	-.0003977	.0001706	-2.33	0.020
distance*	-.0711079	.0257511	-2.76	0.006
inc_high	.0001912	.0004588	0.42	0.677
inc_med*	.001587	.0004594	3.45	0.001
inc_low	-.0001496	.0005859	-0.26	0.798
workers_1*	.0027738	.0004837	5.73	0.000
workers_2*	.0006105	.0001822	3.35	0.001
workers_3*	.0019368	.0005195	3.73	0.000
workers_4*	.0006023	.0002584	2.33	0.020
workers_5	-.0001351	.0001941	-0.70	0.486
workers_6	.0003718	.0005922	0.63	0.530
workers_7	-.0000704	.0001261	-0.56	0.577
workers_8	-.0000193	.0004609	-0.04	0.967
workers_9*	-.0011678	.000412	-2.83	0.005
workers_10	.000583	.0005668	1.03	0.304
workers_11*	.0040927	.0012613	3.24	0.001
workers_12	-.0000193	.0011152	-0.02	0.986
ntype_1	-14.97858	855.0469	-0.02	0.986
ntype_2*	.7528804	.1304474	5.77	0.000
ntype_3	.023307	.0930117	0.25	0.802
ntype_4*	.2374443	.1158595	2.05	0.040
ntype_5	.0686	.1109531	0.62	0.536
ntype_6	-.0861355	.1267239	-0.68	0.497
ntype_8	.0577018	.1101872	0.52	0.601

* = significant at $\alpha \leq 0.04$. n = 627 establishments. LR chi2 (24) = 437.64. Prob > chi2 = 0.0000. Pseudo R2 = 0.0503.

Having estimated the locational choice models, the next step was to re-express the coefficients for the neighborhood type variables (and those for the household and employment variables) into a form suitable for TRANUS. As we discussed in the previous section, the attractor function requires weights (b_k^n) that are greater than or equal to one; further, the only important aspect of the value of a weight is its relative magnitude, i.e., weights of 2 and 10 are equivalent to weights of 10 and 50. We created weights for the coefficients that were statistically significant and positive. To create each relevant independent variable's weight, we calculated the variable's odds ratio $e^{\hat{\beta}}$. For a unit change in an independent variable, holding all other variables constant, the odds change by the factor $e^{\hat{\beta}}$. Because the change is independent of the independent variable's initial value, the odds ratio provides an overall quantitative summary of the variable's influence on locational choice and permits different independent variables' influence to be consistently ranked.

Therefore, to the extent TRANUS allows, our locational choice models inform the structure of our land use-transportation model. The variables with negative coefficients receive weights of zero; because neighborhood types were treated as dummy variables and hence one neighborhood had to serve as the reference, the reference neighborhood type (7, exurbs) receives a weight equal to one; and variables with significantly positive coefficients receive weights equal to their odds ratio.

Conclusions and Next Steps

We have demonstrated an approach to incorporating built-environment factors into models of residential and business establishment choice, which in turn are building blocks for an integrated transportation-land use model with a mobile-source emissions module. Not only is the approach viable, it is founded on microeconomic theory, and it is empirically grounded.

We are using TRANUS, an existing, open-source modeling platform. All of the data we are using are available to the public, and we obtained the data at minimal cost. The data requirements and structure of TRANUS are similar to other spatial input-output models currently in development or implementation, such as PECAS.

Our technical approach is replicable in other areas. Nonetheless, however important analytical tools and data, the success of similar efforts—at least those attempted by university based research teams is very likely to hinge on establishing and sustaining informal research partnerships with stakeholders.

Our immediate plans for exercising the simulation model entail assessing future regional development scenarios, i.e., planned or hypothetical changes in transportation policy and provision and/or land use policy. TRANUS facilitates investigation of land use policies implemented with market-based and regulatory instruments. That capability allows us to gain insight into the leverage of both direct and indirect approaches to achieving smart growth. Due to their sensitivity to the built environment, our model's forecasts will more accurately portray the spatial patterns of activities and of the associated motor vehicle traffic and air pollutant emissions.

Looking further ahead, our current project is creating an infrastructure for extensions in several directions. We could enhance our current model by estimating emissions of airborne toxins and greenhouse gases, explicitly incorporating a larger study area (the entire "travelshed"), and linking the model's outputs to modules representing such additional environmental attributes as vegetative cover and water quality. Our data, the Neighborhood Transect, and our estimated behavioral models are steppingstones to working with additional land use-transportation modeling platforms. Use of multiple platforms would allow us to characterize the model-based component of forecast uncertainty.

Acknowledgments

The project has been made possible with funding from the National Center for Environmental Research (EPA STAR Grant 2004-STAR-B1). Funding has also been provided by North Carolina State University's Institute for Transportation Research and Education.

The other senior members of the transportation-land use research team are Chris Frey (NCSU), Leta Huntsinger (now at Parsons Brinckerhoff), Asad Khattak (Old Dominion University), and Nagui Rouphail (NCSU). Several Ph.D. students have also participated: Tracy Hadden (UNC), Elizabeth Shay (UNC), Bev Wilson (UNC), and Haibo Zhai (NCSU).

References

Ewing, R., and Cervero, R. (2001). "Travel and the built environment." *Transportation Research Record*, 1780, 87-114.

Handy, S. (1996). "Urban form and pedestrian choices: study of Austin neighborhoods." *Transportation Research Record*, 1552, 135-144.

Boarnet, M. G., and Crane, R. (2001). "Travel by design: The influence of urban form on travel." *Oxford University Press*, New York.

Linking urban structure and air quality

H. Martins, [1] A. I. Miranda,[1] and C. Borrego[1]

[1] CESAM, Department of Environment and Planning, University of Aveiro, 3810-193 Aveiro, Portugal, Ph +351 234 370220; Fax +351 234 429290; e-mail: hmartins@dao.ua.pt ; aicm@dao.ua.pt ; borrego@ua.pt

Abstract

The degradation of the urban air quality is today object of concern since a great part of the world population lives in urban areas. While the environmental implications of industrial and transport activities are recognized for decades, the influence of the urban patterns in air quality is still poorly understood. Currently, urban areas are expanding towards rural ones which, together with the segregation of land uses related to daily activities (home, school, work, leisure), lead to an increase of motorized trips and trip distances, resulting in an increase of air pollutants emissions.

This paper carries out the state of the art concerning the link between urban structure and air quality. This link is neither simple nor direct; currently there is a debate on which is the most sustainable urban form: the compact city or the disperse city. Several empirical and modeling studies integrating land use, transport and air quality issues and its main conclusions are presented. Results indicate that compact cities promote a better air quality when compared to the dispersed cities; on the other hand, when the subject is human exposure to pollutants, compact cities present worst results. These studies allow concluding about the influence of the city structure in air pollutants levels, highlighting the importance of air quality as a relevant urban planning indicator.

Introduction

The environmental degradation and, in particular, the deterioration of urban air quality is today an object of effective concern since a large part of the world population resides in urban areas and, therefore, are exposed to high pollution levels.

In urban areas the main atmospheric pollutants emission sources are road traffic, industrial activities and the use of fossil fuels for heating and energy production. Despite the reduction in specific emissions from motorized vehicles and large point sources, achieved through the use of cleaner fuels and more efficient technologies, urban areas still show increasing signs of environmental stress: loss of open spaces, traffic congestion, noise and air quality degradation. Thus, a substantial number of people – around 1.5 billion, corresponding to 25% of the world population – are exposed to excessive gaseous and particulate pollutants concentrations (WHO, 2000). Since future projections point to the increase of population in urban areas, the situation tends to aggravate.

While the environmental implications of industrial and transport activities are recognized for decades, the study of the influence of urban patterns in the environment is still relatively recent, namely in what concerns its influence on air quality. However, it seems clear that in the last decades, urban growth has not followed the concept of sustainable development, with urban areas expanding to rural ones and with the replacement of green spaces by low density urbanizations, while the population of the historic city centres decreases. This tendency, together with the segregation of activity-related land uses (home, school, work and leisure), gives rise to traffic growth resulting in an increase of pressures over the environment, such as atmospheric and particulate pollutants and greenhouse gases (GHG) emissions. As a result of these growth patterns, the environment, which should be preserved for future generations, is compromised by the effects of the excessive consumption of resources.

However, our cities don't have to persist in this development pattern of space waste and automobile dependence in prejudice of walking, cycling and public transports, and in the development of residential, commercial and industrial spaces with no consideration for spatial interdependence. Newton (1997) presents the city as a villain, a victim and white knight with respect to air quality. A villain since its transport and industries consume enormous amounts of energy, contributing significantly to urban air pollution. A victim because its residents, industries and image are affected negatively by the atmospheric pollution, which reduces the quality of life and health as well as the attractiveness of the city to tourists and potential new industries and residents. But the city can also be a white knight since changes in its structure and development may lead to a substantial reduction of traffic congestion and, consequently, energy consumption and air pollutant levels.

Today it's widely accepted by the scientific community, that there is a relation between the city's form, size, density and land use, and its sustainability; however, the consensus about the exact nature of this relation as not yet been reached (Williams et al., 2000). There is a wide variety of studies concerning the effects of land use changes in motorized trips and in urban emissions, but the results are not always consistent, and the exact extension of the cause-effect relationship is not conclusive (PLUME, 2003).
In this paper the link between urban structure and sustainable urban development, namely air quality, is explored through the analysis of the current state of the art. Several modeling studies integrating land use, transport and air quality issues and its main conclusions are presented.

What the theory tells us

Since cities are one of the main natural resources consumers, the largest producers of waste and the centre of most human activities, it's evident that a significant part of the debate over sustainability must have an urban framing; i.e., if the cities are part of the problem, they must be part of the solution.

In response to environmental sustainability issues, urban planners have focused their attention in the types of urban structure that will best serve our growing cities. By urban structure is understood not only the morphologic structures of the city, represented by its key-structures (road and rail networks, ports and airports, telecommunications and social infrastructures), but also the way how residential, industrial, services and recreational land uses are distributed along the city.

The role of urban planning in urban sustainability, more concretely on which urban structure will provide more environmental protection, is now on discussion. In a first glance, the scope of the debate can be summarized by classifying positions in two groups: the "decentrists", in favour of urban de-centralization, defending the dispersed city characterized by low population densities and large area requirements; and the "centrists", who believe in the virtues of high density cities with low area requirements, defending the compact city (Breheny, 1996).

Those in favour of the compact city defend that urban containment, with mixed land use, will reduce the need for motorized trips, therefore reducing traffic emissions, and promoting public transport, walking and cycling. Higher densities will help to make the supply of infrastructures and leisure services economically feasible, also increasing social sustainability. On the other hand, the compact city can become overcrowded loosing urban quality, with less open spaces, traffic congestion and pollution, not representing the type of environment one would choose to live in.

Against the dispersed city defenders the argument is that low densities, and the consequent large area needs and land use segregation, result in a high dependence from motorized vehicles. This argument is strengthened in the current climate change context, not only at the GHG emissions level but also at the level of greater resources consumption (Breheny, 1996). Another aspect to take in to account is the fact that most actual cities follow the dispersed city development pattern, presenting the environmental problems already described.
Given the merits and demerits of each side, a compromise solution seems attractive. Newton (1997) presents six city-types, representing urban systems with different spatial configurations, according to its shape and structure (Figure 1).

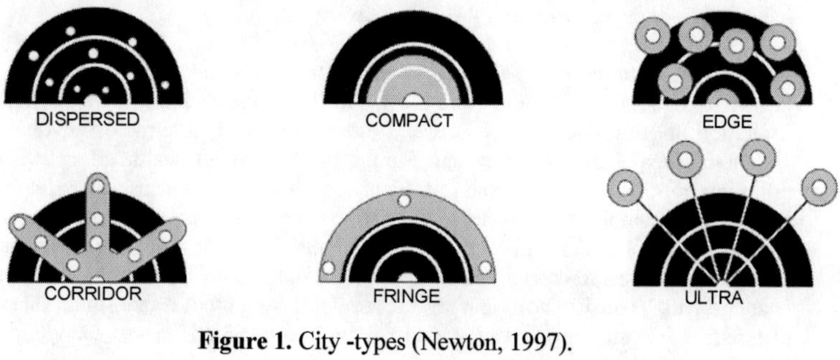

Figure 1. City -types (Newton, 1997).

The dispersed city represents the current trend of many of our contemporary cities, with expansion of urban development at low densities, a well defined city centre and radial structure transport network. The compact city emerges as result of an effort for containing urban expansion, through the increase of population density in the city centre and in the adjacent suburbs. The edge city, or multi-nodal, is constituted by several high development central points (nodes), connected by highways and arterial roads, where labour, commerce and leisure activities are concentrated. The corridor city is characterized by linear corridors originating from the city centre, served by high-quality transport infrastructures, along which growth takes place. In the fringe city, growth is accommodated in the suburbs and rural zones, away from the city centre. In a futurist vision emerges the ultra-city, where the concept of metropolis is replaced by the concept of metropolis-based region, extending some hundred kilometres from its historical origins; high-speed transport and communications provide the basis to this concept.

Today the debate is focused on technical issues and not so much in historical discussions, since the main concern is the attainment of concrete results. With concepts and theories often contentious, the need for extracting knowledge from research and practice is acknowledged (Jenks et al., 1996; Williams et al., 2000; PLUME, 2003; Marquez e Smith, 1999). Therefore there is a need to proceed to the integrated modelling of the subject, including land use, transport, air quality and human exposure aspects.

What modelling tells us

Several empirical and modeling studies integrate land use and transport issues and its relation with urban structure (ECOCITY, 2005; Irving and Moncrieff, 2004; Herala, 2003; PLUME, 2003; EPA 2001; Crane, 2000; Handy, 1996); however, few of them explore the connection to air quality and human exposure. Exposure is the key factor in assessing the risk of adverse health effects, since high pollutants concentrations do not harm people if they are not present, while even low levels become relevant when many people are present (WHO, 1999). Concerning health risk, O_3 (ozone) and PM_{10} (particulate matter with diameter less then 10μm) are particularly important due to their adverse health effects: exposure to high ozone levels can induce changes in lung function and airway inflammation, increasing the number of hospital admissions from respiratory diseases; effects on mortality, respiratory and cardiovascular hospital admissions have been associated with high PM_{10} daily averages.

Conclusions from most of the studies done so far have been harmed by the lack of knowledge about the complex and slow path between an initial action for the limitation of atmospheric emissions and the final benefit in terms of air quality and human exposure (Marquez and Smith, 1999). Human health effects of air pollution are the result of a chain of events going from the release of pollutants leading to an ambient air concentration, over the personal exposure, uptake and resulting internal dose to subsequent health effects (Hertel at al., 2001). Emissions limitation conducts to changes in pollutants atmospheric concentrations, but those changes will have different spatial and temporal magnitudes and signs, due to differences in emissions, weather patterns and population exposed to pollution according to the time of day, day of the week or month of the year, and also according to the population age structure (children, adults and elderly suffer different effects due to their different respiratory frequencies). The need for the introduction of air quality tools in the equation that relates urban structure and environmental sustainability is clear.

Marquez and Smith (1999) developed a framework for linking urban form and air quality, integrating land use, transport and air quality models. In order to evaluate six alternative scenarios, corresponding to the six city types described previously, and its contribution to air quality improvement, Newton (1997) applied the developed framework to the city of Melbourne. Land use and transport models were applied to a time horizon of 20 years; dispersion and air quality models were applied to a typical summer day (with meteorological conditions favourable to the occurrence of photochemical smog – ozone) and a typical winter day (with meteorological conditions favourable to the occurrence of PM_{10} high concentrations). Modelling results show the impact of urban development patterns on human exposure to ozone: the corridor development scenario for 2011 presents a 55% improvement over the 1991 base case, compact and edge scenarios also deliver significant enhancements with 24% and 21% improvement; on the other extreme, the business-as-usual scenario (corresponding to the dispersed city development) results in a 71% increase in human exposure. For the winter episode, the corridor city scenario results in a 14% improvement in human exposure to PM_{10}, while the compact city scenario presents the worst situation, with a 160% aggravation. Despite the low levels of pollutants and CO_2 emissions and fuel consumption, the localization of new residences and working places in the compact city centre, lead to the exposure of a greater number of residents and workers to high dosages of PM_{10}. The study concludes that urban structure does matter, not just for urban air quality, but also for human exposure to pollutants.

More recently, a study conducted by Borrego et al. (2006) concluded, through the application of dispersion and photochemical models, that compact cities with mixed land uses promote a better air quality when compared with dispersed cities with land use segregation. In this study, three cities with distinct urban structures – dispersed, corridor and compact – were idealized; the same number of inhabitants was distributed along each city considering different population densities for different land use classes. Besides emissions from transports, calculated according to different mileages for different land uses, residential, industrial and biogenic emissions were also considered. The corridor city presents the highest per capita and per area emissions, the dispersed city presents the lowest emissions per area and the compact city the lowest emissions per capita. With the objective of characterizing, temporal and spatially, the dispersion of pollutants for each city case, a meteorological and photochemical modelling system was applied to a typical summer day; O_3 and NO_2 (nitrogen dioxide) 2-D concentration fields were analysed, as well the concentrations temporal evolution. Regarding ozone, the dispersed and the compact cities present the highest and lowest concentrations, respectively; the corridor city presents the highest NO_2 concentrations. Figure 2 presents the spatial distribution of ozone concentrations for each city at 2:00 PM: the ozone plume reveals that higher concentrations are reached and larger areas are affected in the dispersed and corridor cities in comparison with the compact city.

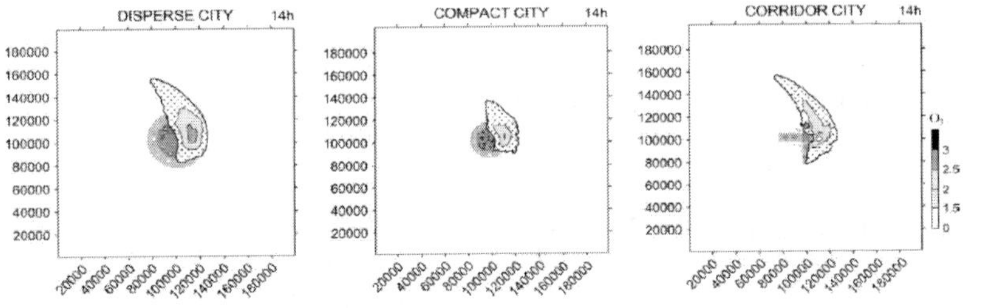

Figure 2. Ozone concentration fields (in relation to the background concentration) at 2:00 PM in the dispersed, compact and corridor cities (Borrego et al., 2006).

A subsequent study, conducted by the same team (Ferreira et al., 2005), investigated the influence of urban structure on human health, estimating the human exposure to atmospheric pollutants for each of the three cities presented in Borrego et al. (2006). Whereas the pollutant concentration refers to an environmental characteristic in a determined time and space, exposure describes the interaction between the environment and the human being. Therefore, in order to estimate exposure, the concentrations estimated by Borrego et al. (2006) were combined with the population distribution and occupation. Results reveal that, although the highest exposure value is found in the corridor city, it's in the compact city that more people are exposed to higher pollution levels (Figure 3), due to the existent high population densities. Thus, the study concludes that air pollution episodes are more critical for human exposure in compact cities.

Figure 3. Total human exposure to ozone (inhab.µg.m^{-3}) in the dispersed, corridor and compact cities (Ferreira et al., 2005).

Civerolo et al. (2007) investigated the potential effects of extensive changes in urban land cover, in the New York City metropolitan region, on surface meteorology and ozone concentrations. They extrapolate urban land cover from present conditions to the year 2050, therefore taking into account climate change: the development of the future land use scenario followed the A2 emissions scenario from the Intergovernmental Panel on Climate Change (IPCC). The study integrates a diverse set of modelling tools from the global scale (general circulation models), to the regional scale and urban scales (meteorology and air quality models); results include high resolution meteorology and air quality. For meteorology, results suggest an increase of afternoon temperatures by more than

0.6°C over the study region. Regarding ozone, future changes in urbanization, may lead to an increase in episode-average O_3 levels by 1-5 ppb and in episode-maximum by more than 6 ppb; however the spatial patterns are heterogeneous, and there are some areas where O_3 concentrations decrease.

Conclusions

The potential impact of land use changes on the environment, namely on air quality, has stimulated research in the understanding of land use change and its main effects. Recent advances in computer technology have allowed the integration of land-use and traffic models with dispersion and air quality models. The studies analysed confirm the importance of urban structure in the city's sustainability. Results indicate that compact cities with mixed land use promote a better air quality when compared to dispersed cities with lower densities; however when considering human exposure to pollutants, compact cities present worst results, due to the high population densities that place more people in areas with higher pollutants concentrations.

Today the great majority of our cities have an urban structure that encourages unsustainable practices, creating a dependency on road-based transport, enhancing GHG and pollutants emissions, and thus contributing to the worsening of air quality, and more broadly, of the quality of life. Therefore there is a need to define more efficient urban spaces, at the energy and environmental levels, rethinking the current organization models and development patterns, not forgetting that air quality should be considered as an important urban planning indicator.

Since urban structure changes take several decades to take place, future work should address the climate change issues, by considering future emissions and future meteorology. Also, longer time-periods must be studied, instead of short time episodes, in order to get a clearer and broader view of the effects of the city's structure on air quality and human exposure.

Acknowledgments

The authors are grateful to the Portuguese Foundation for Science and Technology and the European Social Fund, for the Structural Funds period between 2000-2006, for the Ph.D. Grant of H. Martins (SFRH/BD/13581/2003).

References

Borrego, C., Martins, H., Tchepel, O., Salmim, L., Monteiro, A. and Miranda, A.I. (2006). "How urban structure can affect city sustainability from an air quality perspective." *Journal of Environmental Modelling & Software*, 21, 461-467.

Breheny, M. (1996). "Centrists, Decentrists and Compromisers: Views on the Future of Urban Form." *The Compact City: A sustainable urban form?*, Ed. M. Jenks, E. Burton and K. Williams, Oxford Brookes University, Oxford, UK.

Civerolo, K., Hogrefe, C., Lynn, B., Rosenthale, J., Ku, J-Y, Solecki, W., Cox, J., Small, C., Rosenzweig, C., Goldberg, R., Knowlton, K. and Kinney, P.(2007). "Estimating the effects of increased urbanization on surface meteorology and ozone concentrations in the New York City metropolitan region." *Atmospheric Environment*, 41, 1803-1818.

Crane, R. (2000). "The influence of urban form on travel: an interpretative review." *Journal of Planning Literature*, 15 (1), 3-23.

ECOCITY Project (2005). *Urban development towards appropriate structures for sustainable transport* – Final Report (deliverable 18), EVK4-CT-2001-00056.

EPA (2001). *Our Built and Natural Environments: A Technical review of the interactions between land use, transportation and environmental quality.* EPA, 231-R-01–002, Washington.

Ferreira, J., Martins, H., Miranda, A.I. and Borrego, C. (2005). "Population Exposure to Atmospheric Pollutants: the influence of Urban Structure." *1st Environmental Exposure and Health Conference*, Atlanta, EUA, 5-7 October 2005 – Environmental Exposure and Health. Eds. M..M. Aral, C.A. Brebbia, M.L. Masila and T. Sinks. WIT Press, 13-22.

Handy, S. (1996). "Methodologies for exploring the link between urban form and travel behaviour." *Transportation Research,* D 1 (2), 151-165.

Herala, N. (2003). "Regulating traffic with land use planning." *Sustainable Development*, 11, 91-102.

Hertel, O., De Leeuw, F., Raaschou-Nielsen, O., Jensen, S., Gee, D., Herbarth, O., Pryor, S., Palmgren, F. And Olsen, E. (2001) "Human exposure to outdoor air pollution." *Pure Appl. Chem.*, 73 (6), 933-958.

Irving, P. and Moncrieff, I. (2004). "Managing the environmental impacts of land transport: integrating environmental analysis with urban planning." *Science of the Total Environment,* 334-335, 47-59.

Jenks, M., Burton, E., Williams, K. (Ed.) (1996). *The Compact City: A sustainable urban form?* Oxford Brookes University, Oxford, UK.

Marquez, L. and Smith, N. (1999). "A framework for linking urban form and air quality." *Journal of Environmental Modelling & Software*, 14 (6), 541-548.

Newton, P.W. (Ed.) (1997). *Re-shaping cities for a more sustainable future – Exploring the link between urban form, air quality, energy and emissions.* CSIRO, Australia.

PLUME Project (2003). *Planning and urban mobility in Europe – Synthesis Report, Land use planning measures*, EVK4-CT-2002-20011.

WHO (2000). *Air Quality Guidelines for Europe*, 2nd Edition. World Health Organization. Regional Office for Europe. .WHO Regional Publications, European Series, No. 91, Copenhagen.

WHO (1999) *Monitoring Ambient Air Quality for Health Impact Assessment.* World Health Organization Regional Publications, European Series, No.85, Copenhagen.

Williams, K., Burton, E., Jenks, M. (Ed.), 2000. *Achieving Sustainable Urban Form*, USA, 2000.

Benefits of Air Quality Simulation Models

B. Kim, [1] and J. Rachami[1]

[1]Research and Consulting, Wyle Laboratories, Inc., 522 Papillion Trace, Woodstock, GA 30188; PH (703) 577-8221; FAX (703) 415-4556; email: brian.kim@wylelabs.com

Abstract

As the air quality modeling needs of transportation planning professionals become more complex, the models must follow suit to provide more sophisticated and advanced capabilities. The current CALINE and CAL3QHC-series roadway models promulgated by the Environmental Protection Agency (EPA) provide static environments with time-averaged, aggregate variables that are unsuitable for micro-scale assessments. Therefore, the successor (next generation) models should be based on a simulation approach and leverage new computing technology to provide more realistic and robust modeling environments. The TRaffic Air Quality Simulation Model (TRAQSIM) developed by Wyle Laboratories meets each of these qualities. TRAQSIM's 4-dimensional (4D) environment with virtually no temporal and spatial constraints provides the model with significant growth potential including the possibility of modeling the effects of road grade and facilitating the emissions and dispersion modeling of both particulate matter (PM) and chemically reactive pollutants. As an example of the robustness of a simulation environment, the ability to directly assess input and output uncertainties is presented in this paper. The assessment is conducted for a queue reduction example where Monte Carlo simulations are used to determine whether or not the air quality related to different scenario cases are statistically different from a baseline case. These methods and results provide additional quantitative information to help transportation land use and in planning professionals make informed decisions.

Introduction

With the increasing concerns over health impacts from roadway emissions, it is becoming evident that higher fidelity and more robust and dynamic models need to be developed to better assess air quality impacts from highway planning and congestion mitigation projects (Bailey 2003 and Cadle 2007). This paper summarizes the benefits of simulation models that transportation planning professionals can use to better assess air quality impacts and promote sustainable transportation projects (e.g., projects that reduce congestion without adversely impacting air quality). These benefits are exemplified through the use of the TRaffic Air Quality SImulation Model (TRAQSIM) developed by Wyle Laboratories (Kim 2006 and 2007).

The US EPA currently promulgates the use of CAL3QHC for air quality modeling for interrupted flow traffic scenarios (e.g., intersections) (EPA 1995). CALINE4 is similarly promoted for use by CALTRANS in California (Benson 1989). These models take an averaging approach where vehicles are modeled as part of line sources with emissions homogenously distributed along the lines and a steady-state Gaussian plume model used for modeling dispersion of relatively inert pollutants such as carbon monoxide (CO). The average and steady-state characterization of traffic movements, vehicle emission rates, and atmospheric dispersion methods employed by the models present several shortcomings.

- First, CAL3QHC's use of line sources with evenly distributed emission factors (i.e., uniform density) and the representation of idling (queued) vehicles with equivalent length line sources are homogenous approximations of actual traffic conditions and do not allow accurate modeling of modal activities such as acceleration on departure links. CALINE4 improves upon this through conversion of composite emission factors to modal factors but the factors are still applied at an aggregate level to the roadway segments.
- Second, the use of steady-state plume equations for dispersion modeling does not allow the direct use of time-varying meteorological data, and therefore, precludes the direct modeling of time variances for high-resolution outputs.
- Third, the dispersion effects of vehicle wakes and exhaust gas buoyancy are generically approximated through the use of homogeneous mixing zones that do not reflect the complex interactions between vehicle locations and these dispersive factors.
- Fourth, the models do not account for the effects of different drive cycles which could cause significant deviations from the use of composite emission factors based on a standard cycle such as the Federal Test Procedure (FTP) (EPA 2002).
- Lastly, the approximations used with the Highway Capacity Manual (TRB 1985) and macroscopic queuing theories to predict queue lengths are not suitable for under-saturated conditions (i.e., volume to capacity ratio less than 1), and does not account for queues that may extend into or form under the green phase of the signal cycle.

In response to these limitations and in recognition of computational capabilities now offered by technology, TRAQSIM was developed. The model currently exists in a prototype state and is currently capable of predicting concentrations of CO near roadways and intersections using a traffic simulation framework with modal emissions and Gaussian puff dispersion. These components are tightly integrated such that each vehicle is treated as a discrete moving source that emits CO with varying emission rates by modal activity. Once a puff of pollutant is released during a simulated time-step, it experiences vehicle wake effects, atmospheric rise, and atmospheric turbulence. The basic concept of this model is depicted in Figure 1.

Although the HYROAD model developed under sponsorship by the National Cooperative Highway Research Program (NCHRP) also uses traffic simulation and puff dispersion, it does not integrate these components; rather, they are used in a linear fashion with aggregated variables being passed from one component to the next (Carr 2002). For example, one set of results from the traffic simulation modeling in HYROAD are traffic volumes aggregated into 10-meter roadway segments which are then used by the dispersion model. HYROAD also does not have a modal emissions modeling capability, but rather uses composite emission factors from the EPA's MOBILE-series models. The fact that HYROAD does not consistently and in an integrated fashion use a time-varying methodology means that it falls into the same category as CAL3QHC and CALINE4 in not being able to satisfy the high resolution, micro-scale modeling needs of transportation and environmental professionals. Therefore, HYROAD would not qualify as a true simulation model as defined by TRAQSIM's consistent and integrated time-varying methodology. Time-varying emissions and concentrations would ultimately help improve the assessment of health effects by other models.

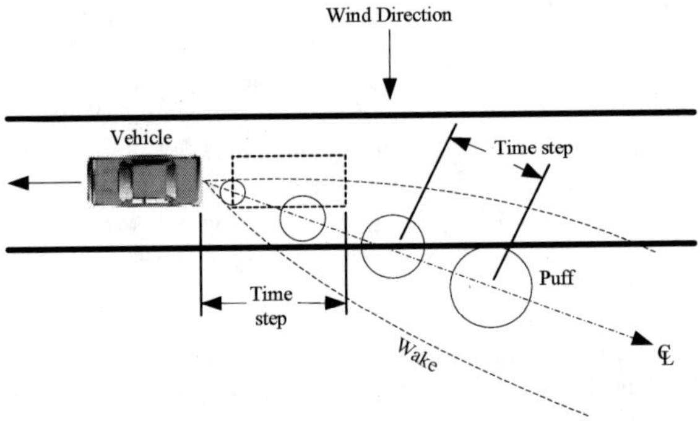

Figure 1. Basic Modeling Concept in TRAQSIM.

Overview Of Simulation Benefits

There are several advantages that simulation models can provide beyond the capabilities of static models. The following sections provide a survey of these advantages.

Improved Cause and Effect Analyses

The benefits of a simulation model become evident when considering the time-varying nature of the model. The core of TRAQSIM is built on the use of a time-step whereby the movements of each vehicle and puff of pollutant are incremented. This allows for true micro-scale modeling of air quality near roadways where hotspots even within a micro-scale environment could be identified. That is, the second-by-second, detailed outputs could be used to isolate the specific interactions between vehicle operations, signal timings, wake effects, etc. to determine the root causes of concentration differences along a roadway rather than simply attributing it to lower speeds or queues in general. As a result, a simulation model would be able to provide greater tools to differentiate the effects of different modeled scenarios. Excerpts from a detailed simulation output are exemplified in tables 1 and 2. The raw data (with full decimal places) is simply presented as is to illustrate the kinds of information preserved from the simulation. This data would normally be further processed to derive the underlying cause and effects.

Capability to use Monitored Data

The simulation framework also allows for the direct use of time-varying monitored data including traffic operations and meteorological data. Second-by-second or even lower resolution data (e.g., in 5, 10, 20, etc. second increments) could allow more accurate modeling of scenarios since actual data are being used. Rather than averaging such data for use in steady-state models like CAL3QHC, the data could be used directly within TRAQSIM with interpolations as necessary depending on the resolution of the data (i.e., interpolation would be necessary if the data was coarser than the time-step used in the simulation). In addition, any spatially varying data could be directly incorporated as well. For example, rather than using one averaged set of wind speed and wind directions, a 4D wind field could be directly used to more accurately advect the puffs of pollutants.

Potential for Chemical Transformations

The 4D environment is ideal for the incorporation of atmospheric chemical transformations. This spatially and temporally unconstrained environment is conducive to the modeling of chemical reactions in a more realistic and accurate way than the steady-state approach used in CALINE4. Although CALINE4 can model NOx reactions to produce estimates of NO_2 concentrations, it relies on steady-state assumptions including uniform mixing within the mixing zone. A more refined chemical transformation module could be potentially developed using existing components from CALPUFF and some of the Eulerian-based grid models such as the Community Multiscale Air Quality (CMAQ) modeling system, Comprehensive Air Quality Model with extensions (CAMx), Urban Airshed Model (UAM), and Regional Modeling System for Aerosols and Deposition (REMSAD) (EPA 2007). Since these grid models generally use plume in grid (PinG) approaches for finer scale modeling, they have already resolved the issues associated with translating Eulerian chemistry components to work under a Lagrangian (e.g., puff) environment. Therefore, the potential exists to incorporate these chemical transformation mechanisms into a model like TRAQSIM to allow higher fidelity modeling of reactive pollutants. Non-reactive gases such as CO and non-volatile particulate matter (PM) would no longer have to be used as surrogates for roughly approximating concentrations of reactive species.

Table 1. Example Excerpt of Detailed Traffic Outputs from a TRAQSIM Run.

Time (sec)	Veh. ID	Speed (m/s)	Accel (m/s^2)	Headspace (m)	Headway (s)	Dist. to Signal (m)	EF (g/s)	...
1	1	16.04326	0	0	0	197	7.29E-02	...
2	1	16.04326	0	0	0	180.9575	7.29E-02	...
3	1	16.04326	0	0	0	164.915	7.29E-02	...
4	1	16.04326	0	0	0	148.8726	7.29E-02	...
5	1	16.04326	0	0	0	132.8301	7.29E-02	...
5	2	15.60466	0	61.16993	3.92017	197	0.1877306	...
6	1	16.04326	0	0	0	116.7876	7.29E-02	...
6	2	15.60466	0	61.60851	3.948277	181.3961	0.1877306	...
7	1	16.04326	0	0	0	100.7451	7.29E-02	...
7	2	15.60466	0	62.0471	3.976384	165.7922	0.1877306	...
8	1	16.04326	0	0	0	84.70263	7.29E-02	...
8	2	15.60466	0	62.48568	4.004491	150.1883	0.1877306	...
9	1	14.169	-1.8743	0	0	68.66014	0.0195211	...
9	2	15.60466	0	62.92427	4.032599	134.5844	0.1877306	...
10	1	12.29475	-1.8743	0	0	54.49182	2.02E-02	...
10	2	15.60466	0	61.48869	3.940598	118.9805	0.1877306	...
...	...	...	...	...	...	...	...	...

High Resolution Modal Emissions

TRAQSIM's second-by-second, vehicle-specific, modal emissions modeling framework also provides higher fidelity modeling because its emission factors are based on the specific properties of a vehicle at a point in time. The use of the vehicle's instantaneous speed and acceleration values to represent power demand in calculating a modal emission factor is more realistic and accurate than the composite emission factors used in CAL3QHC. These modal emission factors provide better spatial distributions of emissions to more properly reflect the operations of a vehicle. This, again, allows for better identification of root causes of concentration differences along a roadway, and hence, a better understanding of the air quality impacts of roadway changes.

Unconstrained Receptor Placements

As a consequence of the spatial and temporally unconstrained modeling framework, TRAQSIM is not limited by the roadway 3 m boundary that constrains CAL3QHC and CALINE4. This allows receptors in TRAQSIM to potentially be placed anywhere including directly on sidewalks (closer than 3 m to the roadway) and even on the roadways themselves (e.g., to model concentrations on crosswalks). This capability is due to the specific modeling of individual vehicle wake effects and atmospheric turbulence so that artificial mixing zones around a roadway are not necessary.

Table 2. Example Excerpt of Detailed Dispersion Outputs from a TRAQSIM Run.

Time (sec)	Puff ID	X (m)	Y (m)	Z (m)	Sigma HT (m)	Sigma ZT (m)	Sigma HR (m)	...
1	1	16.04248	99.18582	0.730085	4.48E-02	0.0459775	6.57E-02	...
2	1	19.02839	98.42526	1.1533	0.1960135	0.1648185	0.1866571	...
2	2	32.08496	99.18582	0.730085	4.48E-02	0.0459775	6.57E-02	...
3	1	21.03682	97.481	1.382393	0.2994951	0.2380762	0.2521122	...
3	2	35.07074	98.42525	1.153285	0.1960072	0.1648138	0.1866529	...
3	3	48.12745	99.18582	0.730085	4.48E-02	0.0459775	6.57E-02	...
4	1	22.06539	96.42856	1.51882	0.3667131	0.2838558	0.2910914	...
4	2	37.07921	97.481	1.382383	0.2994905	0.238073	0.2521095	...
4	3	51.11324	98.42525	1.153287	0.1960083	0.1648146	0.1866535	...
4	4	64.16993	99.18582	0.730085	4.48E-02	0.0459775	6.57E-02	...
5	1	22.69461	95.32423	1.629653	0.424078	0.3220794	0.322758	...
5	2	38.10778	96.42857	1.518812	0.3667089	0.2838529	0.2910891	...
5	3	53.1217	97.481	1.382384	0.299491	0.2380734	0.2521098	...
...	...	...	...	...	...	...	...	...

Facilitated Uncertainty Assessments

The simulation framework also provides a more conducive environment for conducting uncertainty assessments. As these types of assessments often entail simulations using stochastic techniques, the applicable modules already existing within TRAQSIM could be used for these assessments. Although an external "wrapper"-type module could be built to feed input parameter variations to a model like CAL3QHC, it would be awkward and incomplete because of the internal parameters that are not exposed to the typical user. An advanced user (i.e., programmer) would have to figure out how to make the internal variables (coded in FORTRAN®) accessible. In addition, a simulation tends to make these assessments easier due to the conceptually simpler (more straightforward) input parameters. For example, the TRAQSIM user would not have the burden of determining uncertainty estimates for mixing zone widths and saturation flow rates as the CAL3QHC user would. This is due to the fact that in TRAQSIM, atmospheric turbulence caused by each vehicle is individually modeled and the combination of traffic volume, signal timings, vehicle speeds, etc. are assumed to inherently represent the capacity of the modeled roadway. That is, using these data, TRAQSIM "naturally" models a queue through the simulation process. So, while uncertainty assessments could be conducted with both CAL3QHC and TRAQSIM, it would be easier to do so with the latter model.

Uncertainity Assessment Demonstration

The purpose of this demonstration is essentially two-fold: (1) Show the usefulness of uncertainty assessments in evaluating air quality impacts of transportation projects; and (2) more importantly, exemplify the advantages of a simulation framework for conducting such assessments. For this demonstration, a simple and hypothetical, queue reduction scenario was used. The Baseline case modeled in TRAQSIM is illustrated in Figure 2.

This baseline case results in modeled queues at the light between 15 to 20 vehicles. To reduce this potential congestion, two options were considered as indicated in figures 3 and 4. These show a new roadway under Case 1 (build case) which allows traffic volumes to be split between the two roadways. For Case 2 (no build alternative), changes in the speed limit and signal timings are proposed. The closest receptor location (worst case) that is common for all of these cases is 20 m away from the centerline of the existing roadway as shown in Figure 2.

The air quality, uncertainty assessment was based on conducting the following three tasks: (1) Assign uncertainty distributions to each of the model parameters, (2) conduct Monte Carlo simulations to propagate uncertainties through the model, and (3) assess output uncertainties to help determine which alternatives (Case 1 or Case 2) to choose. Ideally, all of the external (input) and internal parameters used as part of the modeling process would need uncertainty estimates. As a surrogate, the main parameters contributing most of the uncertainties to the output results could just be used. However, that would likely require additional assessments to identify those parameters and was considered outside the scope of this demonstration. Therefore, only a few independent parameters were arbitrarily chosen and assigned hypothetical uncertainty distributions as shown in Table 3.

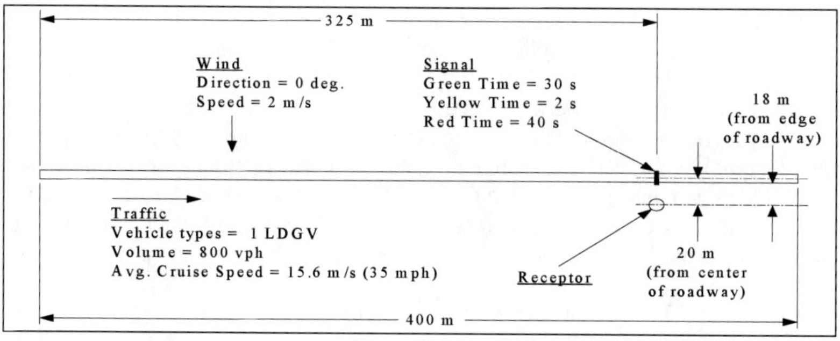

Figure 2. Baseline Case for Queue Reduction Example.

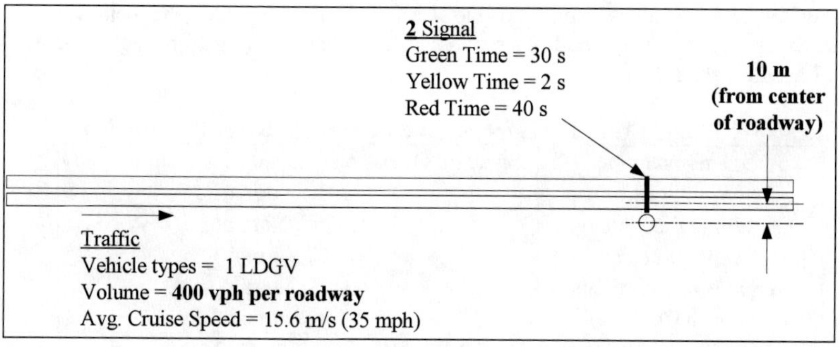

Figure 3. Case 1: Additional Roadway Construction (Build Case).

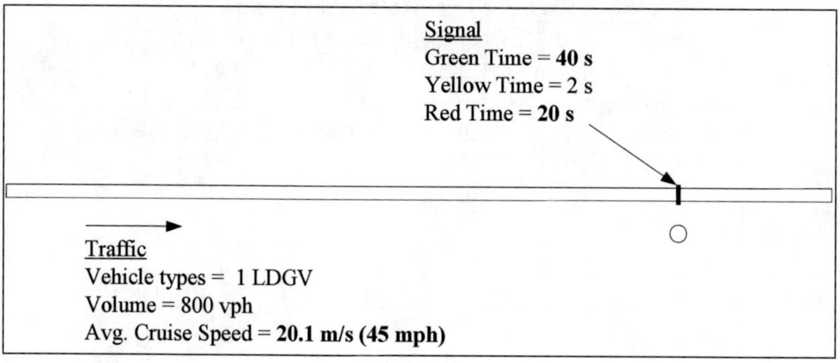

Figure 4. Case 2: Traffic Operational Changes (no build alternative).

Without detailed knowledge of each parameter, the uncertainty distributions were all assumed to be Normal for simplicity. The actual shape of these distributions will need to be determined when conducting more accurate uncertainty assessments. Also, any dependencies between the chosen parameters will need to be taken into account to prevent double counting of uncertainties. The intention of this example is simply to provide a clear illustration of how the existing stochastic components within TRAQSIM could be leveraged to conduct such an analysis. That is, applying these distributions in TRAQSIM was straightforward since the model's simulation framework already allows for stochastic assignment of parameter values.

Using these data, 1000 runs of each case were conducted as part of the overall Monte Carlo simulations. The first step in analyzing the results involved regressing (through multiple regression) the output concentrations on the five model parameters. To do this, each of the parameter and concentration values were standardized (z-scored) to allow for relative comparisons. Table 4 shows an example of the beta values (standardized coefficients) and other summary statistics resulting from regression of the Baseline case's data.

Table 3. Sample Model Parameters and their Uncertainty Distributions.

Parameter	Standard Deviation (σ), and Distribution Type
Wind Speed	± 10%, Normal
Wind Direction	± 5 deg., Normal
Cruise Speed	± 2%, Normal
Internal Emissions Data*	± 10%, Normal
Dispersion Coefficient	± 5%, Normal

*This parameter refers to the underlying emission factor database used in TRAQSIM to derive modal emission factors based on speed and acceleration.

Table 4. Baseline Case Regression Results.

Parameter	Beta (β)	Standard Error	t-value	p (t)
Wind Speed	0.212	0.0258	8.20	0
Wind Direction	-0.463	0.0253	-18.32	0
Cruise Speed	0.000	0.0254	0.02	0.985
Internal Emissions Data	0.263	0.0259	10.15	0
Dispersion Coefficient	-0.208	0.0254	-8.18	0

R^2	F	p (F)
0.37	148.20	0

As indicated by the p-values, all of the parameters appear to provide significant information to the regression with the possible exception of cruise speed. These results are based on the hypothetical uncertainties shown in Table 3 and are somewhat specific to the modeled scenario. As a result, they are presented for demonstration purposes only; other scenarios will likely produce different results.

The low R^2 value also suggests that additional parameters and/or better uncertainty estimates may be used to improve the prediction. However, for demonstration purposes, if these parameters were assumed to represent most of the uncertainties in the modeling process, relative comparisons between the three cases can be conducted. Table 5 and figures 5 and 6 show summary statistics from the simulations.

The results from the Monte Carlo simulations indicate that based on 95% confidence intervals (2 standard deviations), the means of Case 1 and Case 2 are statistically different than the baseline case's mean. That is, in addition to any reductions of the queue, both cases will produce statistically different modeled concentrations of CO. The results of more rigorous t-tests shown in Table 6 help to confirm these findings.

Table 4. Summary Statistics from Monte Carlo Simulations.

Case	Mean Concentration (ppm)	Standard Deviation (σ)
Baseline	3.53	0.86
Case 1	5.34	1.03
Case 2	2.23	0.36

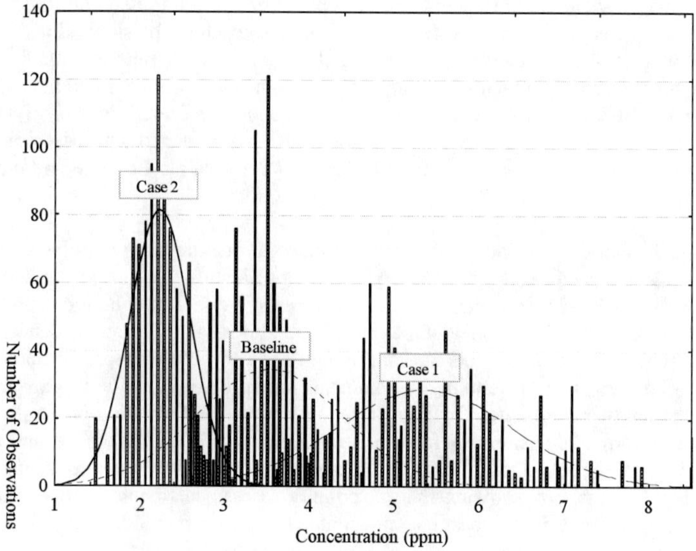

Figure 5. Histogram of Monte Carlo Results.

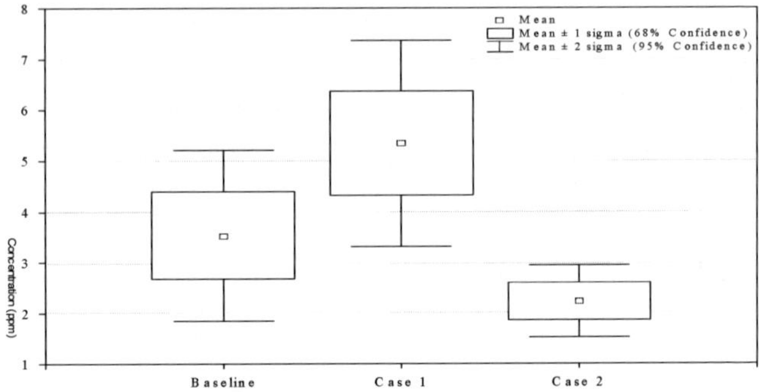

Figure 6. Confidence Intervals of Monte Carlo Results.

Table 5. Results from the Tests of Significance of Sample Means for the Three Cases.

Description	t-value	p-value
Baseline vs. Case 1	-42.7844	0
Baseline vs. Case 2	44.0088	0

These t-tests were conducted for this demonstration based on assuming the samples were independent. The reason for this was because each of the simulations for each case was modeled independently using different sets of random numbers. A refinement to this approach would be to use the same sets of random numbers for the different cases so that Case 1 and Case 2 would be considered dependent samples compared to the Baseline case. This would also allow uncertainty assessments of difference metrics (i.e., concentration differences between the Baseline and the other cases).

Although both Case 1 and Case 2 provide queue reduction and possibly congestion relief, Case 1 produces higher CO concentrations since the traffic on the additional roadway will be closer to the receptor. As a result, Case 2 may be more preferable purely from an air quality standpoint. However, additional information such as future traffic volumes, noise impacts, costs, etc. will need to be considered in making the right choice. The uncertainty assessment just provides additional quantitative information that can be used by transportation land use and planning professionals in making informed decisions. While this example's hypothetical scenarios and results may not reflect actual situations, the demonstration clearly illustrates a benefit of simulation models by showing that uncertainty assessments could be performed by leveraging the built-in, stochastic components of a simulation model.

Summary

This paper provided an overview of the benefits of air quality simulation models using TRAQSIM as an example. The benefits stem from the higher fidelity modeling environment that can generate detailed, time-varying information for each modeling element (e.g., vehicles, puffs of pollutants, etc.). This allows truly micro-scale modeling where the root causes of concentration deviations at receptor locations can be assessed. The spatially and temporally unconstrained modeling environment allows receptors to be placed anywhere including on crosswalks, and allows for the use of detailed traffic and meteorological data as necessary. Also, the 4D environment is conducive to allowing the modeling of chemical reactions, prediction of vehicle-specific modal emissions including the effects of road grade, and uncertainty assessments. The demonstration example showed that uncertainty assessments could be facilitated through the use of the stochastic components already existing within the simulation framework. Ultimately, simulation models allow improved determination of atmospheric concentrations that could be used in other models to better understand health impacts from various transportation projects. Such modeling capabilities and improvements would allow transportation land use and planning professionals to make better decisions.

References

Bailey, C. (2003). "Modeling Mobile Source Air Toxics Hot Spots." U.S. EPA Office of Transportation and Air Quality, Air Toxics Center.

Benson, P. (1989). "CALINE4 – A Dispersion Model For Predicting Air Pollutant Concentrations Near Roadways." State of California Department of Transportation.

Cadle, S. H., Alberto A., Kevin N. B., Fulper C. E., Graze R. R., Minassian F., Murray H. B., Natarajan M., Tennant C. J., and Lawson D. R. (2007). "Real-World Vehicle Emissions: A Summary of the Sixteenth Coordinating Research Council On-Road Vehicle Emissions Workshop." *Journal of the Air & Waste Management Association,* Volume 57, Page 139.

Edward C. L., Johnson R. G., and Ireson R. G. (2002). "HYROAD Model Formulation." *SYSAPP-02/074d.,* National Cooperative Highway Research Program, Transportation Research Board, National Research Council. Systems Applications International, Inc., San Rafael, CA 94903. KLD Associates, Inc., Huntington, New York.

Environmental Protection Agency (EPA) (1995). "User's Guide to CAL3QHC Version 2.0: A modeling Methodology for Predicting Pollutant Concentrations Near Roadway Intersections." *EPA-454/R-92-006 (Revised),*

U.S. Environmental Protection Agency, Office of Air Quality Planning and Standards. Research Triangle Park, N.C.

Environmental Protection Agency (EPA) (2002). "User's Guide to MOBILE6.1 and MOBILE6.2: Mobile Source Emission Factor Model." *EPA Report No. EPA420-R-02-028,* Assessment and Standards Division, Office of Transportation and Air Quality, Environmental Protection Agency. Washington, DC.

Environmental Protection Agency (EPA) (2007). http://www.epa.gov/scram001/aqmindex.htm.

Brian K. Y., Wayson R. L., and Fleming G. G. (2006). "Development of the Traffic Air Quality Simulation Model (TRAQSIM)." *Transportation Research Record, Journal of the Transportation Research Board,* No. 1987, p. 73.

Brian K. Y., Wayson R. L., and Fleming G. G. (2007). "Component Verification and Sensitivity Analyses of the Traffic Air Quality Simulation Model (TRAQSIM)." *A paper presented at the Transportation Research Board's 86th Annual Meeting.*

Transportation Research Board (TRB (1985). "The Highway Capacity Manual." *TRB Special Report 209,* Washington, DC.

Developing a Freight Capacity Model for Land Use Planning and Air Quality Impact Analysis

S. Shackelford[1] and D. Murray[2]

[1]American Transportation Research Institute, 2277 West Highway 36, Suite 302, Roseville, MN 55113, PH (651) 641-6162; email: sshackelford@trucking.org
[2]American Transportation Research Institute, 2277 West Highway 36, Suite 302, Roseville, MN 55113, PH (651) 641-6162; email: dmurray@trucking.org

Abstract

Congestion is a growing concern among those professionals tasked with managing traffic flows and air quality. Congestion is also a growing concern for the trucking industry, which is tasked with delivering goods to retailers, manufacturers and service providers across the country. Congestion is costly for both the public and private sectors due to traffic delays, fuel consumption, pavement and bridge damage, and environmental concerns such as air and noise pollution. Policymakers, planners and motor carriers must make decisions on how to best manage limited capacity to optimize the performance of goods movement on the road system. However, it can be difficult to predict the impact of different decisions on the actual performance of the network. To aid decision makers in this process, the American Transportation Research Institute has developed a freight capacity model which is able to estimate the likely financial, environmental and performance-related impacts of different planning decisions for both motor carriers and public sector planners.

Introduction

Traffic congestion is an increasingly serious concern in the nation's metropolitan centers. Annual mobility reports conducted by the Texas Transportation Institute conclude that congestion is increasing in areas of all population size. Researchers estimated that congestion resulted in 3.7 billion hours of delay time and 2.3 billion gallons of fuel consumption in 2003 (Schrank & Lomax, 2005). The clear reason for the increased congestion is that demand for road usage is increasing at a much faster rate than the supply. According to one source, between 1970 and 2003 the number of licensed drivers increased by 71 percent, the number of registered vehicles increased by 99 percent and the number of vehicle miles traveled increased by 148 percent. In contrast, new road miles have only increased by 6 percent (Young, 2003).

The increased levels of congestion have serious impacts on trucking industry operations and air quality. These large delay times result in higher operational costs and greater amounts of fuel consumption, which in turn lead to higher vehicular emissions and more air pollution.

The trucking industry is a significant user of the highway system. Trucks are an integral part of the supply chain, and truck movements correlate strongly with the state of the economy. In 2004, commercial trucks accounted for more than 13 percent of vehicle miles traveled. In addition, trucks hauled nearly 69 percent of the freight tonnage moved in the U.S. in 2005, resulting in total gross revenue of $623 billion (American Trucking Associations, 2006). As the economy continues to grow, the amount of freight being transported will increase as well.

Two factors are important to consider when discussing freight transportation options: 1) the number of vehicles, and 2) the size of the vehicles. Assuming that economic growth is unimpeded, one or both of these factors will have to increase as freight tonnage increases. In most states, truck sizes and weights are limited to 5-axles or less and operating weights up to 80,000 pounds on the Interstate System. Although longer combination vehicles (LCVs) are permitted to operate in certain states, there has been a freeze since 1991 on allowing states to increase truck size and weight allowances. The current freeze is nearing its expiration. If the freeze is not renewed, presumably states may revisit truck size and weight regulations.

The expanded use of LCVs may provide both operational and environmental benefits by minimizing truck traffic growth. In a study completed by ATRI, it was determined that the use of higher productivity vehicles (HPVs) resulted in emissions reductions per ton-mile (Tunnell, 2004).

To better understand the operational and environmental impacts of HPV and LCV utilization from both a motor carrier and network system perspective, ATRI has developed the ATRI Freight Capacity Model (AFCM). The model transparently integrates user inputs such as fuel efficiency, vehicle miles traveled, operational costs and fleet configurations, and allows users to see the likely impacts on fuel consumption, delay times and emissions.

Impacts of Congestion

Trucking Industry Operations

Increasing levels of congestion have serious implications for the trucking industry. In a survey of trucking company managers, 80 percent perceived traffic congestion to be a serious problem. The number one reason given was because it results in unreliable travel times (Golob & Regan, 2001). Travel time reliability is especially important due to the prevalence of just-in-time delivery schedules. As manufacturing emphasis shifts away from mass production and towards customized production, retailers strive to eliminate unnecessary expenses like excessive inventory and warehousing. As a result, just-in-time delivery has become essential and the expectations placed on freight companies are increasing. Some retailers and shippers now require that deliveries be on time with nearly perfect reliability to comply with the overall operational strategy of the company and assess fines when a truck fails to arrive in the specified window of time.

To further exacerbate the challenge of meeting rigorous customer demands in highly congested conditions, the trucking industry is currently facing a serious driver shortage. Currently, there is a shortage of 20,000 drivers, and the shortage is expected to increase to 111,000 by 2014 if nothing is changed (Global Insight, Inc., 2005). An annual survey conducted by ATRI revealed that the driver shortage was the top concern among motor carriers in 2006 and was the second highest concern in 2005 (American Transportation Research Institute, 2006).

In addition to simply not having enough drivers, motor carriers also have to comply with Hours of Service regulations that limit the amount of time drivers can be on-duty. Drivers do not have the flexibility of working longer hours if circumstances prevent them from reaching their final destination within the allowable driving time period, and motor carriers do not have the option of working drivers overtime to meet tight schedules.

Air Quality

The importance of the link between congestion management and air quality is highlighted by programs such as the Federal Highway Administration's Congestion Mitigation and Air Quality Improvement Program, which was authorized in 1991 to provide funding to develop strategies to reduce congestion and improve air quality. Although environmental impacts are difficult to measure precisely, the projects funded through this program have all shown that improving traffic flow results in lower amounts of Volatile Organic Compounds (VOCs) and nitrous oxides (NO_x) (Federal Highway Administration, 2003). In addition to these pollutants, commercial trucks are also known sources of the smallest category of particulate matter ($PM_{2.5}$).

The impacts of these pollutants vary. VOCs and NO_x react with other chemicals in the atmosphere to form ozone. More than 25 percent of the ozone in the atmosphere comes from automobile emissions. Both ozone and $PM_{2.5}$ have been identified as respiratory irritants and can result in or further aggravate respiratory conditions such as asthma (Centers for Disease Control and Prevention).

Assuming no changes to the status quo, vehicular emissions will increase along with freight tonnage. However, utilizing HPVs can limit the negative environmental impacts by improving the traffic flow and minimizing commercial truck contributions to increasing congestion levels.

The ATRI Freight Capacity Model

The AFCM is a software-based program utilizing a user-friendly interface. The user either accepts the existing default data [primarily generated from government sources], or inputs a set of values and parameters, which are then used to calculate outputs such as fuel consumption, emissions, number of trips and congestion delay time. The most significant feature of the model is that it allows the user to estimate the relative impacts of different scenarios. For example, the AFCM can be used to estimate the effects of increased freight tonnage or different vehicle fleet configurations on related congestion levels, and/or operational and environmental outputs.

The AFCM is designed as a tool for both motor carrier fleet managers and government transportation planners. The user selects the analysis option that best meets his or her needs (either the "Carrier Analysis" or the "Transportation System Analysis"). The user then selects the outputs he or she is interested in assessing. After those selections are made, the user is then prompted to enter the information needed to complete those specific analyses.

There are four major components of the AFCM:

- *Common input data*: This component allows users to input variables common to all levels of analysis. These variables include vehicle configurations, VMTs, fuel consumption and emission rates.
- *Shipment and trip analysis*: This component aids users in determining the estimated savings between proposed and current fleet configurations. The input variables are current and future freight tonnage, number of vehicles, vehicle payloads, vehicle operating weights, and current and proposed fleet configurations. From this information, the model is able to estimate the number of trips needed to haul the specified freight tonnage (for both the current and proposed vehicle fleets) and summarize the estimated cost differences between the new and proposed fleet. Fuel consumption and emissions differences between the two fleets are also estimated.
- *Network analysis*: The network analysis components are used in the transportation system analysis. The input variables are road types, vehicle mix, percentage of road type that is congested, average traffic speed when road is at capacity (degree of saturation = 1) and under free-flow conditions, and the degree of saturation of each road type. Based on this information, the model outputs VMTs, congestion levels and costs for each vehicle type on each road system on a daily, hourly and annual basis. The model is also able to estimate vehicle travel times and delays due to congestion.
- *Cost analysis*: This component allows carriers to assess the economic factors of differences between the current and proposed conditions. The input variables are average societal pollution costs and travel time value. Using this information, the model assesses costs relating to driver delays and externalities (pollution and noise costs).

Default data has been pre-populated into the model for many of the technical inputs that do not vary by user. These defaults include average vehicle emissions, average fuel consumption per vehicle type, average fuel costs, average pollution costs per mile, value of passenger vehicle driver time and vehicle payloads. However, if the default values are changed, a warning will be displayed on the model.

Examples

Motor Carrier

Consider a motor carrier that currently operates 100 trucks. Each of these trucks is a tractor/ single-trailer, 5-axle configuration with a payload weight of 44,800 pounds. The motor carrier is considering the purchase of several tractor/single-trailer, 6-axle configurations and would like to estimate the financial impacts of the investment.

To analyze the financial impacts, the carrier estimates that it will be moving 1.5 million tons of goods and that each trip length is about 500 miles. The motor carrier enters this information into the model and estimates the percentage of freight that will be moved by each of the truck configurations, based on maximum legal gross weights. In this example, 65 percent of the goods will be moved by the 5-axle "trucks" and 35 percent will be moved by the 6-axle "trucks."

The motor carrier then enters the known operational costs associated with the different truck configurations as shown in Table 1.

Table 1 Estimated operating costs for truck configurations.

Truck Types	Per truck cost (buy price)	Per truck cost (sell price)	Estimated maintenance cost per mile	Labor cost per mile	Asset insurance cost per mile	Miscellaneous costs per mile
Tractor-semi 5-axles	$120,000	$95,000	$0.26	$0.53	$0.32	$0.15
Tractor-semi 6-axles	$130,000	$102,000	$0.26	$0.55	$0.33	$0.16

The model then calculates the total operating costs for the current and proposed fleets and displays the percent change for each of the cost categories as shown in Table 2.

Table 2 Total cost summary for current and proposed fleets.

Cost Aspect	**Current**	**Proposed**	**Percent Change**
Fuel	$ 9,952,504	$ 9,728,749	-2%
Asset Acquisition	$ 0	$ 2,935,000	N/A
Maintenance	$ 12,801,100	$ 11,914,370	-7%
Labor	$ 26,094,550	$ 24,563,415	-6%
Asset Insurance	$ 15,755,200	$ 14,802,055	-6%
Miscellaneous	$ 7,385,250	$ 7,011,890	-5%
Total	**$ 71,988,604**	**$ 70,955,479**	**-1%**

This analysis estimates that the carrier will save over $1 million by utilizing 6-axle vehicles. With the exception of asset acquisition, costs are reduced in each of the major cost categories resulting in an overall cost reduction of one percent.

Transportation System Analysis

Now consider a city planner that would like to assess the costs associated with traffic delay related to congestion. The city planner first chooses the "network" to analyze. The network can consist of a road segment, an entire city, or a state. The planner will estimate the vehicle mix and the vehicle miles traveled by each vehicle configuration for the network being analyzed. The values used for this example are displayed in Table 3.

Table 3 Vehicle types and current VMT data.

Vehicle Type	**Annual VMT (Millions)**
Passenger Cars	56,800
Single Unit 2-axles	900
Single Unit 3-axles	8,800
Tractor-semi 5-axles	1,400
Tractor double semi 5-axles	100
Total	**68,000**

The model will then prompt the planner to enter information on the average occupancy and value of travel times for the drivers of each of the vehicle configurations included for the network. In this example, average passenger car occupancy is estimated to be 1.2 people per car and average occupancy for the other vehicles is estimated to be 1. The travel time value per person per vehicle configuration is shown in Table 4. These default rates can be changed to any known value (from the default values), but a pop-up notice references that the default values have been revised.

Table 4 Value of travel time per person per vehicle.

Vehicle Type	**Time Value ($/hour)**
Passenger Cars	$13.50
Single Unit 2-axles	$16.00
Single Unit 3-axles	$16.50
Tractor-semi 5-axles	$20.00
Tractor double semi 5-axles	$20.00

The planner will then be asked to supply information on the current traffic conditions for the network being analyzed. The information the planner provides includes the types of roads included in the network, the posted speed limits, the percent of VMTs that occur during congested conditions and the degree of saturation (= traffic volume / maximum road capacity). The information supplied for this example is shown in Table 5. If a planner is only interested in evaluating a road segment, he or she will only select that road type and input the information relating to that segment into the model.

Table 5 Current traffic conditions.

Road Type	% VMT Occurring Under Congested Conditions	Traffic Speeds (mph)		Degrees of Saturation in Congested Conditions
		Posted	Average Speed when Traffic Volume = Max. Capacity	
Urban Arterials	50	45	25	0.9
Urban Freeway	55	60	40	0.9
CBD	55	40	15	0.9
Urban Local Government Roads	10	40	25	0.75

The model then prompts the planner to supply information on the percent of vehicle miles traveled accruing to each road type for each vehicle configuration as shown in Table 6. The model uses these sets of information to calculate travel time delays and the associated costs.

Table 6 Percentage of VMT accruing to each road type.

Vehicle Type	Urban Arterials	Urban Freeway	CBD	Urban Local Government Roads	Total %
Passenger Cars	17	17	33	33	100
Single Unit 2-axles	20	40	20	20	100
Single Unit 3-axles	38	40	13	9	100
Tractor-semi 5-axles	40	20	20	20	100
Tractor double semi 5-axles	40	20	20	20	100

Using the information provided by the user, the model estimates costs associated with the truck component of the traffic mix. The costs are disaggregated into categories and the results from this example are shown in Table 7.

Table 7 Total cost summary for trucks.

Cost Category	Cost (Millions)
Fuel	$ 2,666
Air Pollution	$ 44
Noise	$ 9
Pavement Damage	$ 412
Bridge Damage	$ 67
Delay	$ 4,748
Total	$ 7,946

In this scenario, the model estimated the costs of congestion for the trucking industry to be 6.98 cents per mile. The model also estimated that the externality costs associated with pollution produced by trucks totaled $53 million.

Discussion

The model's capabilities are much more robust than demonstrated in these examples, but they provide a glimpse into some of the analyses that can be completed. In addition to the analyses shown above, the model is also capable of comparing current and future conditions for the network analysis and can estimate expected costs of freight growth if current conditions remain unchanged.

Model Limitations

It is important to remember that this model is only intended to be a policy and planning analysis tool. The accuracy of the model is dependent upon the quality of the data entered. The quality of the data is high for variables such as VMT, tonnage, operating weights and forecasts, but may be lower for variables such as vehicle-specific emissions data. Better vehicle emissions data will result in a more accurate model.

The model is limited in these areas:

- The AFCM does not consider partially loaded trips. It assumes all vehicles are fully loaded at the weight levels that are inputted. However, the model does account for back-hauls at the prescribed input levels.
- The AFCM is not able to simulate complex network scenarios. For example, the model cannot account for multiple shipment drop-off points.
- The model does not account for the relationship between emissions and fuel consumption as the vehicle load changes.

Although addressing the first two limitations may result in a more accurate assessment of the output variables, requiring additional information will further complicate the user entry interface. It is important to balance the capability and the user-friendliness of the model. Additional data on fuel efficiency and emissions can be incorporated into the model as it becomes available.

Conclusions

This model provides users with a tool to better understand and manage freight logistics and related system impacts. Motor carrier managers and government planners can estimate environmental and economic impacts of operational or policy decisions before implementing or advocating for changes to the current system. The model is not intended to be a predictive tool, but to help users assess potential impacts of changes to current systems and/or fleets. The usefulness of the model will only continue to increase with the additional availability of data.

References

"Top Industry Issues." *American Transportation Research Institute*, 2006.

American Trucking Associations, American Trucking Trends, 2006.

Centers for Disease Control and Prevention, Respiratory Health and Air Pollution, Department of Health and Human Services. Available online: http://www.cdc.gov/healthyplaces/healthtopics/airpollution.htm

"CMAQ: Advancing Mobility and Air Quality." *Federal Highway Administration*, Office of Natural Environment, May 2003.

"The U.S. Truck Driver Shortage: Analysis and Forecast." *Global Insight, Inc., American Trucking Associations*, Alexandria, VA, 2005.

Golob, T. & A. Regan, A. (2001). "Impacts of Highway Congestion on Freight Operations: Perceptions of Trucking Industry Managers," *Transportation Research Part A*, Vol. 35, 2001, pp. 577-599.

Schrank, D. & T. Lomax, T. (2005). "The 2005 Urban Mobility Report." Texas A&M University, Texas Transportation Institute.

Tunnell, M., (2004). "Energy and Emissions Impacts of Operating Higher Productivity Vehicles." *American Transportation Research Institute.*

Young, D. (2003). "New Measure Will Meet Transportation Needs," *Roll Call.*

Clean Air Act General Conformity Study for the Port of Baltimore

R. P. Newman[1] and N. K. Brown[2]

[1]P.E., DEE, Vice President, EA Engineering, Science, and Technology Inc.
[2]Principal Planner, Maryland Port Administration

1.0 Introduction

This report presents an analysis of the air quality impacts from the construction of the proposed Masonville Dredged Material Containment Facility (DMCF). The project entails the construction of a disposal site for dredged material in the Middle Branch of the Patapsco River, at Masonville, Baltimore City, Maryland. The Masonville DMCF project on completion will provide a disposal site to accommodate dredged material generated by various dredging projects that will occur over the next 20 years in the Baltimore Harbor area.

The goal of this study was to develop air emission estimates for the different types of equipment that will be utilized in the construction phase of the Masonville DMCF project. Criteria air pollutant emissions that will result from both direct and indirect sources relative to the project were calculated to address the requirements of General Conformity of the Clean Air Act. Under the General Conformity regulations, an emissions analysis is required to determine the total direct and indirect emissions for each criteria pollutant within the project limits. Emissions over certain prescribed threshold must be mitigated

2.0 General Conformity – Regulatory Background

The U.S. Environmental Protection Agency (EPA) promulgated the General Conformity Rule on November 30, 1993 to implement the conformity provision of Title I, Section 176(c)(1) of the Federal Clean Air Act (CAA). Section 176(c)(1) requires that the Federal Government does not engage, support, or provide financial assistance for licensing or permitting, or approving any activity that does not conform to an approved CAA implementation plan.

Title I, Section 176(c)(1), of the CAA defines conformity as the upholding of "an implementation plan's purpose of eliminating or reducing the severity and number of violations of the National Ambient Air Quality Standards (NAAQS) and achieving attainment of such standards." Conforming activities or actions should not, through additional air pollutant emissions:

- cause or contribute to new violations of any NAAQS in any area;
- increase the frequency or severity of any existing violation of any NAAQS; or
- delay timely attainment of any NAAQS or interim emission reductions.

The General Conformity Rule establishes conformity in coordination with, and as part of, the National Environmental Policy Act (NEPA) process. The rule takes into account air pollutant emissions associated with actions that are federally funded, licensed, permitted, or approved, and ensures emissions do not contribute to air quality degradation, thus preventing the achievement of state and federal air quality goals. Succinctly, General Conformity refers to the process of evaluating plans, programs, and projects to determine and demonstrate that they meet the requirements of the CAA and applicable State Implementation Plan (SIP).

Conformity determination is a two-step process: (1) applicability analysis, and (2) conformity analysis. Applicability analysis is achieved by comparing the project's annual emissions to "*de minimis*" pollutant thresholds outlined in the conformity rule. The more severe the "non-attainment" status of a region, the smaller the corresponding *de minimis* threshold is set. Federal actions are assumed to conform to the most recent federally approved SIP if total direct and indirect emissions caused by the federal action are less than the *de minimis* thresholds. Such thresholds were originally established for Ozone (NOx and VOC precursors), carbon monoxide, nitrogen dioxide, sulfur dioxide, particulate matter-10, and lead. Particulate matter 2.5 (with precursors) was added in 2006

Pursuant to the General Conformity Rule, a federal agency must make a General Conformity Determination for all federal actions in non-attainment or maintenance areas where the total of direct and indirect emissions of a non-attainment pollutant or its precursors exceeds levels established by the regulations.

Both the Masonville DMCF site and Masonville Cove are located in the Baltimore region for air quality planning purposes. The Baltimore region was in severe non-attainment for 1-hour ozone but was reclassified in June 2005 as in "moderate" non-attainment for the new 8-hour ozone standard. The region is in non-attainment status for particulate matter 2.5 (PM2.5) per EPA final rule of January 5, 2005. On July 17 2006, EPA published a direct final rule (71 FR 40420) establishing a 100 tons per year (TPY) *de minimis* levels for both PM2.5 and each precursor (SO2, NOx, VOCs, and ammonia).

3.0 Methodology For Determining General Conformity

3.1 Masonville DMCF Construction Overview

The Masonville DMCF project is divided into eight construction phases identified as Crews A, B1, B3, C, C1, D, E, and F. Each Crew has a specific job responsibility to be completed within an assigned time frame. Crew A will be responsible for demolition of Pier 3 and Pier 1 decks; Crew B1 will perform the pre-dredging of the Masonville DMCF site; Crew B3 represents 25% of dredging activities in Seagirt Channel that will be incorporated in the Masonville DMCF project; Crews C and C1 will be responsible for the construction of the DMCF Dike and Spillway; and Crews D, E, and F will, engage on the Cofferdam Retention Structure construction, Phase 2 Storm Drain Relocation, and the construction of Trail, Mitigation and Education center, respectively. Table 3.1(a) and 3.1(b) below present summaries of the project schedules and activity distribution that were used in the emissions calculations.

Table 3.1(a). Project Schedule

Crew	Activity	Start Date	Completion Date
A	Pier 1 & 3 Demolition	8/12/06	12/15/06
B1	Predredging (Masonville)	6/2/07	8/1/07
B3	Seagirt Deepening (materials to Masonville)	8/1/07	10/9/07
C	Dike construction	7/1/08	9/12/08
C1	Spillway Construction	8/10/08	1/24/09
D	Cofferdam Retention	1/5/07	2/12/08
E	Phase 2 Storm Drain Relocation	10/14/06	2/10/07
F	Mitigation (Education Center and Trail Construction)	1/26/07	4/25/09

Table 3.1(b) - Activities Distribution

Crew	2006	2007	2008	2009
A	100%	0%	0%	0%
B1	0%	100%	0%	0%
B3	0%	100%	0%	0%
C	0%	0%	100%	0%
C1	0%	0%	85%	15%
D	0%	90%	10%	0%
E	66%	34%	0%	0%
F	0%	41%	45%	14%

3.2 Emission Types and Sources

As stated earlier, a conformity determination is required where a federal action causes the total of direct and indirect emissions to equal or exceed the prescribed emissions de minimus levels. The direct and indirect exhaust emissions from onshore and offshore sources that will be used in the construction activity must be determined. The following sections discuss all the emissions sources (direct and indirect) involved in the proposed Masonville DMCF project. Table 3.2 below presents a list of various direct equipments (onshore and offshore) and their corresponding horsepower capacities that will be utilized by the Crews during the DMCF construction.

Table 3.2 Equipment List for Direct Emission Sources

Source Type	Equipment Type	Population	Capacity (HP)
Marine	Cranes	3	1,800
	Hopper Dredges	5	7,500
	Hydraulic Dredge	1	10,000
	Tugboats	12	12,900
Marine Total		**21**	**32,200**
Nonroad	Cranes	3	2100
	Dozers	4	1,800
	Excavators	4	1,200
	Pumps	2	200
	Pumps and Pile hammers	12	1,200
	Trucks	19	5,700
	Unloaders	2	4,000
Nonroad Total		**46**	**16,200**
Grand Total		**67**	**48,400**

The direct and indirect emission sources for the proposed Masonville DMCF project will consist of marine and land-based sources.

The marine sources include two types of dredges (hydraulic and hopper) that will be utilized to dredge material from the main channel, and dredge supporting equipment (tugboats and cranes). The land-based emission sources include both off-road and on-road equipment. The off-road equipment comprises heavy equipment that will be used in the construction and maintenance of the disposal sites (dozers, unloaders, cranes, excavators, off-road trucks, welders, pile hammers, and pumps). Refer to Table 3.2 above for a list of direct emissions sources. The on-road emission sources, identified as indirect emission sources, will be comprised of employee vehicles and delivery trucks. The marine emission sources and off-road equipment all consist primarily of diesel-powered engines. The on-road sources are combinations of gasoline and diesel-powered vehicles.

3.3 Air Emission Models

In order to determine the air emission quantities from the sources, background information (engine horsepower, hours of operation, and fuel source) of the different equipment types was researched. Once this information was obtained, different engine load factors and emission factors were determined. Depending on an emission source, (marine, land-based off-road equipment, or on-road vehicle) EPA has published guidelines that determine the appropriate engine load factors and emission factors. The EPA guidelines and models are discussed below for various direct and indirect emissions sources that will be used in the project.

3.3.1 Emissions from Marine Diesel Engines

The marine emission sources are comprised of two types of dredges along with the associated support equipment. EPA currently has an extensive compilation of air emission factors for various types of equipment (Compilation of Air Emission Factors, AP-42). The latest EPA technical report for developing load factors and emission factors for large compression-ignition marine diesel engines is given in the *Analysis of Commercial Marine Vessels Emissions and Fuel Consumption Data*; EPA 420-R-00-002, published February 2000. The technical report is a compilation of engine and fuel usage test data from various types of marine vessels, including bulk carriers, container ships, dredges, tankers, and tugboats. This report was employed in the determination of the load factors and emission factors for the various pieces of marine equipment that will be operational during construction of the DMCF project. The load factors for the marine equipment shown in Table 5-2 of the EPA technical report *Analysis of Commercial Marine Vessels Emissions and Fuel Consumption Data* are based on the suggested operating mode of the vessel. These load factors are given for the corresponding operating mode (cruise, slow cruise, maneuvering, and hoteling) for the different types of vessels. Detailed emission factors were determined through a regression analysis of the representative test data published by EPA. Emission factor algorithms were determined for the different pollutants and also for fuel consumption, which is used to determine the SO_2 emission factor. The sulfur content for the fuel consumption regression for SO_2 was set to 0.3 percent, which is the national sulfur content average for nonroad diesel. The marine engine emission factor and fuel consumption algorithms are presented in Table 5-1 of the EPA technical report *Analysis of Commercial Marine Vessels Emissions and Fuel Consumption Data*. The emission factor and fuel consumption rate algorithms are applicable to all engine sizes since the emissions data showed no statistically significant difference across engine sizes. All the equipment required for dredging, transport, and placement of the dredged material is accounted for in this emissions modeling analysis.

3.3.2 Emissions from Off-road and On-road Sources

a). Nonroad Emissions Model:

The off-road land-based emissions were calculated using the new EPA's computer-based model known as National Mobile Inventory Model (NMIM). This computer application was developed by EPA to help estimate current and future emission inventories for on-road motor vehicles and nonroad equipment. NMIM uses current versions of MOBILE6 and NONROAD models to calculate emission inventories based on multiple input scenarios entered into the system. NMIM comprises a Java framework, graphical and command line user interfaces, the MOBILE6 and NONROAD models, a national county database, and post-processing and aggregation capabilities. The NMIM model estimates emissions for six different exhaust pollutants: hydrocarbons (HC), NOx, CO, carbon dioxide (CO_2), SOx, and PM. HC are reported as either total hydrocarbons (THC), total organic gases (TOG), non-methane organic gases (NMOG), non-methane hydrocarbons (NMHC), or volatile organic compounds (VOCs). The NONROAD option of the NMIM model contains several different sets of data files that are used to specify the options for a model "run". These data files provide the necessary information to calculate and allocate the emissions estimates. The NMIM database contains information on load factors, emission factors, activity, geographic location (region, state, and counties), equipment source classification codes, and temporal allocation. The data files can be modified to reflect the project conditions by selecting the Fleet option and importing a comma delimited (csv) text files containing user–specified and/or project-specific information like the equipment source classification code (SCC), equipment capacity (HPmax), model year, technology type, equipment population, annual hours of use, and monthly activity distribution.

The NMIM Graphical User Interface was utilized for this project. The input options modified for the purpose of the Masonville DMCF project emissions estimate are discussed in the analysis section of this report. Information on the latest draft of the NMIM model is available at EPA's website (http://www.epa.gov/otaq/nonrdmdl.htm).

b). Mobile Source Emissions Model:

The indirect emissions sources for the project will be as a result of employee vehicles and other over-the-road vehicles utilized during the construction process. EPA has developed a mobile source emissions model, MOBILE6.2, to calculate emissions from different vehicle types. MOBILE6.2 is a model that calculates emissions, in grams per mile, for different vehicle types under various operating conditions. The user specifies various input options on vehicle types, quantity, and operating conditions. The model will calculate the emission quantities for HC, CO, NOx, CO_2, PM, and toxics for each type of vehicle. The emission quantities are then multiplied by the number of miles traveled to determine the final quantities. The input options utilized to run the model are discussed in the analysis section of this report.

3.4 Masonville DMCF Project Emissions

3.4.1 Marine Emissions

The first step in determining the marine emission quantities was to develop a list of all the marine equipment (dredges and support) that would be utilized on the Masonville DMCF project and the engine operating characteristics (horsepower and fuel type). The marine operations comprised the following equipment: one hydraulic dredge, three hopper dredges, twelve tugboats, and three cranes. Each of these equipment types was analyzed based on its capacity.

Emission rates, reported in tons of pollutant emitted per hour of operation (tons/hr) for each engine, were calculated for each of the six criteria air pollutants: CO, NOx, PM2.5, PM10, SOx, and VOC. The emission rates were derived from the formula:

Emission Rate (tons/hr) = Engine Horsepower ×Engine Load Factor ×Emission Factor

As stated earlier, the EPA technical report: *Analysis of Commercial Marine Vessels Emissions and Fuel Consumption Data* was utilized in determining the appropriate load factors and emission factors used for the marine equipment emissions calculations. Load factors and emission factors for different marine equipment were determined, and the project schedule was used to estimate the annual hours of operation for the each piece of equipment. The Masonville DMCF project will be completed by a total of eight different construction groups (Crews) as presented in Table 3.1(a) above.

Applying the project schedule, the annual hours of operation were developed for each piece of marine equipment over the construction period. Emission amounts for each of the six pollutants were then calculated based on the following formula:

Emission Amount (tons/year) = Emission Rate (tons/hour) ×Working Hours (hours/year)

3.4.2 Off-Road Emissions

The criteria air pollutants emissions from the off-road land-based equipment were determined using the new EPA NMIM model (March 2006). The equipment types, engine parameters, hours of operation, and geographic location were derived from the project estimates. The off-road equipment was categorized by contracts (Crews) along with the corresponding engine capacities (horsepowers), equipment population, and contract schedules (hours). Based on individual Crew project schedule, and various Crew activities per project year, the hours of operation for each Crew were then distributed to the corresponding years to determine annual operating amounts (see Table 3.1(b) above).

In order to quantify emissions, the NMIM model will require certain input parameters such as fuel type; temporal information; geographic region (state and county); equipment source classification codes (SCC), equipment technology type, equipment population, and monthly activity distribution ratio. The model contains a database of emission factors which is a function equipment SCC, equipment technology type, fuel type and metrological data of the geographical region.

To properly characterize nonroad emissions that will result from the project, some of the Model's input parameters were modified to accurately reflect the estimated project schedule. For each Crew, the NMIM model was executed with a properly formatted input data file.

3.4.3 On-Road Emissions (Mobile Sources)

The indirect emissions for the Masonville DMCF project result from mobile sources (employee vehicles and other on-road vehicles) utilized during the construction of the project. The construction phase on-road vehicle emissions were estimated using the EPA Mobile6.2 emission factor model and activity data representing the employee and delivery vehicle miles traveled (VMT) for all four years of construction and all eight construction crews. CO, NOx, VOC, SO_2, PM10, and PM2.5 emissions were calculated for employees' vehicles and the delivery trucks. Mobile6.2 calculates emission factors in grams per mile for different vehicle types and operating conditions. These emission factors were multiplied by the VMT to calculate the total emissions.

The emission factors were estimated from the execution of Mobile6.2 for the summer and winter seasons of calendar years 2006, 2007, 2008, and 2009. Required local input data such minimum and maximum temperature, fuel Reid's vapor pressure (RVP), fuel sulfur content, and state inspection and maintenance (I/M) control program data were obtained through literature search and approved by the Maryland Department of the Environment.. Data concerning VMT mix fractions and registration distribution used Mobile6.2 default values that represent national average data.

The Mobile6.2 emission factors were further reduced to calculate composite emission factors. In doing so, it was assumed that employee vehicles were composed of light-duty gasoline vehicles (LDGV) and light-duty gasoline trucks. The delivery vehicles were assumed to be composed of heavy-duty gasoline vehicles (HDGV), light-duty diesel vehicles (LDDV), light-duty diesel trucks (LDDT) and heavy-duty diesel vehicles (HDDV).

To account for the monthly fleet turnover, the emission factors for each month were estimated by using EPA recommended mathematical interpolation between the two MOBILE6.2 results by weighting the January and July results relative to 6 months.

The activity data represented the VMT by vehicle class. The total mileage in a given crew was obtained by multiplying the number of employees by the mileage driven per employee per day and by the duration of the crew in days. In doing so, it was assumed that each employee would drive its own vehicle, take two trips on the construction site, and drive 1 mile per trip.

For the delivery vehicles, a maximum of four deliveries per construction day was assumed. It was also assumed that each delivery vehicle would take two trips on the construction site and drive 1 mile per trip.

4. General Conformity Determination

4.1 Emissions Results

The criteria air pollutants emissions presented in the table below represent the estimated total of direct and indirect emissions that will occur during the proposed Masonville DMCF construction. The analysis of these emissions is to address the requirements of General Conformity of the Clean Air Act. The emissions for the marine and land-based equipment were determined as discussed in Section 3.4. The calculated emissions were then totaled on an annual basis.

Table 4-1 and Table 4-2 present total emissions summaries from the Masonville DMCF construction project. The resulting emissions was first distributed per Crew (Table 4-1), and then distributed per calendar year in which the contract year is defined (Table 4-2). Each contract year emissions were compared to the General Conformity Threshold requirements. From Table 4-2, it is observed that NOx emissions will exceed the Federal Conformity limits (bold print) in the second and third years of the project (2007 and 2008). All other pollutants (PM2.5 and VOC) are below the Federal Conformity limits in all years and therefore no further analysis is required for those pollutants. The tables below present the summaries of emissions for the Masonville DMCF construction project:

Table 4.1. Emissions Summary for the Masonville DMCF Project (tons)

	CO	NOx	PM2.5	PM10	SO2	VOC
CREW A	2.164	10.955	0.323	0.332	1.803	0.327
CREW B1	26.585	113.007	3.773	3.897	16.820	3.477
CREW B3	8.371	53.574	1.348	1.348	9.085	0.940
CREW C	18.242	114.344	2.916	2.923	19.239	2.094
CREW C1	2.841	6.057	0.312	0.340	0.132	0.529
CREW D	12.582	37.202	1.854	1.929	7.654	2.025
CREW E	0.843	1.595	0.083	0.092	0.251	0.154
CREW F	7.200	16.593	0.846	0.920	0.973	1.600
TOTAL	**78.83**	**353.33**	**11.46**	**11.78**	**55.96**	**11.15**

Table 4.2. Total Annual Emissions Compared to the General Conformity (GC) Threshold (tons)

Pollutant	GC Threshold	2006	2007	2008	2009	TOTAL
CO	**NA**	2.72	49.55	25.12	1.44	78.83
NOx	**100**	12.00	**207.47**	**130.60**	3.26	**353.33**
PM2.5	**100**	0.38	7.17	3.74	0.17	11.46
PM10	**NA**	0.39	7.39	3.81	0.18	11.78
Sox	**100**	1.97	33.28	20.55	0.16	55.96
VOC	**50**	0.43	6.95	3.46	0.31	11.15

4.2 Emissions Summary

Based on results as shown in Table 4-2 above, NOx emissions are observed to be above the Federal Conformity limits for the second and third years of the proposed Masonville DMCF construction project. Due to this exceedance, according to the General Conformity ruling (40 CFR 93.158 (a)(2) and (a)(4)), it will be required to mitigate or offset the NOx emissions that will result from construction activities to zero for those years (2007 and 2008). The project areas (Baltimore City and Anne Arundel County) are both designated as a moderate nonattainment for ozone (the pollutant of concern for the project). As such, the *de minimis* threshold for these regions is 100 tons of NO_x per year. NOx is an ozone precursor. Therefore, general conformity mitigation will be required for NO_x for the second and third year of the project (2007 and 2008).

5. Masonville Air Pollutant Mitigation Plan

As demonstrated in the previous chapter, the Masonville project will require the mitigation of air emissions for the construction years of 2007 and 2008. NOx emissions have been projected to be 207.5 tpy for 2007 and 130.6 tpy for 2008.

The initial approach for developing an emissions mitigation program was to identify emission reduction opportunities for NOx at the Maryland Port Administration facilities. Several possibilities were examined, such as the electrification of gantry cranes at the Dundalk Marine Terminal, however, none were deemed cost effective or practical in terms of the Masonville schedule (pursuant to USEPA guidance emission reduction "projects" should occur at the same time new emissions are generated).

The next option examined was to consider the use of securing permanent emission reduction credits (ERC). Several options in this regard were pursued. The first was to obtain NOx credits for the necessary 2-year period from a utility that has secured the necessary credits, however does not need them until the 2009/2010 timeframe. This option became the preferred alternative and a contract was executed with Sempra Energy Inc. in May, 2007. The approach generated considerable debate since the credits were "leased" rather than wholly purchased. Sempra required the credits returned by 2010, the date of startup of their Maryland power plant project. The approach was eventually approved by both the state and federal regulatory agencies.

The MDE approved a 1:1 ratio for offsetting credits. An equipment monitoring program has put in-place in order to document activity and the associated emissions.

On April 13, 2007, the Baltimore District, U.S. Army Corp of Engineers issued an initial approval of the conformity study, findings, and mitigation plan. This ruling will be subject to the approval of the Final Environmental Impact Statement that should occur in the summer of 2007.

Recommended Practices for NEPA Air Toxics Analysis of Highway Projects

D. A. Ernst and E. L. Carr

Abstract

The Clean Air Act Amendments of 1990 designated 188 air toxic pollutants as Hazardous Air Pollutants (HAPs). Certain of these HAPs are characteristic of motor vehicles and have become known as mobile source air toxics (MSAT). Public concern about evaluating MSATs in the National Environmental Policy Act (NEPA) process has prompted transportation agencies to begin addressing air toxic emissions as part of their environmental review procedures. This study developed best-practices guidance for MSAT assessments under NEPA. Current approaches used by transportation agencies to assess and communicate MSAT emissions and health impacts were compiled and reviewed. A tiered methodology was developed which guides the analyst in identifying the appropriate level of analysis for a transportation project, using typically available information and potential level of exposure based on project characteristics. Five potential levels of analysis are suggested based on both technical and policy considerations. Details are presented on how to conduct the MSAT assessment as well as on the amount of information that should be included at each level of analysis. Recommendations are provided on how best to communicate the findings of the MSAT assessment as part of environmental documents and in public review.

1. Introduction

The Clean Air Act Amendments of 1990 designated 188 air toxic pollutants as Hazardous Air Pollutants (HAPs) and required the U.S. Environmental Protection Agency (EPA) to undertake a number of study and regulatory activities to reduce HAP emissions. Public concern about evaluating these toxic air pollutants in the National Environmental Policy Act process (NEPA, 42 U.S.C. 4321 et seq.) increased in the 1990s, partly due to several pivotal agency studies (Houk and Claggett 2006). Chief among these studies and rules were the EPA National Air Toxics Assessment (EPA 2000), the California South Coast Air Quality Management District MATES II study (SCAQMD 2000), and the EPA Mobile Source Air Toxics Rule (EPA 2001, the "MSAT Rule"). The National Air Toxics Assessment (NATA) was a nationwide modeling and risk assessment exercise that estimated the cancer and non-cancer risk from air toxics for each census tract in the U.S. The NATA estimated that every county in the U.S. experiences an overall cancer risk of greater than 10^{-5} or 10 in 1 million (i.e., 10 cancer cases per million population over a lifetime) from all sources. The MATES II study, which estimated mobile source-related risks in the California South Coast region, assigned 90% of the total cancer risk to mobile sources with 70% of the total risk assigned to diesel particulate matter from mobile sources.

The NEPA process requires that major Federal actions that "significantly affect the quality of the human environment" undergo assessment of their environmental impact. A history of regulatory actions for non-mobile sources of HAPs has led the public to ask that assessments be made for MSATs when part of a Federal action. As concern about MSATs has mounted, the Federal Highway Administration (FHWA) and state departments of transportation have increasingly received requests for MSAT analysis in agency-funded environmental impact statements (EISs). The issue of air toxics has been raised with several major highway projects around the country, resulting in lengthy deliberations and in some cases, litigation (FHWA 2004, FHWA 2006). With the recent availability of motor vehicle air toxic emission factors from EPA's MOBILE6.2 model, the feasibility of assessing air toxics under NEPA needs to be evaluated. If air toxic assessments are feasible, the NEPA process can then be used to disclose the potential impact, analyze alternatives and possibilities for mitigation, and inform the public about the impacts of air toxic emissions from a proposed project.

The EPA MSAT Rule identified 21 hazardous air pollutants as mobile source air toxics. EPA identified six of the 21 pollutants as of greatest concern due to their high relative emissions and risk and because state agencies have indicated that they are major mobile source pollutants of concern. These six pollutants have become known as the "priority MSATs"[1] and are:

- Acetaldehyde
- Acrolein
- Benzene
- 1,3-Butadiene
- Formaldehyde
- Diesel particulate matter and diesel exhaust organic gases (DPM+DEOG)

Information on recommended practices for MSAT analysis in NEPA is needed to:

- Help practitioners to communicate MSAT project level-impacts in NEPA documents in ways consistent with the limitations and uncertainties of current modeling tools and in the absence of National Ambient Air Quality Standards (NAAQS);
- Promote consistency among MSAT evaluation methods so that the relative impacts of roadway projects and programs can be compared and;
- Assure that the quality of MSAT analysis for NEPA documents is sufficient to meet statutory and regulatory requirements, to support agency decision-making, and to adequately inform the public about the air quality impacts of projects in the NEPA context.

---

[1] These priority MSATs are subject to change based on improved understanding of ambient levels and health effects. In particular, EPA's proposed MSAT rule on Control of Hazardous Air Pollutants from Mobile Sources (71 FR 60:15813-15814, March 29, 2006) discusses the MSATs which pose the greatest risk at current levels based on updated information, and includes naphthalene.

MSAT Issues in NEPA Studies for Transportation Projects

Relevant Pollutants:

Most projects focus on priority MSATs as they represent the bulk of total health risk. All of the priority MSATs can cause respiratory health effects, and all except acrolein are EPA-designated probable or known carcinogens. Benzene, a known carcinogen, and DPM are viewed as especially harmful.

Proximity to Emission Sources:

Proximity to transportation facilities, typically roadways, has been established as a primary factor leading to community exposure and potentially increased risk. Numerous studies in recent years have found adverse health impacts that seem to be linked to proximity to a roadway, including increased incidence of asthma and cancer. (Houk and Claggett 2006; for one summary of studies, see Johns Hopkins 2004). The public health community perspective is best reflected in the Johns Hopkins summary document which states "In conclusion, a substantial and growing body of evidence from epidemiologic studies indicates that residence in close proximity to roadways with high traffic density is associated with increased risk of a broad spectrum of health outcomes in adults and children. The scientific evidence is stronger for the health outcomes of mortality, lung function, and lung cancer in adults, and for respiratory symptoms including asthma/wheezing and lung function in children. The interpretation of study results for asthma medication or health care use, cancer in children, and atopy are less consistent."

More recent studies support a finding of increased risk from exposure in proximity to transportation facilities. For example, two recent studies, Gauderman *et al.* (2005) and McConnell *et al.* (2006), both observed a statistically significant association of increasing childhood asthma rates with decreasing distance to freeways in several California towns. The weight of the current evidence indicates that it is reasonable to use proximity to a transportation project facility as a screening tool in NEPA evaluations of MSATs. Future work may investigate distance-based project screening criteria.

Thresholds of Impact:

In NEPA air quality studies of criteria pollutants, the threshold for "significant" impacts is commonly taken to be an ambient concentration standard. Section 109 of the Clean Air Act states that NAAQS are to be set at levels that, "allowing an adequate margin of safety, are requisite to protect the public health." In contrast to criteria pollutants, no NAAQS have been established for HAPs or for MSATs in particular (lead is designated as both a criteria pollutant and a HAP). The large uncertainties in the scientific understanding of health effects of HAPs as well as in the methodologies for estimating HAP emissions and concentrations continue to prevent the establishment of NAAQS or single-number standards for MSATs under NEPA. Instead of single-number concentration standards as thresholds, quantification of health impacts relies on health risk assessment which is subject to large uncertainties of its own. The definition of the degree of project impact at which risk becomes excessive—the impact criterion—is a social and policy decision that is best addressed by each community.

Value of Emissions Inventory:

Many NEPA analyses have estimated emissions but not the resulting concentrations or health risk. An emissions inventory analysis provides information on total emissions levels that can be used to satisfy the NEPA purpose of comparing project alternatives. This approach depends on the assumption that potential impacts of the alternatives are adequately represented by the aggregate emissions. The uncertainties involved in estimating emissions alone are considerably less than those involved in estimating risk. By only estimating emissions, the typical NEPA analysis will use resources to improve the emission inventory to support better relative aggregate comparison of the project alternatives for NEPA decision making, but at the cost of not estimating the absolute risk magnitude comparison.

3. Health Risk Assessment Concepts and Their Application to Transportation Agencies

Public agencies with air quality responsibilities have developed a number of approaches to HAP analysis. Many human health risk assessments have used the four-step human health risk assessment (HRA) process as detailed in EPA (1989) and elsewhere. This guidance is part of a set of EPA manuals for use in the remedial investigation/feasibility study process for Superfund hazardous waste sites. The HRA process was adapted from well-established chemical risk assessment principles and procedures. The HRA procedures and principles are valid for mobile sources and could be applied to MSATs as California air quality agencies and others have. The HRA process comprises four steps:

(1) Data Collection (Hazard Identification): identifies the substances that may present a hazard due to the project and, for air quality studies, characterizes their emission rates.
(2) Exposure Assessment: estimates the magnitude, frequency, and duration of potential human exposures. Pollutant concentrations in the project region are estimated using dispersion modeling. Demographic and spatial data for the project region are used to assess exposure to the estimated concentrations by location and over time. For most NEPA air quality studies the inhalation pathway is the only appreciable route of exposure.
(3) Toxicity (Dose-Response) Assessment: considers the types of adverse effects associated with the substances identified, the relationship between magnitude of exposure and adverse effects, and related uncertainties such as the weight of evidence of a substance's carcinogenicity in humans. NEPA air quality studies almost always rely on published agency data for this step rather than conducting new research.
(4) Risk Characterization: combines the outputs of the exposure and toxicity assessments to define the project-related risk in qualitative and quantitative terms. The potential for adverse effects to occur is characterized in terms of cancer risks (probabilities) and noncancer hazard quotients (HQ). The hazard quotient is an index that expresses potential exposure levels relative to threshold levels at which health effects are not known to occur. The HQs are aggregated to derive a Hazard Index (HI).

Toxicity Information for Priority MSAT Analysis:

It is helpful to consider HRA risk metrics and criteria along two dimensions: cancer versus non-cancer, and acute versus chronic. Acute refers to health impacts of short-term exposure, typically

24 hours or less. Chronic refers to longer-term exposures, such as an annual average, up to a lifetime.

Cancer Risk: To measure risks from developing cancer, many risk assessments have used the metric Cancer Risk (CR) or Excess Lifetime Cancer Risk (ELCR). This metric represents the probability that an individual will develop cancer in his or her lifetime as a result of exposure to the substance in question. EPA generally considers risks (CR or ELCR) of less than 10^{-6} (one in a million) to be acceptable and acts to reduce cancer risks greater than 10^{-4} (1 in 10,000). The calculation of CR/ELCR is dependent on a Unit Risk Factor (URF) as an input parameter, defined as the probability that a person will get cancer from exposure to the source over 70 years, per 1 microgram per cubic meter ($\mu g/m^3$) concentration of the pollutant of interest. In a NEPA project MSAT analysis, the URFs normally are taken as a given and the CR/ELCR is derived as the ratio of a modeled concentration (in $\mu g/m^3$) to the URF.

Non-Cancer Risk: To measure non-cancer risks, many risk assessments have used the metrics Hazard Quotient (HQ) and Hazard Index (HI). HQs are fractional indices that can be derived for various pathways or target human health/organ systems. An HQ is the ratio of an exposure concentration for a given compound to the RfC for that compound. The sum of HQs for several compounds is an HI. An HQ or HI less than one indicates that no adverse health effects are expected, while an HQ or HI in excess of one indicates that adverse effects are possible. HQ and HI estimates cannot be interpreted as a probability of adverse health effects. The calculation of HQ and HI is dependent on a Reference Concentration (RfC) or California Reference Exposure Level (REL) as an input parameter for toxicity. In a NEPA project MSAT analysis, the RfCs/RELs normally are a given and each HQ is derived as the ratio of a modeled concentration (in $\mu g/m^3$) to the RfC/REL.

Exposure Information for Priority MSAT Analysis:

For those MSAT analyses that derive quantitative estimates of risk, the basis for the exposure calculation is ambient concentrations output from modeling. Various metrics of exposure can be selected according to policy choice, characteristics of the source and the study area. In modeling terms this determines the selection of the receptor locations to be used in the dispersion modeling. A single location of maximum modeled concentration, sometimes referred to as the Point of Maximum Impact (PMI), is often chosen. When combined with conservative (tending to overstate impact) assumptions a worst-case risk estimate is created that can be used in a screening process. If all worst-case results fall within acceptable levels of risk, then adverse impacts may be assumed not to occur. If worst-case screening results fall above the acceptable risk levels then a more refined analysis is needed. The PMI approach cannot be used in refined analysis because the PMI bears no relationship to the actual exposure patterns that lead to MSAT health impacts.

More refined analyses may use known locations of sensitive populations—where the impact of exposure might be relatively great (i.e., high response per dose in a given population), population-weighted ambient concentrations, estimates of Reasonable Maximal Exposure (RME)—hypothetical maximum exposure expected from the project, or identification of a Maximum Exposed Individual Resident (MEIR) and Maximum Exposed Individual Worker (MEIW)—hypothetical individual who might receive maximum exposure.

Existing or Background Concentrations:

The most current assessment of nationwide MSAT concentrations is available through EPA's NATA (http://www.epa.gov/ttn/atw/nata1999/ tables.html). This assessment modeled 1999 outdoor air concentrations at county-level resolution. For those MSATs not modeled in NATA, observed concentrations from EPA's AirData Reports (http://www.epa.gov/air/data/geosel.html) can be used to develop estimates of background level concentrations. This was done as part of the NCHRP study (ICF 2007). While EPA does not recommended that the data be used to estimate a specific MSAT concentration at a census tract the NATA data may be used to estimate the distribution of concentrations within the county for each MSAT; combining this with information about the project setting (urban, suburban, rural, population density) provides an approach to estimate the likely existing background MSAT concentration at the project location.

4. Recommended Procedures for Analyzing MSAT

Recommendations have been developed on how to select and apply the best available models and associated techniques for MSAT impact assessment in the NEPA process. The approach uses both policy and technical considerations to determine the need and appropriateness for conducting a MSAT assessment. The policy questions identify the appropriate level of analysis and the technical questions address the technical feasibility of the desired policy-level analysis. Based on these considerations five analysis levels were identified with each level of analysis balancing the need for increased information due to the project's potential for risk versus the increased level of effort to conduct the analysis.The full set of questions that determine the five levels of analyses for air toxic assessment under NEPA is presented in flowchart form in Figure 1 and elaborated below. The analysis at each level will focus on the six priority MSATs.

Level 1—No Air Toxic Risk Assessment

To reach this level of assessment the project must be categorically excluded under FHWA's NEPA regulations (23 CFR 771.117(c)), in which case no analysis or discussion of MSATs is needed. However, the project should document how it qualifies for the categorical exclusion or exemption, and the basis for the determination that no meaningful air-related impacts would occur.

Level 2—Qualitative Air Toxic Assessment

To qualify for this level of assessment a project must not exceed recommended screening thresholds. The NCHRP study (ICF 2007) provides a detailed discussion on the development of these health risk-based screening thresholds from analysis of the key risk drivers[2]. The recommended screening thresholds are:

- Freeway volume of 125,000 AADT in the design year;
- Arterial volume of 100,000 AADT in the design year;
- Interchange volume of 40,000 AADT in the design year;

[2] A variety of transportation settings (freeway, arterials, and intersections) in combination with a number of lane configuration and speeds were modeled to determine the level at which traffic volumes in 2008 were sufficient to reach a one in a million cancer risk level for benzene or for non-cancer health effects the diesel reference concentration level.

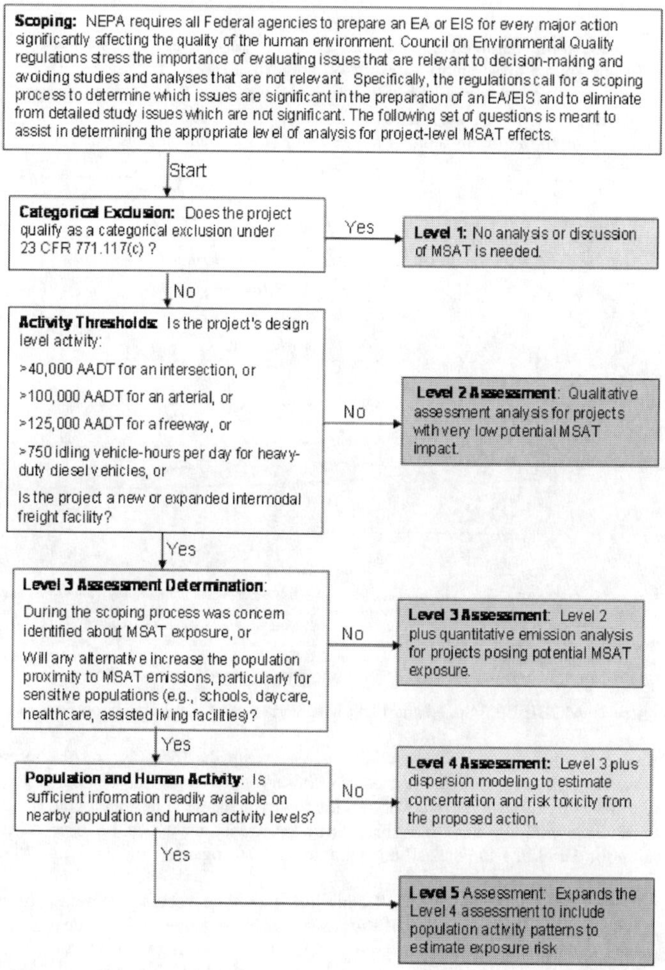

Figure 1. Tiered Decision Tree For Mobile Source Air Toxic Analysis

Heavy duty diesel vehicle activity of 750 idling vehicle-hours per day at a single location. Qualitatively describe how each project alternative would affect traffic volume, speed, and vehicle mix. Using summary information based on the latest emission factor model and studies, discuss how these three parameters would likely affect MSAT emissions for each project alternative. In one source of summary information done for FHWA (Tang et al. 2003), priority MSAT emission rates were determined as a function of speeds and vehicle mix. Figure 2 shows the results for freeways which are based on calendar year 2005 using MOBILE6 national defaults and a maximum/minimum summer temperature of 88° F and 100° F, respectively.

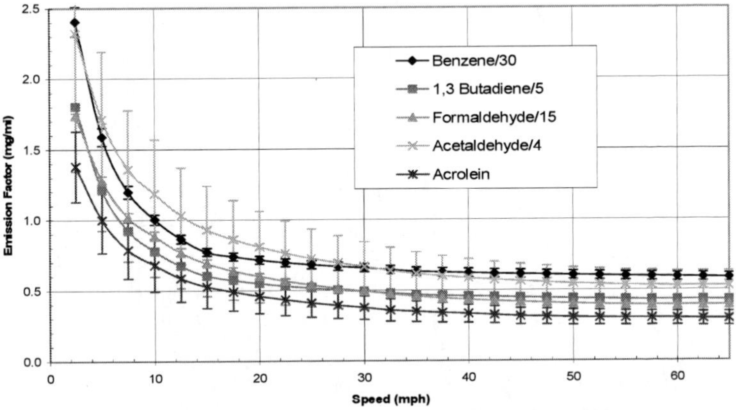

Notes: Data are for MOBILE6.2 Freeway facility type. Ranges are based on fleet mix of 2% and 15% heavy duty diesel vehicles (HDDV). Error bars show change relative to default fleet (8.5% HDDV) for 2% HDDV and 15% HDDV. Benzene emission factor is highest with 2% HDDV. All other MSATs are higher with greater HDDV percentage except butadiene which is higher with 15% HDDV at low speeds and then crosses at 47.5 mph to have higher emissions with lower (2%) HDDV at higher speeds. To display MSATs on same graph the emission factors have been scaled; the true emission factor is equal to the value in the graph multiplied by the denominator (in legend). For example, the true benzene emission rate at 10 mph is (1.0*30) = 30 mg/mi. Source: Adapted from Tang et al. 2003.

Figure 2. MOBILE6.2 Speed and Fleet Mix Effects on MSAT Emission Factors

If an alternative would change the relative distance between the roadway and the location of exposed individuals then Figure 3 can be used as a conservative estimate of the potential increase or decrease in MSAT exposure. Figure 3 shows the relative change in pollutant concentration as a function of downwind distance. The black carbon measurement can be used as a close approximation for diesel PM and the carbon monoxide as a surrogate for the gas-phase MSATs.

Likely secular (non-project trends) in MSAT emissions should be discussed to define differences between the existing conditions and the NEPA No-Action Alternative in future years, and to set the context for the project impacts. Figure 4 illustrates national trends in priority MSAT emissions and VMT. Figure 4 or similar data can be used in the Level 2 discussion of trends.

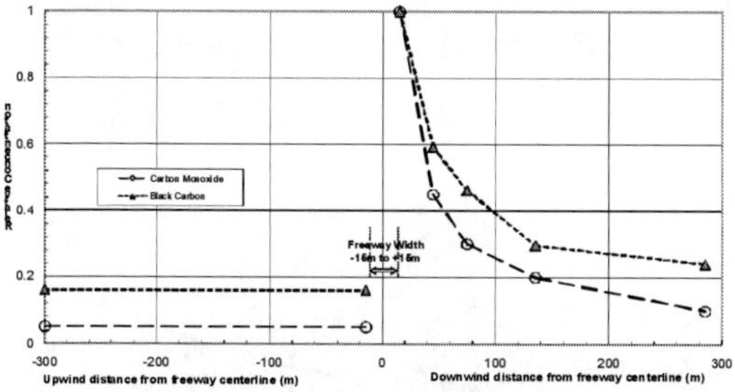

Notes: Adapted from Zhu et al. (2002) Figure 8. Measurements collected during the daytime (May-July 2001) adjacent to the 30-m wide I-405 freeway. Traffic density ranged from 140 to 250 vehicles per minute. Wind direction was consistently perpendicular to the freeway at 1-2 m/s. Less than 5% of vehicles were heavy-duty diesel trucks.

Figure 3. Relative Carbon Monoxide and Black Carbon Concentrations vs. Downwind Distance

Level 3—Quantitative Air Toxic Emissions Assessment

The Level 3 assessment will follow the same procedures as outlined in the Level 2 analysis, but also will include an emissions inventory of project-related MSATs. This will provide decision makers with the ability to compare the relative differences in total emissions among the various alternatives. The most current emission factor models (currently MOBILE6.2 and EMFAC2007) should be used with project-specific traffic activity data. The projects that fall into this category are those that would exceed the activity level screening thresholds, thus demonstrating potential for increased risk due to MSAT emissions, and for which concern about MSAT exposure was expressed during the scoping process.

Level 4—Quantitative Air Toxic Concentrations and Risk Assessment

This assessment will follow the same procedures as outlined in the Level 3 analysis, but will involve air quality modeling of ambient pollutant concentrations to provide the reviewer with a better perspective on the relative impact of the increased air toxic risk of the project relative to existing air toxic risk. The analysis will use the available information on the project alternatives in conjunction with the best available emission factors to quantitatively estimate the impact on MSAT emissions, and then conduct dispersion modeling and assess the cumulative air toxic risk for the proposed action. The projects that fall into this category have not only shown a high potential for MSAT emissions to concentrate at high enough levels to be of potential concern, and have been raised as a public concern during the scoping process, but also would increase proximity to MSAT exposure (i.e., decrease source-receptor distances)[3]. In order to determine the cancer risk from the project the modeled air concentrations at each receptor location for each priority MSAT must be multiplied by their individual unit risk level. While each MSAT does not target the same organ, a conservative approach is to assume that they do so that the sum of the risks from each MSAT is the maximum increased cancer risk. The receptor location which has the highest risk should be identified as the maximum individual cancer risk.

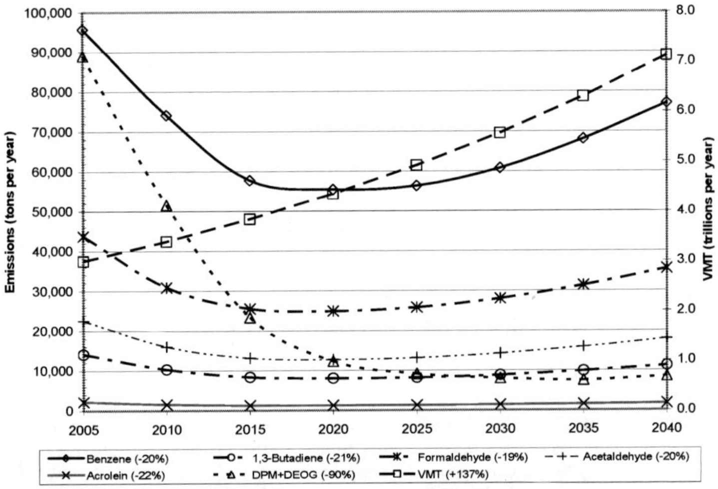

Notes: Data are for on-road mobile sources. Emissions were generated with MOBILE6.2. Oxygenates held constant at 50% market share of E-10. Constant VMT growth of 2.5%/yr. "DPM+DEOG" is the sum of SO_4, elemental carbon, and organic carbon from diesel vehicles. Gasoline benzene content = 0.97% (national average) for 2005 and 2010, 0.62% after 2010. Diesel sulfur content = 300 ppm for 2005, 15 ppm after 2005. Fuel RVP, aromatics, and olefins held constant at 8.5 psi, 15% and 15%, respectively. Temperature range 45°-65° F. No refueling; tailpipe emissions only. Otherwise, MOBILE6.2 defaults. MOBILE6.2 includes no emission controls beyond the EPA Tier II light-duty standards and 2007 heavy-duty standards, and does not forecast potential future controls. Thus, this is likely a worst-case view of future MSAT emissions.

Figure 4. U.S. Annual Vehicle Miles Traveled (VMT)vs. MSAT Emissions (2005–2040)

Level 5—Quantitative Exposure and Population Risk Characterization

This assessment will follow the same procedures as outlined in the Level 4 analysis, but will add an additional component following the air dispersion modeling that incorporates exposure assessment techniques. This analysis will provide the reviewer more information as to the population exposed to the increased risk. The analysis will still use the available information on

[3] The relative increase in proximity serves as an indicator that the potential for increased exposure is more likely than under a no-action alternative thus warranting a more refined analysis.

the proposed action and its alternatives in conjunction with the best available emission factors to quantitatively estimate the impact for MSAT emissions and then conduct dispersion modeling. The projects that fall into this category have not only shown a high potential for MSAT emissions to concentrate at high enough levels to be of potential concern, but have been raised as a public concern during the scoping process, as well as having sufficient local information available on nearby population and human activity levels to conduct the exposure assessment.

Exposure models combine information about the geographic pattern of pollutant concentrations (typically from an air dispersion model) with information about population activities. In exposure model applications the receptor is a person, population subgroup, or specific area. The simplest models estimate inhalation exposure using ambient concentrations and residential population distributions, with the implicit assumption that the populace is outside at home at all times[4]. More refined models combine concentration data with time-activity data. Activity profiles specify a schedule of movements among specified locations (e.g., indoors at home, at work or school, outdoors at a neighborhood park) and activities (e.g., sleeping, walking the dog) for an individual over a period of time. Risk characterization combines the results of the dose-response assessment with the exposure assessment to estimate the likelihood of adverse effects occurring. This information on time-activity and duration of exposure is an important factor in reducing uncertainty as duration coupled with concentration plays an important role in assessing overall exposure. Further discussion on available exposure models, their interconnection with air quality models, and their implications on reducing uncertainty are discussed in the NCHRP study which also provides recommendations for models for use in refined MSAT analysis.

5. Uncertainty in MSAT Analysis

The health risk assessment literature contains extensive coverage of the uncertainties involved in human health risk assessment. A number of discussions of uncertainties for MSATs in particular have appeared in agency guidance documents. Some recent EISs have included discussions of uncertainties to comply with the Council on Environmental Quality regulations requiring discussion of incomplete and unavailable information (40 CFR 1502.22). (For further detail see Claggett and Houk 2006, Claggett and Miller 2005, FAA 2001, FAA 2003, FAA2005, and FHWA 2006.) Based on review of this information and the current state of the science, the following conclusions may be drawn concerning the treatment of uncertainty in MSAT analyses for NEPA:

- Transportation projects vary widely in the need for and usefulness of MSAT analysis. Uncertainty is a substantive issue for larger projects.
- Use of health risk assessment, with its attendant uncertainties, is warranted for some of the larger projects[5].

[4] Exposure to MSATs indoors or in vehicles can exceed exposure outdoors under some circumstances. For examples from several studies see Winer 2005. Because the NAAQS and EPA guidance are based on outdoor exposure, NEPA studies have not taken indoor or in-vehicle exposure into account.

[5] The NCHRP study (ICF 2007) identified projects of sufficient size and activity level, based on health risk screening, that warrant further assessment.

- The public demand for information in NEPA documents may exceed the level of analysis in which the agency has confidence for purposes of alternative selection and impact assessment. Thus, agencies may need to educate the public about uncertainties in the analysis to forestall comment that seeks to stop the project rather than guide the selection of alternatives under NEPA.
- A large uncertainty range in MSAT results does not automatically invalidate their use in comparing alternatives. Although the uncertainties associated with risk assessment often are larger than the estimated differences in travel activity due to the project, both the MSAT analysis and the travel forecast assume implicitly that all of the factors not included in the modeling – i.e., most of the real world – will remain constant. In other words, all of these analyses are done on the basis of "all other things being equal". If this condition is accepted for the project analysis then relative (not absolute) differences among alternatives, when calculated by consistent methodology, are generally valid for purposes of ranking alternatives in NEPA.[6]
- Many MSAT analyses show a decline in emissions over time regardless of the project, and the difference between future alternatives is typically much less than the overall secular reduction. A large overall reduction does not necessarily guarantee that the impacts of the project itself will be insignificant. Thus, this result does not relieve the agency of its responsibilities under NEPA to characterize the MSAT differences among the project alternatives if MSAT impacts are significant, even in the presence of uncertainty (40 CFR 1502.22. See also the supplementary discussion of this regulation at 51 FR 80:15620, April 25, 1986.) Even if the agency judges MSAT impacts not to be significant the agency should, as is consistent with the intent of NEPA, compare the MSAT impacts of the project alternatives and disclose the results for purposes of agency decision-making and public information.

6. Communication of MSAT Analysis Results and Health Impacts

Considerable literature exists on how to communicate health risk assessment results to the public (see http://www.epa.gov/ecocommunity/bib.htm); the points below are limited to risk analyses of priority MSATs for transportation projects under NEPA. These points also apply in concept to MSAT analyses that do not include a full HRA.

- Describe the project emission sources, the relevant MSATs, and the types of cancer and non-cancer health risks they pose.
- Define clearly the criteria for a significant impact of the project in the NEPA context.
- Explain and reference the information on toxicity, exposure, and dose-response that the transportation agency takes as given for purposes of the analysis.
- Identify and explain any health studies undertaken in the project area that are relevant to the MSAT analysis.
- For each impact metric, provide a clear definition and show a comparison of the results for each project alternative and the selected criterion of significance.

---

[6] Consistent methodology does not necessarily increase reliability of results in the statistical sense, but does improve the predictive ability of the analysis to compare alternatives to each other for NEPA decision-making purposes.

- Compare project impacts to other exposure information, such as regional or county-level MSAT emission inventories and measured MSAT concentrations.
- Show MSAT assessment results in easily-understood graphic formats where possible.
- Provide a careful discussion of sources of uncertainty. Include the level of confidence in toxicity information, modeled concentrations, and exposure estimates. The uncertainty discussion must adequately support the agency's decision on what level of MSAT analysis to prepare.
- Provide benchmarks to assist the reader in putting the project-related risks into perspective. Commonly-used benchmarks include everyday risks and sources of hazard such as the probability of having an auto accident or being injured in a fall at home. Also, the NATA summaries (EPA 2000) provide information on overall cancer risk and cancer risk from radon for comparison purposes.

7. Conclusions

This study and the NCHRP report enables transportation agencies to effectively develop an approach to evaluate and communicate the impacts of MSATs emitted from surface transportation sources. The study provides a suggested approach that:

- Provides a health-based screening procedure that balances the level of detail, analytic rigor, and resource requirements with the likely magnitude and significance of project impacts;
- Helps transportation analysts choose procedures for applying air quality and emissions models and other technical methods in the analysis of MSAT impacts for NEPA;
- Provides advice for communicating MSAT impacts in the NEPA documents that is consistent with the limitations and uncertainties with current assessment tools;
- Promotes consistency among MSAT evaluation methods so that the relative impacts of roadway projects and programs can be compared;
- Assures that the quality of MSAT analysis for NEPA documents is sufficient to meet statutory and regulatory requirements, to support agency decision-making, and to adequately inform the public about the air toxic impacts of projects in the NEPA context.

Acknowledgement

This work was sponsored by the American Association of State Highway and Transportation Officials, in cooperation with the Federal Highway Administration, and was conducted in the National Cooperative Highway Research Program, which is administered by the Transportation Research Board of the National Research Council.

References

Michael, C., and Houk J., (2006). *FHWA Workshop on Project-Level Mobile Source Air Toxics.* Phoenix, AZ. March 7, 2006.

Michael, C., and Miller, T. (2005). "A Method for Evaluating Mobile Source Air Toxic Emissions Among Transportation Project Alternatives." Paper presented at *Air and Waste Management Association 98th Annual Conference and Exhibition,* Minneapolis, MN. June.

U.S. Environmental Protection Agency, Office of Emergency and Remedial Response. (1989). "Risk Assessment Guidance for Superfund, Volume I, Human Health Evaluation Manual (Part A), Interim Final." *EPA/540/1-89/002,* Washington, D.C. December. http://www.epa.gov/oswer/riskassessment/ragsa/index.htm

U.S. Environmental Protection Agency. (2001). "1996 NATA National Scale Air Toxics Assessment." EPA 2001,
http://www.epa.gov/ttn/atw/nata/natsaov.html

U.S. Environmental Protection Agency. (2001). "Control of Emissions of Hazardous Air Pollutants from Mobile Sources." *EPA 2001, 40 CFR Parts 80 and 86, Final Rule.* 66 FR 61:17230-17273. March.

U.S. Department of Transportation, Federal Aviation Administration. (2001). "LAX Master Plan EIS/EIR, Technical Report14a. Human Health Risk Assessment." Los Angeles, CA. January. http://www.laxmasterplan.org/docs/draft_eir_NE/T14a_LR.pdf.

U.S. Department of Transportation, Federal Aviation Administration, Office of Environment and Energy. (2003). "Select Resource Materials and Annotated Bibliography On The Topic of Hazardous Air Pollutants (HAPs) Associated With Aircraft, Airports, and Aviation." *Technical Directive Memorandum D01-010.* Washington, DC. July.

U.S. Department of Transportation, Federal Aviation Administration.(2004). "O'Hare Modernization Final Environmental Impact Statement, Appendix I, Hazardous Air Pollutant Discussion." Chicago, IL. July.
ftp://public-ftp.agl.faa.gov/ORD FEIS/Appendix I.pdf.

U.S. Department of Transportation, Federal Highway Administration.(2005) "Transportation-Related Air Toxics: Case Study Materials Related to US 95 in Nevada." *Prepared by Sonoma Technology, Inc.,* undated (*circa* 2004). http://www.fhwa.dot.gov/environment/airtoxic/casesty2.htm.

U.S. Department of Transportation, Federal Highway Administration. (2006). "Interim Guidance on Air Toxic Analysis in NEPA Documents." Memorandum from Cynthia Burbank to Division Administrators. February.
http://www.fhwa.dot.gov/environment/airtoxic/020306guidmem.htm.

James, G. J., Avol, E., Fred, L., Nino, K., Frank, G., John, P., Rob, M. (2005). *"Childhood Asthma and Exposure to Traffic and Nitrogen Dioxide." Epidemiology,* 16(6):737-743, November.

Jeffrey, H., and Claggett, M. (2006). *"Mobile Source Air Toxics Analysis of FHWA Projects." Paper presented at Air and Waste Management Association 99th Annual Conference and Exhibition,* New Orleans, LA. June.

ICF International LLC. *Final Report (2007)-NCHRP25-25 (Task 18) Analyzing, Documenting, and Communicating the Impacts of Mobile Source Air Toxic Emissions in the NEPA Process.* Prepared for National Cooperative Highway Research Program. Washington, DC. March.

Johns Hopkins School of Public Health. "Traffic Proximity and Density Related Health Effects: Summary of Recent Research". *Workshop on Traffic, Health, and Infrastructure Planning.* Baltimore, Maryland. February. http://www.jhsph.edu/RiskSciences/Research/TrafficProximity.html

Rob, M., Berhane, K. (2006). *" Traffic, Susceptibility, and Childhood Asthma." Environmental Health Perspectives* 114(5): 766-72, May.

South Coast Air Quality Management District. "Multiple Air Toxics Exposure Study in the South Coast Basin: MATES II." Final report (and appendices). Diamond Bar, CA. March. http://www.aqmd.gov/matesiidf/matestoc.htm.

Tang, T., Claggett, M., Byun, J., Roberts, M., Granell, J., and Aspy, D. (2003). "MOBILE6.2 Air Toxic Trend and Sensitivity Analysis." *U.S. Department of Transportation, Federal Highway Administration, Resource Center.*

Winer, A. (2005) "Transportation-Related Air Pollutant Exposure: Implications for Regional Policies and Public Health." Paper presented at *Transportation/Land-Use/Environment Symposium,* Lake Arrowhead, CA. October.

Zhu, Y., Hinds, W., Kim, S., and Sioutas, C. (2002). "Concentration and Size Distribution of Ultrafine Particles Near a Major Highway." *Journal of the Air and Waste Management Association,* 52:1032-1042. September.

Traffic Emission Impacts on Air Quality near Large Roadways

R. Baldauf,[1] K. Black,[2] V. Martinez,[2] M. Gaber,[2] E. Thomas,[1] and D. Costa[1]

[1] U.S. Environmental Protection Agency, Office of Research and Development, Research Triangle Park, North Carolina, USA
[2] Federal Highway Administration, Washington, D.C., USA

Abstract

A growing number of epidemiological studies conducted throughout the world have identified an increase in occurrence of adverse health effects for populations living near major roads. In addition, several air quality studies have identified increased concentration levels of certain pollutants near high volume roadways. However, questions remain on the spatial extent of elevated concentrations near the road, and the factors affecting concentration variability, including traffic activity (volume, fleet mix, speed, etc.), environmental conditions, and land development. In the U.S., federal agencies have been under increasing pressure to evaluate potential air quality and health impacts related to transportation systems.

This presentation provides a summary of health effect and air quality studies that have been conducted near large roadways. The presentation also describes research plans being proposed by the Federal Highway Administration (FHWA) and the U.S. Environmental Protection Agency (EPA) to address future research projects related to the FHWA's US95 settlement agreement. These studies will measure criteria and air toxic pollutant concentrations near highways, including benzene, 1,3-butadiene, acetaldehyde, acrolein, formaldehyde, and particulate matter. In addition, meteorological and highway vehicle data will be collected to identify the relationship of these factors to the ambient air concentrations. The study will be conducted at three different geographic sites in the United States to investigate regional (climate, fuel, vehicle fleet differences), seasonal, and land use effects on near road pollutant impacts. Preliminary results from a pilot study conducted in Raleigh, North Carolina during the summer of 2006 will also be presented. This information will be presented to provide information on the impacts of vehicle emissions on near road air quality, as well as the development of control strategies to mitigate the influence of vehicle emissions on the near road microenvironment.

Introduction

In recent years, a large number of epidemiological studies have examined associations between living near major roads and different adverse health endpoints. These studies indicate that populations living, working or going to school near major roads may be at increased risk for a number of adverse health effects. Air quality monitoring studies have measured elevated concentrations of pollutants emitted directly by motor vehicles near large roadways, when compared to overall urban background concentrations. These elevated concentrations generally occur within a few hundred meters of the road, although this distance may vary depending on traffic patterns, environmental conditions, topography, and the presence of near roadway urban structures.

Many of the health studies examining near road populations employed non-specific metrics to represent exposures to traffic generated pollutants. Typical exposure metrics include distance to the nearest road, traffic volume on the nearest roads (usually as the average annual daily traffic (AADT) count), summary measure of total traffic density within a specified area around the residence/school, and measured outdoor concentrations (typically nitrogen dioxide (NO_2) although some studies have used particulate matter (PM) and black carbon (BC) measurements). A few studies have relied upon general air quality dispersion modelling of a limited number of traffic-generated chemicals or exposure assignments based on land use regression.

The following sections summarize studies evaluating health risks to near road populations and air quality variability in the near road microenvironment. These discussions focus on work conducted in the United States; however, a significant amount of work has been conducted in Europe and Asia as well.

Health Studies

A wide range of adverse health impacts have been reported for populations living near large roadways. Studies have ranged over time periods with fleets containing varying motor vehicle emission control technologies. Respiratory effects have been associated with residential and school proximity to major roads (Brauer et al., 2002; Brunekreef et al.,1997; Buckeridge et al., 2002; Delfino, 2002; English et al., 1999; Garshick et al., 2003; Gauderman et al., 2005; Heinrich et al., 2005; Janssen et al., 2003; Kim et al., 2002; McConnell et al., 2006; Varner, 2001). Adverse outcomes identified in these studies include chronic bronchitis, hospital admissions for respiratory causes, emergency room visits for asthma, acute bronchitis, upper and lower respiratory symptoms, and exacerbation of asthma. In addition, studies in Europe, Asia and North America have found increased risk of respiratory symptoms such as wheeze, cough, chronic phlegm production, and shortness of breath in children and adults with increased residential proximity to roadways and/or associated with increased local traffic density. Most of these studies were cross-sectional and relied on questionnaire assessments of health outcomes, in combination with simple exposure indicators. Although this section focuses on studies conducted in the United States, European studies reach similar conclusions.

Premature mortality from all-cause and cardiopulmonary effects for near road populations has been investigated in Europe and North America (Finkelstein, 2003; Finkelstein et al., 2004; Hoek et al., 2002; Jerrett, et al., 2005; Laden et al., 2000). These studies found increased mortality effects for populations living near a major road, typically with a residence within 100 meters of a large highway or within 50 meters of a major arterial roadway. No difference in response was found among those with pre-existing respiratory illness.

Studies on cardiovascular effects have focused on both short-term variations in exposure, as well as long-term residential history. There are also stressors in the roadway microenvironment (e.g., noise, anxiety) that may have an impact on cardiovascular effects that could not be accounted for in these studies. A study from Germany found that heart attack victims were more likely to be in a car, in transit, or on a bicycle in the hour prior to the event (Peters et al., 2004). This study accounted for potential confounding factors such as smoking, gender and time of day. Although the study did not specifically account for stress, similar results for varying modes of transportation may suggest that stress is not a predominant cause of the effects. A study of healthy young police officers in the United States found in-vehicle $PM_{2.5}$ concentrations to be associated with altered heart rate variability, an indicator of cardiac stress. In-vehicle pollutant concentrations were also associated with increased concentrations of factors in the blood associated with long-term cardiac risk (Riediker et al., 2004). Studies examining birth outcomes in populations living near major roads have found evidence of low birth weight, preterm birth, reduced head circumference, and heart defects among children of mothers living in proximity to heavy traffic (Gehring et al., 2002; Mannes et al., 2005; Miyake et al., 2002; Ritz and Yu, 1999; Ritz et al., 2000; Wilhelm and Ritz, 2003).

Several studies suggest that residential proximity to major roads increases cancer risk for adults and children although evidence is equivocal. Several studies suggest the potential for elevated risk of childhood leukemia associated with living near roads with higher traffic volumes, with the strongest associations for roads with at least 10,000 vehicles per day (Harrison et al., 1999; Pearson et al., 2000; Knox et al., 1997). However, other studies do not show positive associations for cancer and roadway proximity.

Air Quality Studies

Motor vehicles influence the temporal and spatial patterns of regulated gases, particulate matter (PM), and air toxic pollutant concentrations within urban areas. Emissions from motor vehicle operations may lead to elevated concentrations of particular air pollutants near major roads. Air quality monitoring studies have detected elevated levels of carbon monoxide (CO); nitric oxides (NO_x); nitrogen dioxide (NO_2); coarse ($PM_{10-2.5}$), fine ($PM_{2.5}$), and ultrafine ($PM_{0.1}$) particle mass;

black carbon (BC), polycyclic aromatic hydrocarbons (PAHs), and benzene near large roadways, as compared to overall urban background levels (Duci et al., 2003; Gilbert et al., 2003; Giugliano et al., 2005; Harrison et al., 2003; Kim et al., 2004; Reponen et al., 2003; Sapkota and Buckley, 2003; Weisel et al., 2005). Several recent studies have focused on PM measurements, identifying elevated ultrafine particle number concentrations in proximity to the roadway (Zhu et al, 2002a; Zhu et al., 2002b; Reponen et al., 2003). Other studies have noted that PM concentrations in the ultrafine, fine, and coarse modes may be elevated near major roads (Hutchins et al., 2000; Kittelson et al., 2004; Zhang et al., 2004). These studies suggest that elevated concentrations occur within several hundred meters of the road, although the extent of elevated concentrations will likely vary depending on traffic conditions, environmental conditions, topography, and structures present near the road. In addition to air pollutants, the presence of non-specific stressors (i.e. noise, socioeconomic status, etc.) may be increased for populations living near roads, which adds complexity to the assessment of near road air quality concentrations to adverse health effects.

Federal Research Plans

The Federal Highway Administration (FHWA) and Environmental Protection Agency (EPA) plan to collaborate on several studies addressing the near road issue. In June of 2005, the FHWA entered into an agreement with the Sierra Club to study the air quality concentrations of several air toxic compounds emitted by motor vehicles in proximity to roadways. These studies complement research needs identified by the FHWA and EPA to assess the relationship of traffic activity and environmental conditions on near road air quality. These parties have agreed to combine resources to implement a comprehensive evaluation of near road air quality.

The studies will measure pollutant concentrations of criteria gases, particulate matter, and air toxics at several distances away from a major roadway, as well as at a background site on the opposite side of the road. The air toxics identified for this study include benzene, 1,3-butadiene, acetaldehyde, acrolein, formaldehyde, and particulate matter based on the pollutants identified in EPA's Mobile Source Air Toxics (MSAT) rule of 2001 as national and regional risk drivers. In addition to the air quality data, highway vehicle activity information will be collected to identify the volume, speed, and vehicle fleet mixes present on the roadway to associate with the ambient air pollutant concentrations. Meteorological monitoring for wind speed, direction, temperature and humidity will also be conducted to determine the effect of varying environmental conditions on the air quality concentrations.

Measurements will be collected for one year in three cities in the United States. The three cities represent different geographic sites to evaluate seasonal, climatic, and traffic fleet/fuel variability that may affect near road pollutant concentrations and gradients. The first city will be Las Vegas, Nevada, followed by a year long study in Detroit, Michigan. The third study will be conducted on the east coast, with Raleigh, North Carolina as the tentative location. As part of the FHWA settlement, a protocol was developed outlining the methods and time periods to evaluate the traffic, meteorology, and air quality parameters to be assessed. In addition, the EPA conducted a pilot study in Raleigh, North Carolina to evaluate these methods, as well as additional monitoring techniques. Objectives of this study included the assessment of the relationship between vehicle activity and environmental conditions to air pollutant concentrations in the near road microenvironment, as well as the evaluation of existing emissions and dispersion models to characterize near road air quality. These models can be used by urban planners, transportation and environmental agencies, and health professionals to assess the impacts of traffic emissions on air quality and, ultimately, public health.

Disclaimer

This article has been reviewed by the Office of Research & Development, U.S. Environmental Protection Agency, and approved for publication. Approval does not signify that the contents necessarily reflect the views and policies of the Agency nor does mention of trade names or commercial products constitute endorsement or recommendation for use.

References

Brauer, M., Hoek, G., Van-Vliet, P., Meliefste, K., Fishcer, P.H., Wijga, A., Koopman, L.P., Neijens, H.J., Gerritsen, J., Kerkhof, M., Heinrich, J., Bellander, T., Brunekreef, B. (2002). "Air pollution from traffic and the development of respiratory infections and asthmatic and allergic symptoms in children." *American Journal of Respiratory and Critical Care Medicine.* 166(8), 1092-1098.

Brunekreef, B., Janssen, N.A.H., de Hartog, J., Harssema, H., Knape, M., van Vliet, P. (1997). "Air pollution from truck traffic and lung function in children living near motorways." *Epidemiology,* 8(3), 298-303.

Buckeridge, D.L., Glazier, R., Harvey, B.J., Escobar, M., Amrhein, C., Frank, J. (2002). "Effect of motor vehicle emissions on respiratory health in an urban area." *Environmental Health Perspectives,* 110(3), 293-300.

Delfino, R.J. (2002). "Epidemiologic evidence for asthma and exposure to air toxics: linkages between occupational, indoor, and community air pollution research." *Environmental Health Perspectives*, 110(Supp 4), 573-589.

Duci, A., Chaloulakou, A., Spyrellis, N. (2003). "Exposure to carbon monoxide in the Athens urban area during commuting." *The Science of the total environment,* 309, 47-58.

English, P., Neutra, R., Scalf, R., Sullivan, M., Waller, L., Zhu, L. (1999). "Examining associations between childhood asthma and traffic flow using a geographic information system." *Environmental Health Perspectives*, 107(9), 761-767.

Finkelstein, M.M. (2003). "Mortality and indicators of traffic-related air pollution." Lancet 361(9355): 430; author reply 430.

Finkelstein, M.M., Jerrett, M., Sears, M.R. (2004). "Traffic air pollution and mortality rate advancement periods." *American Journal of Epidemiology*, 160, 173-177.

Garshick, E., Laden, F., Hart, J.E., Caron, A. (2003). "Residence near a major road and respiratory symptoms in U.S. Veterans." *Epidemiology*, 14(6), 728-736.

Gauderman, W.J., Avol, E., Lurmann, F., Kuenzli, N., Gilliland, F., Peters, J., McConnell, R. (2005). "Childhood asthma and exposure to traffic and nitrogen dioxide." *Epidemiology*, 16(6), 737-743.

Gehring, U., Cyrys, J., Sedlmeir, G., Brunedreef, B., Belander, T., Fischer, T., Bauer, C.P., Reinhardt, D., Wichmann, H.E., Heinrich, J. (2002). "Traffic-related air pollution and respiratory health during the first 2 yrs of life." *European Respiratory Journal*, 19(4), 690-698.

Gilbert, N.L., Woodhouse, S., Stieb, D.M., Brook, J.M. (2003). "Ambient nitrogen dioxide and distance from a major highway." *The Science of the total environment*, 312(1-3), 43-46.

Giugliano, M., Lonati, G., Butelli, P., Romele, L., Tardivo, R., Grosso, M. (2005). "Fine particulate (PM2.5-PM1) at urban sites with different traffic exposure." *Atmospheric Environment*, 39: 2421-2431.

Harrison, R.M., Leung, P.L., Somervaille, L. (1999). "Analysis of incidence of childhood cancer in the West Midlands of the United Kingdom in relation to proximity of main roads and petrol stations." *Occupational and Environmental Medicine,* 56, 774-780.

Harrison, R.M., Tilling, R., Callen Romero, M.S., Harrad, S., Jarvis, K. (2003). "A study of trace metals and polycyclic aromatic hydrocarbons in the roadside environment." *Atmospheric Environment*, 37, 2391-2402.

Heinrich, J., Topp, R., Gehring, U., Thefeld, W. (2005). "Traffic at residential address, respiratory health, and atopy in adults: the National German Health Survey 1998." *Environmental Research*, 98, 240-249.

Hitchins, J., Morawska, L., Wolff, L., Gilbert, D. (2000). "Concentration of submicrometer particles from vehicle emissions near a major road." *Atmospheric Environment*, 34, 51–59.

Hoek, G., Brunekreef, B., Goldbohm, S., Fischer, P., van den Brandt, P.A. (2002). "Association between mortality and indicators of traffic-related air pollution in the Netherlands: a cohort study." Lancet 360(9341), 1203-1209.

Janssen, N.A.H., Brunekreef, B., van Vliet, P., Aarts, F., Meliefste, K., Harssema, H., Fishcer, P. (2003). "The relationship between air pollution from heavy traffic and allergic sensitization, bronchial hyper responsiveness and respiratory symptoms in Dutch school children." *Environmental Health Perspectives*, 111(12), 1512-1518.

Jerrett, M., Burnett, R., Pope, C.A., Krewski, D., Bruce, K., Thurston, G., Shi, Y., Finklestein, N., Calle, E.E., Thun, M. J. (2005). "Spatial analysis of air pollution and mortality in Los Angeles." *Epidemiology,* 16(6).

Kim, J.J., Smorodinsky, S., Ostro, B., Lipsett, M., Singer, B.C., Hogdson, A.T. (2002). "Traffic-related air pollution and respiratory health: the East Bay Children's Respiratory Health Study." *Epidemiology*, 13(4), S100.

Kim, J.J., Smorodinsky, S., Lipsett, M., Singer, B.C., Hogdson, A.T., Ostro, B. (2004). "Traffic-related Air Pollution Near Busy Roads: The East Bay Children's Respiratory Health Study." *American Journal of Respiratory and Critical Care Medicine*, 170(5), 520-526.

Kittelson, D.B., Watts, W.F., Johnson, J.P. (2004). "Nanoparticle emissions on Minnesota highways." *Atmospheric Environment*, 38(1), 9-19.

Knox, E.G., Gilman, E.A. (1997). "Hazard proximities of childhood cancers in Great Britain from 1953-1980." *Journal of Epidemiology and Community Health*, 51, 151-159.

Laden, F., Neas, L.M., Dockery, D.W., Schwartz, J. (2000). "Association of fine particulate matter from different sources with daily mortality in six U.S. cities." *Environmental Health Perspectives*, 108, 941-947.

Mannes, T., Jalaludin, B., Morgan, G., Lincoln, D., Sheppeard, V., Corbett, S. (2005). Impact of ambient air pollution on birth weight in Sydney, Australia. Occupational and Environmental Medicine 62, 524-530.

McConnell, R., Berhane, K., Yao, L., Jerrett, M., Lurmann, F., Gilliland, F., Kuenzli, N., Gauderman, J., Avol, E., Thomas, D., Peters, J. (2006). "Traffic, Susceptibility, and Childhood Asthma." *Environmental Health Perspectives*, 114(5), 766-772.

Miyake, Y., Yura, A., Iki, M. (2002). "Relationship between distance from major roads and adolescent health in Japan." *Journal of Epidemiology*, 12(6), 418-423.

Pearson, R.L., Wachtel, H., Ebi, L. (2000). "Distance-weighted traffic density in proximity to a home is a risk factor for leukemia and other childhood cancers." *Journal of the Air & Waste Management Association*, 50, 175-180.

Peters, A., von Klot, S., Heier, M., Trentinaglia, I., Hormann, A., Wichmann, E., Lowel, H. (2004). "Exposure to traffic and the onset of myocardial infarction." *New England Journal of Medicine*, 351, 1721-1730.

Reponen, T., Grinshpun, S.A., Trakumas, S., Martuzevicius, D., Wang, Z.M., LeMasters, G., Lockey, J.E., Biswas, P. (2003). "Concentration gradient patterns of aerosol particles near interstate highways in the Greater Cincinnati airshed." *Journal of Environmental Monitoring*, 5(4), 557-562.

Riediker, M., Cascio, W.E., Griggs, T.R., Herbst, M.C., Bromber, P.A., Neas, L., Williams, R., Devlin, R.B. (2004). "Particulate matter exposure in cars is associated with cardiovascular effects in healthy, young men." *American Journal of Respiratory and Critical Care Medicine* 169(8), 934-940.

Ritz, B., Yu, F. (1999). "The effect of ambient carbon monoxide on low birth weight among children born in southern California between 1989 and 1993." *Environmental Health Perspectives*, 107, 17-25.

Ritz, B., Yu, F., Chapa, G., Fruin, S. (2000). "Effect of air pollution on preterm birth among children born in Southern California between 1989 and 1993." *Epidemiology*, 11(5), 502-511.

Sapkota, A., Buckley, T.J. (2003). "The mobile source effect on curbside 1,3-butadiene, benzene, and particle-bound polycyclic aromatic hydrocarbons assessed at a tollbooth." *Journal of the Air & Waste Management Association* 53, 740-748.

Varner, A.E. (2001). "Automobile traffic, atmospheric pollution, and childhood asthma." *Journal of the American Medical Association*, 285(21), 2712-2713.

Wilhelm, M., Ritz, B. (2003). "Residential proximity to traffic and adverse birth outcomes in Los Angeles County, California, 1994-1996." *Environmental Health Perspectives*, 111, 207-216.

Zhang, K.M.; Wexler, A.S.; Zhu, Y.F.; Hinds, W.C.; Sioutas, C. (2004). "Evolution of particle number distribution near roadways. Part II: the 'Road-to-Ambient' process." *Atmos Environ*, 38: 6655-6665.

Zhu, Y., Hinds, W.C., Kim, S.K., Shen, S., Sioutas, C. (2002). "Study of ultrafine particles near a major highway with heavy-duty diesel traffic." *Atmospheric Environment*, 36, 4323-4335.

Zhu, Y., Hinds, W.C., Kim, S.K., Sioutas, C. (2002). "Concentration and size distribution of ultrafine particles near a major highway." *Journal of the Air & Waste Management Association*, 52, 1032-1042.

Hazardous Air Pollutants Benefits Assessment: Houston-Galveston Case Study

J. H. Wilson Jr.,[1] M. A. Mullen,[2] D. McKenzie,[2] K. Thesing,[2]
A. Bollman,[2] H. A. Roman[3] and J. A. Craig[4]

[1]E.H. Pechan & Associates, Inc., 5528-B Hempstead Way, Springfield, VA 22151; PH (703) 813-6700; email: jim.wilson@pechan.com
[2]E.H. Pechan & Associates, Inc., 5528-B Hempstead Way, Springfield, VA 22151.
[3]Industrial Economics, Inc., Cambridge, MA.
[4]U.S. Environmental Protection Agency, Office of Air and Radiation, Washington, DC.

Introduction

Section 812 of the Clean Air Act Amendments of 1990 (CAAA) requires the U.S. Environmental Protection Agency (EPA) to perform periodic, comprehensive analyses of the total costs and total benefits of programs implemented pursuant to the Clean Air Act (CAA). The first analysis required was a retrospective analysis, addressing the original CAA and covering the period 1970 to 1990. The retrospective was completed in 1997. Section 812 also requires performance of prospective cost-benefit analyses, the first of which was completed in 1999. The prospective analyses address the incremental costs and benefits of the CAAA. The first prospective covered implementation of the CAAA over the period 1990 to 2010.

EPA's Office of Air and Radiation (OAR) is now working on a second prospective study, looking at the period from 1990 to 2020. The analytical plan was reviewed by a statutorily-mandated outside peer review group, the EPA Science Advisory Board Advisory Council for Clean Air Compliance Analysis (SAB Council), and the SAB Council provided comments, which have been incorporated into the technical analysis planning.

The purpose of this paper is to describe the development of a benzene emissions inventory in the Houston, Texas, area as part of a case study for the second prospective study. After the first prospective 812 study, the SAB Council encouraged EPA to include a hazardous air pollutant (HAP) benefits case study in future efforts to help address limitations in our ability to estimate benefits associated with HAP controls under the CAA.

They advised EPA to select a well-studied, representative HAP for which to perform a prototype 812 analysis. The SAB recommended benzene in part because of the wealth of available national ambient concentration data from monitors. The SAB felt that an 812 analysis using the available benzene data would:

- Identify limitations and gaps in the database;
- Provide an estimate of the uncertainties in the analyses and perhaps provide a reasonable lower bound on potential health benefits from control; and
- Provide a scientific basis for deciding whether there is merit in pursuing a greater ability to assess the benefits of air toxics.

EPA's response to these SAB comments was to undertake a metropolitan scale analysis of the benefits of Clean Air Act Amendments of 1990 (CAAA) controls on benzene emissions. This scale allows both a more rigorous analytical effort and the opportunity to build on previous EPA modeling efforts for benzene.

The Houston-Galveston area was selected for the case study. The geographic boundary of the study area includes Brazoria, Galveston, and Harris Counties. According to the 1999 EPA National Emission Inventory (NEI) emission estimates for the Houston-Galveston metropolitan area, these three counties contained 99 percent of the metropolitan area's point source benzene emissions.

As with the main criteria pollutant portion of the section 812 analysis, the analysis years are 1990, 2000, 2010, and 2020. Both *with-* and *without-CAAA* scenarios are evaluated for 2000, 2010, and 2020. The base year is 1990 because that is the year when the CAAA were enacted.

This paper describes the methods for developing emissions inventories for the case study and is organized by four major sectors: (1) on-road vehicles, (2) nonroad engines/vehicles, (3) point sources, and (4) nonpoint sources. (Nonpoint sources were formerly called area sources, or area wide sources.) The emission estimation methods and results for each sector are described in separate sections. The nonroad engines/vehicles sector typically includes emissions from commercial marine vessels, aircraft, and locomotives. In this paper, those sources are included in the nonpoint source analysis, and the non-road sector discussion is limited to the engines and vehicles that are included in EPA's NONROAD model.

Figure 1 provides a map of the study area, with the county boundaries outlined for Brazoria, Galveston, and Harris counties.

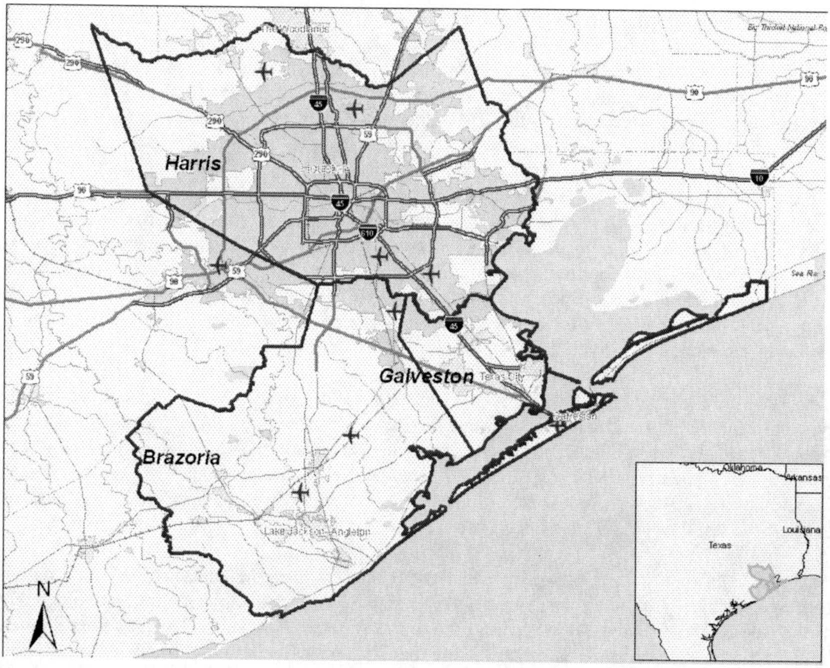

Figure 1. Houston-Galveston Benzene Case Study Area

Methods

- **Onroad Vehicles**

Onroad vehicle emissions were estimated for the 8-county Houston metropolitan area, though ultimately only emissions data for Harris, Brazoria, and Galveston counties were used in the case study analysis. Because the ultimate purpose of this emission inventory was for use in the AERMOD modeling system, it was determined that link-level (i.e., roadway segment) emissions data by hour of the day for each season was the appropriate level-of-detail for this important source type. Vehicle miles traveled (VMT) information was based on data files prepared for the Houston area by the Houston-Galveston Area Council and further processed by the Texas Transportation Institute (TTI). However, adjustments to these VMT estimates were needed to meet the temporal needs of this project. Emission factors were computed using EPA's MOBILE6.2 emission factor model. Where available, local input data for the Houston area, as provided by the Texas Commission on Environmental Quality (TCEQ), were used in MOBILE6.2 input file development. The sections below describe VMT data file adjustments and MOBILE6.2 input file preparation.

TTI provided link-level hourly VMT data for the years 2002, 2009, and 2012. The data included the eight counties in the Houston nonattainment area. Because the data were developed for ozone modeling, the link-level data files had been developed specific to an August/September episode. The data were supplied for four day types: weekday, Friday, Saturday, and Sunday. Several adjustments were made to the VMT data to make it appropriate for use in the Section 812 benzene study.

First, the VMT data were adjusted from the specific modeling period to the four seasons. TTI provided adjustment factors to be used with the data set to adjust the VMT from the episode-specific time period to each of the four seasons. These seasonal adjustment factors were provided for each of the four day types, but did not differ by year. Four new seasonal data sets were prepared from each of the episode-specific VMT data sets for 2002, 2009 and 2012 by multiplying these adjustment factors by the VMT in the episode-specific VMT data sets. These new data sets were still at the link and hourly level.

Next, the VMT data were allocated by vehicle type. TTI provided VMT fractions by vehicle type corresponding to each data set. The VMT fractions differed by time of day (AM peak, PM peak, overnight, and midday) and road class (freeways, arterials, and collectors). These data were matched appropriately to the hour and functional class of each link VMT and then each hourly, link-level VMT record was multiplied by each of the corresponding VMT mix fractions to break the link-level VMT data out by vehicle type.

After the VMT data were allocated by vehicle type, the seasonal VMT data sets were adjusted to combine the VMT from the four-day types into a single VMT value representing an average day of the week. This was done by weighting the four different day type VMT values from a given link, hour, vehicle type, and season by the number of that type of day that occurs in that season and year. For example, in winter 2002 there were 53 weekdays (excluding Fridays), 12 Fridays, 12 Saturdays, and 13 Sundays, for a total of 90 winter days. Thus, the following equation was used to calculate the VMT from a given link, vehicle type, and hour in winter 2002 for an average day:

$$VMT(avgday) = (VMT(wkd)*53 + VMT(Fri)*12 + VMT(Sat)*12 + VMT(Sun)*13)/90$$

The hourly, link-level speed values were similarly weighted to obtain hourly, link-level speeds for an average day.

The final step in preparing VMT that could be used to calculate on-road emissions for Houston was to adjust the VMT to the study years. To obtain the 2000 VMT, the 2002 and 2009 link-level VMT data were matched by link, hour, vehicle type, and season. The average annual growth rate from 2002 to 2009 was then calculated from each set of matching records. The 2002 VMT data were then extrapolated back by two years using this average annual growth rate to estimate 2000 VMT. The link network from the 2002 data set was assumed to be the same as the 2000 link network. Thus, links included in the 2009 data file, but not in the 2002 data file were not included in the 2000 data file. Similarly, the 2010 VMT was calculated using the 2009 and 2012 link-level data files. Again, the average annual growth rate was calculated between these two years for the VMT values matched on each link, hour, vehicle type, and season. One year of this average annual growth rate was applied to the 2009 link-level VMT data set to obtain a 2010 link-level VMT data set. The 2010 link network was assumed to be the same as the 2009 link network. The 2020 VMT data set was calculated based on extrapolating the 2002 and 2012 VMT data. The average annual growth rate was calculated between these two years for the VMT values matched on each link, hour, vehicle type, and season. Eight years of this average annual growth rate were applied to the 2012 link-level VMT data set to obtain a 2020 link-level VMT data set. The 2012 link network was used in 2020.

VMT data for 1990 were not available at the link level. However, TCEQ provided weekday (Monday through Friday) VMT summaries for the 8-county Houston area, based on data included in rate-of-progress plans. A comparable data summary was prepared from the 2002 episode-specific VMT data, after weighting the 2002 weekday VMT (multiplied by 4/5) and the 2002 Friday VMT (multiplied by 1/5) to estimate 2002 average weekday VMT. Backcasting factors were then calculated from 2002 to 1990 by county and functional road class (e.g., 1990 VMT divided by 2002 VMT). These backcasting factors were then matched to the 2002 episode-specific VMT data and multiplied by the corresponding 2002 VMT values to obtain a 1990 episode-specific hourly, link-level VMT file for each of the four day types in the 2002 link-level VMT file. From this point, the adjustments applied to the 1990 VMT were similar to those applied in the other years. First, the VMT were allocated to the four seasons, using seasonal VMT adjustment factors. Next, the VMT were allocated by vehicle type, using 1990 vehicle mix fractions by time of day and facility type, specific to 1990, which were also provided by TCEQ. These vehicle-type specific VMT data were then aggregated to obtain an average day VMT for each hour, link, season, and vehicle type. The 2002 link network was also used for 1990.

In this analysis, speeds were calculated by weighting them by VMT when the VMT data were adjusted. If resources permitted, it would have been preferable to recalculate the link speeds based on the VMT volumes in the seasonal, average day VMT files.

- **Emission Factor Inputs**

Registration Distributions

The default MOBILE6.2 registration distribution of vehicles by age was used in 1990. In 2000, for both the *with-CAAA* and the *without-CAAA* scenarios, three different registration distribution files were used. One file was specific to Harris County. Brazoria, Fort Bend, Galveston, and Montgomery Counties were all modeled with the same urban 2002 registration data file. Chambers, Liberty, and Waller Counties were modeled with a 2002 rural registration distribution file. In both 2010 scenarios, the registration distributions varied for each county. All of these registration distribution files were downloaded from TCEQ's web site. The 2010 registration distributions were also used in 2020.

Diesel Sales Fractions

The default MOBILE6.2 diesel sales fractions were used in 1990. For both scenarios in 2000, Statewide diesel sales fraction estimates were used in all counties. Similarly in 2010 and 2020,

Statewide diesel sales fraction estimates were applied in all counties in both scenarios. The diesel sales data were available from TCEQ's web site.

Inspection and Maintenance (I/M) and Anti-Tampering Program (ATP) Inputs

There was no I/M program operating in the Houston area in 1990. However, Harris County did have an ATP implemented in 1990. Therefore, only the Harris County ATP was modeled in 1990. Since the *without-CAAA* scenarios for 2000 and 2010 represent emissions without CAAA controls, only the Harris County ATP was modeled in those scenarios, with no I/M programs in any county. In the 2000 *with-CAAA* scenario, an I/M program was modeled in all eight Houston area counties. However, the program inputs for Harris County differed from those of the remaining 7 counties. In addition, only Harris County was modeled with an ATP in the 2000 *with-CAAA* scenario. By 2010 in the *with-CAAA* scenario, all eight counties were modeled with both I/M programs and ATPs. However, the program inputs varied for Harris County, the urban counties, and the rural counties. Again, all I/M program data were available from the TCEQ web site. The 2020 data used were the same as the 2010 data.

Temperature Inputs

Since hourly emissions were the desired end product, hourly temperature input data were used. For both 2000 scenarios, hourly temperature and humidity inputs by month were used. These were obtained from the national county database associated with EPA's National Mobile Inventory Model (NMIM). For the 2010 and 2020 scenarios, long-term average hourly temperature and humidity inputs for each month were obtained from the same source. Both the 2000 and long-term average temperature and humidity data sets varied by county. For 1990, an hourly temperature data set for Harris County from the TCEQ web site was used to develop 1990 hourly temperature inputs by month. This same data set was applied to all counties. The MOBILE6.2 default humidity inputs were used in 1990.

Fuel Inputs

Monthly fuel inputs, based on 2000 fuel data, from the NMIM national county database were used as the basis of the 2000 and 2010 *with-CAAA* fuel inputs. The 2010 and 2020 fuel data were the same as the 2000 fuel data, but with gasoline sulfur levels adjusted to account for the low sulfur gasoline provisions that began taking effect in 2004. For 1990, and the 2000 and 2010 *without-CAAA* scenarios, July and January 1990 historical fuel data, used in EPA's 1990 NEI, were applied. The July fuel data was applied in months May through September, while the January data were applied in the other (winter season) months. The fuel inputs for all years included Reid vapor pressure (RVP), as well as information on the oxygen content of the fuel, the benzene content, and additional properties needed to model toxic air pollutants in MOBILE6.2.

Development of MOBILE6.2 Input Files

Using the inputs discussed above, MOBILE6.2 input files were prepared for 1990, and the *without-CAAA* and *with-CAAA* 2000, 2010, and 2020 scenarios. MOBILE6.2 scenarios were developed by hour, using the hourly temperature and humidity inputs. In all years and scenarios, a set of emission factors were developed for the MOBILE6.2 freeway and arterial road categories in 5 mile per hour increments. Additionally, a single scenario representing ramps and one representing local roads were also modeled. (Emission factors for these two road categories do not vary by speed in MOBILE6.2.) In addition to modeling the 1990 fuel characteristics and the 1990 Harris County ATP, the 2000, 2010, and 2020 *without-CAAA* input files included the "NO CAAA" command to exclude the effects of national CAAA programs on the emission factors. For the *with-CAAA* scenarios, this command was removed and the actual I/M programs and ATPs discussed above and the fuel programs appropriate to the year being modeled were included.

- **Point Sources**

EPA's 1990 NEI for HAPs point source file, which was recently redeveloped by EPA, was used to estimate 1990 benzene emissions in the study area, and was used as the base year file for estimating *without-CAAA* scenario emissions for 2000, 2010, and 2020. The EPA 2002 NEI was used to estimate point source benzene emissions for the study area for the 2000 *with-CAAA* scenario, and as the base year emissions file for preparing 2010 and 2020 *with-CAAA* scenario emission estimates. The NEI covers all criteria air pollutants and HAPs for the United States. While the Texas Air Quality Study (TexAQS) study database received strong consideration as the base year data set, the TexAQS study provided hourly emission estimates for an August-September 2000 Houston area modeling episode. This database would have provided detailed information about the emissions during this period, but it would also have required adjustments to allow it to be used to represent emission conditions during other time periods. In addition, as a modeling database, it did not contain some of the control device information that assists in making emission forecasts to future years. Furthermore, the TCEQ submittal for the draft 2002 NEI was improved from previous years for many compounds because of the TCEQ work on species profile improvements. These species profile improvements occurred as part of the recent efforts in the Houston-Galveston area to develop improved point source emission estimates for ozone modeling. Individual point sources were surveyed in order to develop the improved compound-specific emission estimates.

- **Nonpoint Sources**

Nonpoint (area source) emissions were projected for both the *with-CAAA* scenario (2010 and 2020) and the *without-CAAA* scenario (2000, 2010, and 2020). The draft 2002 NEI was used as the base for the *with-CAAA* scenario, while the 1990 NEI for HAPs inventory was used as the base for the *without-CAAA* scenario. This newly revised inventory (released in November 2005) is available from the EPA. This inventory used as a baseline the 1990 NTI and a number of revisions have been made to it. The primary revisions included converting previously county-level emissions to point source level data wherever feasible. Additional estimates were developed for missing MACT source categories and HAPs, so that the baseline inventory was more comparable to the 1999 NEI and 2002 NEI.

- **Off-Road Engines/Vehicles**

Nonroad benzene emission estimates for the Houston area were developed by multiplying benzene speciation factors by VOC emissions output from EPA's NONROAD2004 model. The benzene speciation factors and fuel data inputs were obtained from EPA's National Mobile Inventory Model (NMIM), which was used to develop the final 2002 NEI nonroad estimates. NONROAD model equipment categories include: recreational vehicles, farm and construction machinery, lawn and garden equipment, aircraft and rail support equipment, and other industrial and commercial applications. NONROAD2004 (EPA, 2004a) incorporates all Federal engine standards, with the exception of the large spark-ignition evaporative standards. VOC reductions from this standard were applied outside of the NONROAD model.

A region-specific NONROAD growth file was used for the Houston-area model runs. Input files were prepared for Brazoria, Galveston, and Harris counties to reflect the appropriate temperature and fuel inputs for the *with-CAAA* scenario runs. In addition, fleet emission rate inputs were modified to remove the effect of CAAA-related standards for the *without-CAAA* runs. Using county-specific input files, NONROAD model runs were performed to generate seasonal emission estimates for each scenario year. Seasonal emissions were then summed to estimate annual emissions at the county and SCC level for each scenario/year.

Results

Table 1 summarizes the three-county benzene emission totals by year and scenario for each vehicle type. While this table shows that some benzene emission reductions would occur through 2010 without the additional provisions of the CAAA, the reductions obtained through the CAAA control measures are significant. From 1990, with the CAAA programs in place, on-road vehicle benzene emissions in Houston are reduced by 68 percent in 2000, by 86 percent in 2010, and by 88 percent in 2020. While the reductions from 1990 to 2020 with the CAAA programs in place are similar to the reductions achieved between 1990 and 2010, an 86 percent reduction in on-road benzene emissions is achieved between the *with-CAAA* and the *without-CAAA* scenarios in 2020, compared to a 77 percent reduction between emissions in these two scenarios in 2010. Thus, despite a significant increase in VMT from 2010 to 2020, benzene emissions continue on a downward trend through 2020 in the *with-CAAA* scenario.

Based on an interim set of runs for 2000 and 2010 that were performed with the MOBILE6.2 "NO CAAA" command, but with the fuel and I/M programs and ATP specific to either 2000 or 2010, it was determined that in 2000, about 96 percent of the benzene emission reductions that occur between the 2000 *without-CAAA* and the 2000 *with-CAAA* scenarios are attributable to changes in fuel program parameters and the I/M program, with the remaining reductions due to the phasing in of more stringent emission standards. By 2010, a greater portion of the emission reductions (29 percent) is attributable to greater phase-in of the emission standards.

Table 1. Annual On-road Benzene Emissions from Brazoria, Galveston, and Harris Counties (emissions in tpy)

Vehicle	1990	2000 *without-CAAA*	2000 *with-CAAA*	2010 *without-CAAA*	2010 *with-CAAA*	2020 *without-CAAA*	2020 *with-CAAA*
LDGV	1,335.09	889.04	421.80	831.73	180.82	1,133.47	148.76
LDGT1	642.58	504.70	254.44	474.18	103.16	645.71	90.05
LDGT2	205.73	88.13	46.27	104.57	22.96	160.55	25.91
HDGV	115.44	33.71	16.95	16.94	6.34	18.90	4.06
MC	1.95	2.82	1.57	2.53	1.54	3.25	1.97
LDDV	2.85	0.48	0.47	0.16	0.07	0.13	0.02
LDDT	3.83	0.60	0.55	0.52	0.22	0.62	0.12
HDDV2b	1.87	1.20	0.78	1.81	0.69	2.87	0.83
LHDDV	2.94	1.72	1.22	1.93	0.87	2.56	0.76
MHDDV	4.97	2.68	2.43	2.57	1.61	3.35	1.26
HHDDV	57.17	15.58	14.68	11.38	8.74	14.50	7.37
BUS	1.15	1.01	0.88	1.09	0.63	1.66	0.66
Total	**2,375.58**	**1,541.66**	**762.04**	**1,449.40**	**327.64**	**1,987.58**	**281.78**

LDGV = light-duty gasoline vehicle
LDGT1 = light-duty gasoline truck 1 (< 6,000 lbs GVWR)
LDGT2 = light-duty gasoline truck 2 (< 6,000-8,500 lbs GVWR)
HDGV = heavy-duty gasoline vehicle
MC = motorcycle
LDDV = light-duty diesel vehicle
LDDT = light-duty diesel truck
HDDV2b = heavy-duty diesel vehicle (8,501-10,000 lbs GVWR)
LHDDV = light heavy-duty diesel vehicle (10,001-19,500 lbs GVWR)
MHDDV = medium heavy-duty diesel vehicle (19,501-33,000 lbs GVWR)
HHDDV = high heavy-duty diesel vehicle (33,001 and above lbs GVWR)

It should be noted that EPA has recently determined that cold start hydrocarbon emissions have been underestimated in MOBILE6 at temperatures below 75 degrees Fahrenheit, based on engine certification data collected primarily from Tier 1 and more recent vehicles to comply with the carbon monoxide cold temperature standards. This would likely lead to a similar underestimation of benzene at cold temperatures. As a result, the benzene estimates from this analysis are likely to have been underestimated in the *with-CAAA* scenario. The effect in the Houston area would not be as significant as it would be in colder areas of the country.

A Mobile Source Air Toxics rule was proposed by EPA since this analysis was performed. The potential benzene emission reductions from this rule were not accounted for in this analysis. This rule would be expected to further reduce benzene emissions in the *with-CAAA* scenario.

Table 2 summarizes the Houston-Galveston study area benzene emission results by sector for the scenarios, sectors, and years evaluated. This table shows that there have been significant CAAA-attributable benzene emission reductions in the Houston-Galveston study area since 1990. A large fraction of these CAAA benefits occurred in the period between 1990 and 2000.

For the point and nonpoint source sectors, the benzene emission reductions during this 1990 to 2000 period are largely attributable to Federal maximum achievable control technology (MACT) emission standards, and local volatile organic compound (VOC) measures in the 1-hour ozone attainment plan. The most significant benzene emission reductions in these sectors are from the chemical manufacturing and petroleum refining industries. The emissions path for point and nonpoint benzene sources in the *with-CAAA* scenario is stable or slightly increasing through the 2000 to 2020 period.

Table 2 also shows that there have been significant CAAA-associated benefits in reducing on-road vehicle emissions in the 1990 to 2000 period. In the Houston-Galveston area, these benefits are primarily attributable to the Federal reformulated gasoline program as well as Federal emission standards that reduced exhaust and evaporative VOCs and benzene emissions. Tier 2 emission standards and associated requirements that lower the sulfur content of gasoline are a significant factor in achieving further benzene emission reductions from onroad vehicles between 2000 and 2020.

Table 2. Houston-Galveston Benzene Emissions Summary (tons per year [tpy])

Sector	1990	2000 *without-CAAA*	2000 *with-CAAA*	2010 *without-CAAA*	2010 *with-CAAA*	2020 *without-CAAA*	2020 *with-CAAA*
Point/Nonpoint	5,409	6,532	1,230	6,699	1,258	7,702	1,440
Nonroad	740	900	567	1,127	354	1,351	360
Onroad vehicles	2,375	1,541	762	1,449	328	1,988	282
Total	8,524	8,973	2,559	9,275	1,940	11,041	2,082

Federal nonroad emission standards also produce benzene emission reductions in the Houston-Galveston area. These emission reductions are less than those observed for on-road vehicles in part because the 1990 emissions are much lower, and because many nonroad emission standards were set after the Tier 1 on-road vehicle emission standards took place. Therefore, the biggest differences between *with-* and *without-CAAA* benzene emissions in the study area for nonroad engines/vehicles occur in 2010 and 2020.

The other important observation from Table 2 is that the relative importance of the four major sectors is changing with time as control programs reduce nonroad and on-road benzene emissions post-2000. On-road vehicles account for 28 percent of the 1990 benzene emissions in the study area. This sector's contribution to total benzene emissions stays about the same in 2000. However, on-road vehicle contributions drop to 17 percent in 2010 and 13.5 percent in 2020 under the *with-CAAA* scenario.

Acknowledgments

This work was sponsored by the U.S. Environmental Protection Agency, Office of Air and Radiation, under Contract EP-D-04-006 with Industrial Economics, Inc. It has not been subject to agency review, and, therefore, does not necessarily reflect the views of the agency, and no official endorsement should be inferred. Citations of product, company, or trade names do not constitute endorsement by EPA and are provided only for the purpose of better describing information in this article.

References

U.S. Environmental Protection Agency, (2004), "NONROAD2004 Model," [computer software], U.S. Environmental Protection Agency, Office of Transportation and Air Quality, Ann Arbor, MI, May 2004. http://www.epa.gov/otaq/nonrdmdl.htm. (>May, 2004).

Pechan, E. H., and Associates, Inc., and Industrial Economics Inc., (2006), "Emission Projections for the Clean Air Act Second Section 812 Prospective Analysis, Draft Report," prepared for James DeMocker, Office of Air and Radiation, U.S. Environmental Protection Agency, Washington, DC. http://www.epa.gov/oar/sect812/ (>June, 2006).

Pechan, E. H., and Associates, Inc., and Industrial Economics Inc., "Section 812 Clean Air Act Cost Benefit Study, Air Toxics Study, Benzene Emission Reductions in Houston," prepared for Jeneva A. Craig, Office of Air and Radiation, U.S. Environmental Protection Agency, Washington, DC.

$PM_{2.5}$ Source Apportionment – Case Study of Hickory, N.C.

J. Tippett, L. C. Lane and J. S. Lane[1]

[1]AICP, GISP, Director of Planning,The Louis Berger Group, Inc.,1513 Walnut Street, Suite 250, Cary, North Carolina 27511; PH (919) 467-3885; e-mail: slane@louisberger.com

Abstract

The City of Hickory and Western Piedmont Council of Governments are located in the foothills of North Carolina's Appalachian Mountain range. Hickory is home to one of the State's largest remaining concentrations of furniture and textile industries. Like much of the State and Southeast Region of the U.S., the Hickory Metro area has struggled to find a new economic identity after losing 25,000 jobs since January, 2001. Hickory and the whole of Catawba County were designated as a non-attainment area for $PM_{2.5}$ (fine particulate matter) by the USEPA in December, 2004. Previously, Hickory had also been designated as non-attainment for ozone, which led to the development of an Early Action Compact (EAC) that adopted 14 local control measures to address ozone formation.

This paper and presentation relates the concerns that the Unifour has expressed about the relationships between $PM_{2.5}$ monitoring methodologies, localized land uses, and the non-attainment area designation process. Land use and transportation issues have played a large role during the history of the $PM_{2.5}$ assessment process. The issue is compounded because the only $PM_{2.5}$ monitoring site is affected by out-of-state transport; is located in a low-income / minority community; and is in the heart of the remaining manufacturing sector of the City served by an active diesel rail line. The paper presents the concepts that Hickory and the Unifour Area have devised and implemented that may help other small- and medium-sized communities understand, assess, and mitigate for fine particulates.

Introduction

During the fall of 2006 and early winter of 2007, the Western Piedmont Council of Governments (WPCOG) sponsored a study to determine the sources of and control strategies for fine particulate matter, known commonly as $PM_{2.5}$. The concern motivating the study was prompted by Catawba County being designated as a non-conforming area for the pollutant by the U.S. Environmental Protection Agency (USEPA). The USEPA is responsible, based in part upon recommendations from each state's governor, for designating areas not in compliance with $PM_{2.5}$ pollution standards established by the National Ambient Air Quality Standards (NAAQS) under Section 107(d) of the Clean Air Act and later amendments.

Although $PM_{2.5}$ has not been regulated for as long a time period as "coarse" particulates (PM10), it is considered to be an even graver threat to human health since the finer particles are more readily absorbed deeper in lung tissue. The health effects of being exposed to high levels of $PM_{2.5}$ are serious, and include decreased lung function, irregular heart function including heart attacks, and exacerbating pre-existing asthma conditions. The formation and transportation of $PM_{2.5}$ is still under considerable study, however, it is known that $PM_{2.5}$ has both direct sources and indirect sources. Indirect sources can be generated from fuel combustion working in conjunction with sunlight and water vapor. The direct sources of $PM_{2.5}$ pollution are many and varied: wood smoke from residential or commercial combustion; automobile exhaust in the form of oxides of nitrogen; coal-fired power plants; small engines; open burning of trash or construction debris; and dust from agricultural operations or open areas. Actually, the USEPA first promulgated $PM_{2.5}$ standards in July, 1997. However, as monitoring equipment was not in place, and since three years' worth of data must be collected to make a determination of attainment, designations of non-attainment areas did not occur until 2004.

Once an area has been designated as non-attainment with regard to USEPA standards for a controlled pollutant, the area's local and state governments typically respond to have the designation overturned or lessened (geographically in size or in severity of the designation), or, if it is clear that the designation cannot be ameliorated, they must work to develop and implement a plan to bring the area back into compliance with the national standard. In the current case, the first action was to have the initial USEPA-recommended non-compliance area reduced from four counties to just one: Catawba. In the following excerpt of a letter from North Carolina Governor Michael Easley responding to USEPA's initial designation, the other major concern of a non-attainment designation – economic impact – is clearly seen:

"In its letter of June 29, 2004, EPA has provided flawed analysis to support far-reaching $PM_{2.5}$ nonattainment designations surrounding two isolated, non-attaining monitors in Hickory and Lexington, North Carolina. According to North Carolina's analysis, which is included in the attached letter from Secretary of Environment Bill Ross, these broad designations will not help solve the non-attainment problem at these two monitors. In fact, they are unlikely to have an appreciable effect on North Carolina's efforts to improve air quality.

These excessive non-attainment designations will, however, have a significant dampening effect on economic development efforts in the Triad and further west in the Hickory/Morganton/Lenoir area. These two areas of our state have been hit particularly hard by manufacturing job losses associated with unfair federal trade policies. Both areas are turning a corner now, but they can ill afford non-attainment designations that can undermine their ability to bring jobs to their communities – particularly when there is no beneficial effect."

The technical data employed by North Carolina was sufficient for USEPA to reduce the non-attainment area from four counties to one county. Two other counties to the east in North Carolina also fell into non-attainment.

In the Hickory/Catawba County area, the caution expressed by the N.C. Governor fell especially close to a particular industry that of the furniture manufacturing and finishing businesses that lie scattered throughout the non-attainment area. Furniture manufacturing and national trade has been a source of income for the timber-rich region of middle and western North Carolina since at least the 1880's when sawmill owners began to realize that greater profits could be made by making furniture closer to the mill rather than shipping the raw input across state lines to finished and resold elsewhere. Along with the furniture trade other suppliers of dowels, hinges, varnishes, glues, casters, drawer pulls and so forth also flourished. By 1921, North Carolina had a permanent furniture exposition in High Point, and Hickory and the surrounding area also have similar exhibitions to show the latest designs and styles. Local artisans produced high quality and unique furniture products that sold from coast-to-coast – and could be found in stylish London homes as well. A hundred years later what had started out as a valuable tool to generate much-needed income during the Reconstruction Era had translated into North Carolina in the early 1980's acquiring the moniker "The Furniture Capital of the World." The heart of this "capital" was located squarely in Catawba County. In 2004, of the 15 largest furniture manufacturers in North Carolina, eight were located in Catawba County, directly contributing over one billion dollars in annual revenue and 17,000 employees.

However, the 1990's and early 2000's have not been kind to the furniture industry, as the area has been witness to a number of downsizings and plant closures due to international trade policies and pricing structures that have made competition all but impossible. This has produced a strong sensitivity in the region to any potential threat to the remaining businesses that still employ many people in the area and indirectly support many other suppliers, shop owners, and retail businesses. Since any new or expanded business that might contribute significantly to $PM_{2.5}$ pollution has to undergo a "new source review," this is seen – to borrow a word from the N.C. Governor – as a "dampening" factor on industry. Perhaps just as importantly, no government or business official wants their area to be perceived as having "bad air" for the sake of its residents as well as placing it at a competitive disadvantage relative to others in a regional market.

The Importance of the Monitoring Site Location

This sensitivity to the new standard in Catawba County expressed itself most clearly by the local governments' strong belief that the siting of the Hickory monitoring device, known as the Hickory Water Tower monitoring site, was inappropriate. Several actions challenging the validity of the location and/or data readings taken at the site took place between the times that the Catawba County designation took affect and the WPCOG hired a consultant to study the sources and control measures of $PM_{2.5}$.

The first of these challenges was to ask NCDAQ to place a second monitor within approximately 1 mile of the water tower site to determine if the second monitor (located at the Hickory Rescue Squad) would show lower readings than the Water Tower location. For each quarter between July 2004 and June 2006 the rescue squad monitor $PM_{2.5}$ levels were lower than the water tower monitor readings, indicating that localized events were impacting the monitor at the water tower. In December 2005 the UAQC asked for an EPA audit of both of the rescue squad and water tower monitors. The audit was completed in March 2006. The EPA auditors concluded that both sites "technically" met the EPA citing criteria. They also concluded, however, that local impacts could adversely be affecting the results at the water tower monitor.

Based on the conclusions of the audit and the two monitor comparisons, the UAQC sent out an RFP in May 2006 to determine the source apportionment of the water tower $PM_{2.5}$ monitor. In July 2006, the Louis Berger Group, Inc. was asked to conduct the study.

It bears mentioning that both North Carolina and the Western Piedmont governments have applied as much energy and regulatory efforts to control both ozone and $PM_{2.5}$ pollution as any place in the country of a comparable size. Smokestack controls; enhanced inspection / maintenance in 40 of North Carolina's 100 counties; intensive modeling to refine predictions, sources, and control measure effectiveness; public awareness campaigns; and participation in regulating interstate transport of pollutants have all occurred since the late 1990's thanks to actions taken by the State of North Carolina. The local governments in the Western Piedmont have also exhibited good leadership to address this problem by adopting an Early Action Compact for ozone (one of the precursors of which, NOx, is also a precursor emission to fine particulate matter) that outlines 14 local control actions to be undertaken to reduce pollution and bring the area back into compliance with federal standards. These measures includes bans on open site burns, regional biking/walking plans, local government conservation programs, improving traffic operations, and increasing public transit and ridesharing options.

The location of the Hickory Water Tower monitoring site is shown in Figure 1. The location of the site is within a historically high concentration of urban land uses, including manufacturing and small homes owned or rented by middle- and low-income families. An active rail line (Figure 2) with approximately 10 freight operations daily runs generally east-west on the north side of the monitoring site.

Numerous roadways, some of which are interstate freeways and U.S.-numbered highways, are near the site as well. Most noticeable is the collocation of the water tower itself on the same site as the monitoring equipment. The entire site is surrounded by a chain-link fence with padlocked entry.

From the earliest days when officials knew that Catawba County would probably not be in attainment with the federal $PM_{2.5}$ standard, questions were raised about the siting of the monitor. Some were concerned that the presence of the water tower and its supports could influence site-level wind patterns or humidity, thus altering the monitor readings.

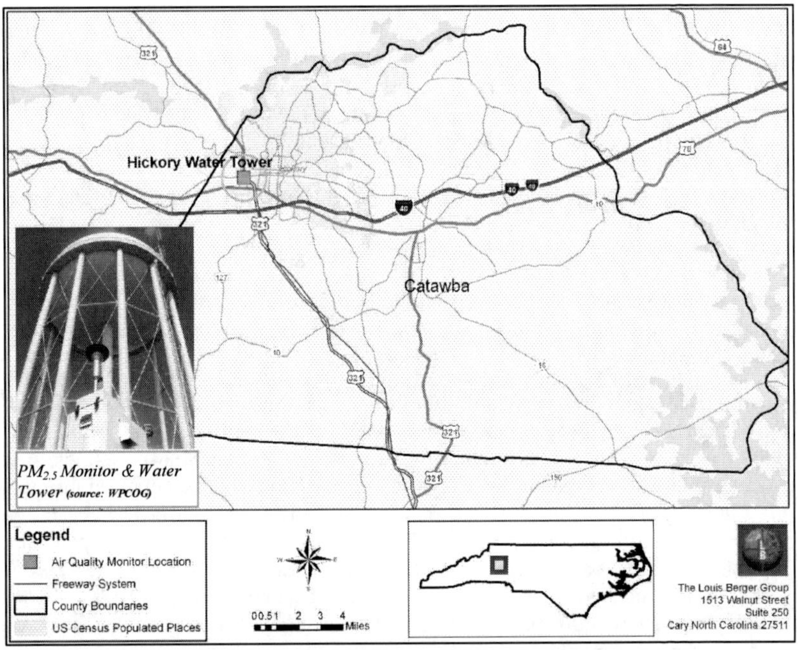

Figure 1. Location of Catawba County, City of Hickory, and Monitoring Site

Others felt that the nearby presence of several electrical substations could "magnetize" some particles or in some other way increase the $PM_{2.5}$ readings at the monitor. Many felt that the location of the monitor in the heart of an industrial sector of the only city in the County (there are several smaller towns) with its concomitant traffic, manufacturing effluent, and other potential sources of $PM_{2.5}$ simply was not

Figure 2. Nearby Active Rail Line with Water Tower Site in Background

representative of the whole of Catawba County, which is generally rural and without many large manufacturers in any great concentration. While the other suppositions about magnetized particles and tower-induced humidity changes have not had much corroborating evidence, the following section summarizes some of the findings that were discussed during the investigation of the Water Tower monitor site.

Population Density

The Hickory Water Tower monitor site is not representative of the rest of Catawba County in terms of the population density of the surrounding area compared to the population density of Catawba County as a whole (Figures 3A and 3B). The population density is considerably higher (more than two standard deviations) than that of most of the rest of the County.

Figures 3A and 3B. Comparison of Population Density of Adjacent Area of Water Tower Monitor Site and Catawba County

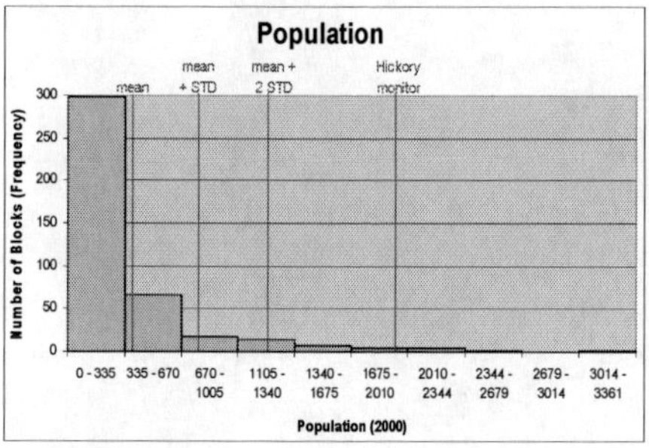

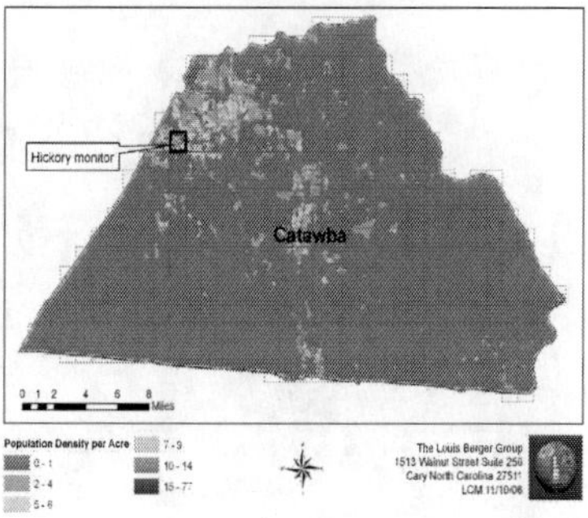

Employment Density

The Hickory Water Tower monitor site is not representative of the rest of Catawba County in terms of the employment density or industrial employment density of the surrounding area compared to the population density of Catawba County as a whole (Figures 4A and 4B). The employment and industrial employment density is considerably higher (more than two standard deviations) than that of most of the rest of the County.

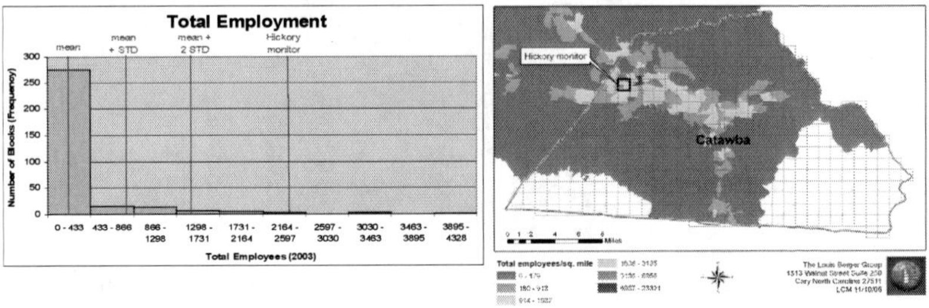

Figures 4A and 4B. Comparison of Employment Density of Adjacent Area of Water Tower Monitor Site and Catawba County

Transportation Network Density

The Hickory Water Tower monitor site is not representative of the rest of Catawba County in terms of the density of the transportation network (streets, rail lines) density of the surrounding area compared to the transportation network density of Catawba County as a whole (Figures 5A and 5B). The transportation network density is considerably higher (more than two standard deviations) than that of most of the rest of the County. These statements are true whether considering only local streets or considering all of the transportation facilities in Catawba County.

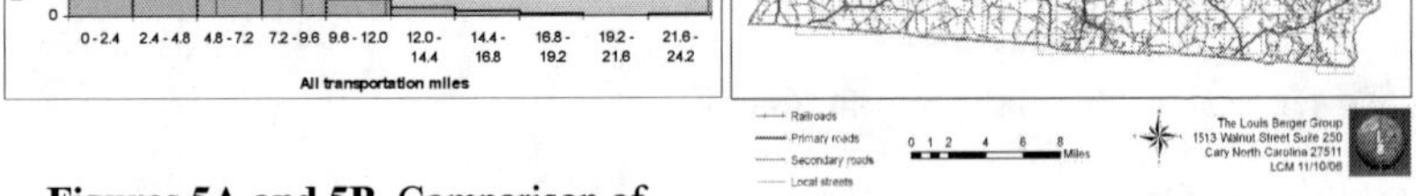

Figures 5A and 5B. Comparison of Transportation Network Density of Adjacent Area of Water Tower Monitor Site and Catawba County

Elevation

The source apportionment study also examined elevation, which can play a role in determining monitor source readings as well. Again the Hickory Water Tower monitor site is relatively higher in elevation as compared to the rest of Catawba County or three other, comparable monitor sites that were used as peer research sites (Figure 6). The monitor site is in a small township within Hickory, the Town of Long View, North Carolina. Long View is so-named because of the distance that one can see when standing in the Town. The monitor is located at the top of a ridgeline as well.

Figure 6. Elevation Map of Hickory Water Tower Monitor Site
(Note: Lighter colors represent higher elevations)

Siting Criteria Used by USEPA

The USEPA has several criteria (40 CFR Part 58.12 Appendix E, Section 8) that it uses to locate monitoring stations. These, along with their applicability to the Hickory Water Tower monitor site, are described in Table 1. None of the criteria were violated at the Hickory Water Tower monitoring site according to a review of the sites performed in March, 2006, but none of the specific criteria deal with densities of population and employment as they relate to the general location of a monitoring station.

USEPA has also offered guidance on the general geographic location of a monitor, which depends upon its purpose and scale of coverage (e.g., microscopic, neighborhood, urban, regional, etc.). An excerpt from this guidance also provides some indications about where to locate a monitor:

"In some cases, the physical location of a site is determined from joint consideration of both the basic monitoring objective and the type of monitoring site desired, or required by this appendix. For example, to determine $PM_{2.5}$ concentrations which are typical over a geographic area having relatively high $PM_{2.5}$ concentrations, a neighborhood scale site is more appropriate. Such a site would likely be located in a

residential or commercial area having a high overall $PM_{2.5}$ emission density but not in the immediate vicinity of any single dominant source. Note that in this example, the desired scale of representativeness was an important factor in determining the physical location of the monitoring site.

...Spatial Scales. (a) To clarify the nature of the link between general monitoring objectives, site types, and the physical location of a particular monitor, the concept of spatial scale of representativeness is defined. The goal in locating monitors is to correctly match the spatial scale represented by the sample of monitored air with the spatial scale most appropriate for the monitoring site type, air pollutant to be measured, and the monitoring objective.

...Neighborhood scale—Defines concentrations within some extended area of the city that has relatively uniform land use with dimensions in the 0.5 to 4.0 kilometers range. (40 CFR Part 58.12 Appendix D, Section 1.2)"

Table 1. EPA Monitor Siting Criteria

Criteria	Requirements
Vertical Probe Placement	2 -7 meters above ground for micro scale
	2-15 meters above ground for other scales
Obstructions on Roof	2 meters from walls, parapets, penthouses, etc.
Spacing from Trees	Should be 20 meters from drip line of trees.
	Must be 10 meters from drip line if trees are an obstruction
Obstacle Distance	2 x height differential (street canyon sites exempt)
Unrestricted Airflow	At least 270 including the pre-dominant wind direction
Furnace or Incinerator Flues	Recommended that none are in the vicinity.
Distance between Collocated Monitors	1 to 4 meters
Spacing from Station to Road	Depends on volume of traffic; ranges from 10 to 250 meters
Paving	Area should be paved or have vegetative ground cover

Figure 7. A Rural Area of Catawba County

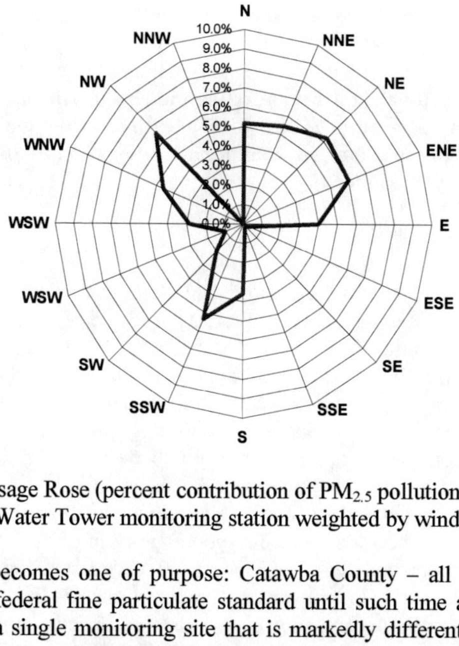

Figure 8. Dosage Rose (percent contribution of $PM_{2.5}$ pollution impacting the Hickory Water Tower monitoring station weighted by wind direction)

The issue then becomes one of purpose: Catawba County – all of it – will be in violation of the federal fine particulate standard until such time as (a) the level of $PM_{2.5}$ impact at a single monitoring site that is markedly different from 90% of the remainder of the non-attainment area decreases below the set threshold levels; and/or (b) the threshold levels or regulations change. Not as unlikely a scenario as it seems, since the 24-hour standard of 65 u/m3 (read as micrograms per cubic meter) tightened to 35 u/m3 during the course of the Hickory/Catawba study in late 2006. It is also possible that the monitor could be relocated, but the area would remain in non-attainment until three more years of data could be collected. Since the monitor is "just over" the federal thresholds for both annual (actually, a three-year average) and hourly standards for fine particulates, and since it is extremely difficult to prove under the current siting guidelines that the monitor has never been representative of anything other than a small wedge of the county, this is unlikely to occur. Furthermore, the area has taken a stand that as long as some of its residents are affected by higher-than-safe levels of pollution, there is a problem present that should be dealt with given fiscal, policy, and environmental limitations.

It is worthwhile to note that North Carolina law now requires smokestack controls on the Marshall Steam Station, located 25 miles to the southeast of the Hickory Water Tower monitoring site, but still inside of Catawba County. This single point source produces more than 90% of the estimated $PM_{2.5}$ emissions in the entire county, but, according to modeling done by the consultant under this study, less than 6% of the $PM_{2.5}$ reaches this area because of wind conditions that almost preclude any wind coming from the southeast. These and other controls applied by both local governments in the Western Piedmont and North Carolina have started to show real improvements in ozone formation, for example. However, localized problems may not be seriously improved by these controls. An additional consideration is that over two-thirds of the $PM_{2.5}$ "problem" at the Hickory Water Tower monitoring site is coming from other states or outside the non-attainment area of Catawba County. There is no drylands problem that is affecting the Hickory monitoring site, and the number of low-wind days is quite high (over one-third of the monitor-height wind readings at all four of the examined monitoring locations had 0°/360° wind readings indicative of low-wind conditions).

Conclusions

Since other areas of the country are also not in attainment of the USEPA standards for $PM_{2.5}$ pollution (or another criteria pollutant), it would seem logical to reconsider the general siting criteria specified in the Federal Register guidance, especially Appendix D to Part 58 - Network Design Criteria for Ambient Air Quality Monitoring. The revised guidance should be formulated considering the relationship between localized pollution sources and monitoring site purpose, including more specific spatial information on the relationship between siting neighborhood and urban scale monitoring stations and tying this back to designation areas. If this approach had been applied to the Catawba monitoring site, it is likely that more than one county would have been designated as non-attainment of the $PM_{2.5}$ standard, but that only an industrial and high-traffic corridor would now be affected by the designation. This might help pinpoint sources that the local governments could actually influence without damaging county-level economic opportunities, avoid expending valuable resources on contesting the viability of the non-attainment designation, and help frame specific control measures that have the best chance of working given the localized conditions.

Acknowledgements

The authors accept all responsibility for errors or omissions in this paper, but would like to credit the following agencies with their invaluable assistance:

- Unifour Air Quality Committee
- Unifour Air Quality Oversight Committee
- Western Piedmont Council of Governments and Staff, especially Taylor Dellinger
- City of Hickory, North Carolina
- Catawba County, North Carolina
- North Carolina Department Environment and Natural Resources Division of Air Quality
- North Carolina Department of Transportation Division 12

References

USEPA, Fine Particle Designations http://www.epa.gov/pmdesignations/faq.htm#0 (>January 28, 2007).

Letter from Honorable Michael F. Easley, Governor, North Carolina to Michael Leavitt, Administrator, US Environmental Protection Agency, USEPA, dated September 9, 2004.

Powell, W. S. (1989), "North Carolina Through Four Centuries." *Chapel Hill: The University of North Carolina Press 1989, pp. 410-412.*

Powell, W. S. (2006), "Encyclopedia of North Carolina." *Chapel Hill: The University of North Carolina Press 2006.*

Letter from R. Douglas Neeley, Chief, Air Toxics and Monitoring Branch, USEPA Region 4 to Hoke Kimball, Chief, Ambient Air Monitoring Section, N.C. Department of Environment and Natural Resources, dated May 10, 2006.

Creating an Evidence Toolbox to Assist Qualitative $PM_{2.5}$ and PM_{10} (Particulate Matter) Hot-Spot Analyses

J. A. Frazier[1], D. Szekeres[2] and C. M. Reitz[3]

[1]Michael Baker Jr., Inc., 1304 Concourse Drive, Suite 200, Linthicum, MD 21090; PH (410) 689-3400; FAX (410) 689-3401; email: jfrazier@mbakercorp.com
[2]Michael Baker Jr., Inc., 4431 N. Front Street, Second Floor, Harrisburg, PA 17110; PH (717) 221-2000; FAX (717) 234-7618; email: dszekeres@mbakercorp.com
[3]Michael Baker Jr., Inc., 1304 Concourse Drive, Suite 200, Linthicum, MD 21090; PH (410) 689-3400; FAX (410) 689-3401; email: creitz@mbakercorp.com

Abstract

A recently published Environmental Protection Agency (EPA) final rulemaking (40 CFR 93.116) established transportation conformity criteria and procedures for identifying transportation projects that must be analyzed for local air quality impacts in $PM_{2.5}$ and PM_{10} nonattainment and maintenance areas. A qualitative hot-spot analysis is required for "projects of air quality concern" as defined in EPA's final rule. PM hot-spot analyses assess projects found to be of "air quality concern" through the interagency consultation screening process, in combination with changes in background air quality concentrations, to determine if new or worsened future violations will result from their implementation.

This paper presents suggestions for the development and utilization of various tools based on existing studies, computations and data, which can be compiled to form an evidence toolbox for use in qualitative PM hot-spot analyses. The evidence toolbox is designed for use by transportation officials and interagency consultation groups (ICG) in order to demonstrate whether a transportation project may or may not have any adverse affect on future PM emissions.

Introduction

On March 10, 2006, the Environmental Protection Agency (EPA) published a Final Rule (updating 40 CFR 93.116) that establishes transportation conformity criteria and procedures for determining which transportation projects must be analyzed for potential local air quality impacts in $PM_{2.5}$ and PM_{10} (PM) nonattainment and maintenance areas (identified in

Figure 1and
Figure 2). Also in March 2006, EPA and the Federal Highway Administration (FHWA) issued a joint guidance document (EPA420-B-06-902) entitled "Transportation Conformity Guidance for Qualitative Hot-spot Analyses in $PM_{2.5}$ and PM_{10} Nonattainment and Maintenance Areas," that outlines how state and local agencies can meet the hot-spot analysis requirements.

The project-level, PM hot-spot conformity requirements apply to all non-exempt, federal, transportation projects which are located in PM nonattainment or maintenance areas, and require FHWA/Federal Transit Administration (FTA) approval or authorization.

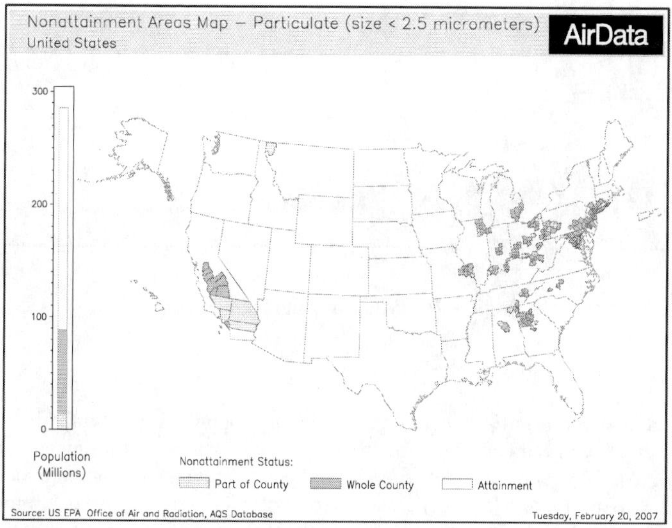

Figure 1: $PM_{2.5}$ Nonattainment Areas

This paper focuses on addressing the challenges of performing detailed qualitative PM hot-spot analyses through the development and utilization of various evaluation options, which can be combined to form an evidence toolbox. The evidence toolbox is designed to assist transportation officials and interagency consultation groups (ICG) with demonstrating whether a transportation project located in a PM nonattainment or maintenance area will have any adverse affect on future PM emissions and possibly require the implementation of mitigation measures. These suggested toolbox items are based, in large part, on the March, 2006 guidance. However, in some cases, the items suggest providing more detailed information than is outlined by the guidance in order to provide a more complete picture of the air quality issues associated with a specific project and its location.

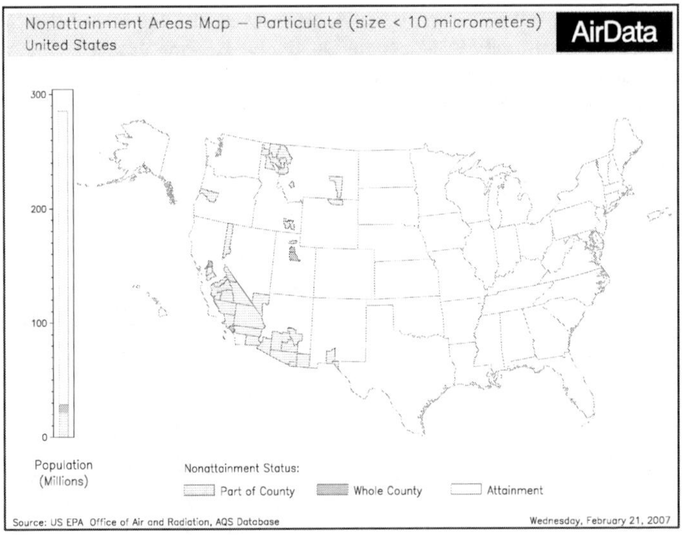

Figure 2: PM_{10} Nonattainment Areas

Arriving at a qualitative analysis

Transportation conformity is required for federally supported transportation projects in areas that have been designated by the EPA as nonattainment (not meeting one or more NAAQS) or maintenance (previously were in violation but are currently meeting one or more NAAQS) areas. Transportation conformity determinations related to the updating of transportation plans (Plan) and transportation improvement programs (TIP) are regional analyses, typically stand-alone documents that are submitted for federal approval in conjunction with the submittal of an updated Plan and/or TIP.

A PM hot-spot analysis, the focus of this paper, is prepared for required projects when a project-level conformity determination is executed. This determination is typically completed as part of the National Environmental Policy Act (NEPA) process (Figure 3), even though it may also be part of a conforming Plan or TIP. The NEPA regulations require that all actions sponsored, funded, permitted or approved by federal agencies undergo planning to ensure that environmental considerations, such as impacts on air quality, are given proper weight during the project decision-making process.

Through the interagency consultation (ICG) process, and as part of the project-level transportation conformity determination process, a project is identified either as a "project of air quality concern" or a "project not of air quality concern." Projects found to be "projects not of air quality concern" must be identified as such and should be accompanied by a reasonable explanation supporting this conclusion. Those projects identified as "projects of air quality concern" must have relevant, qualitative, hot-spot analysis documentation and a determination prior to the next federal (FHWA/FTA) action to adopt, accept, approve or fund the project.

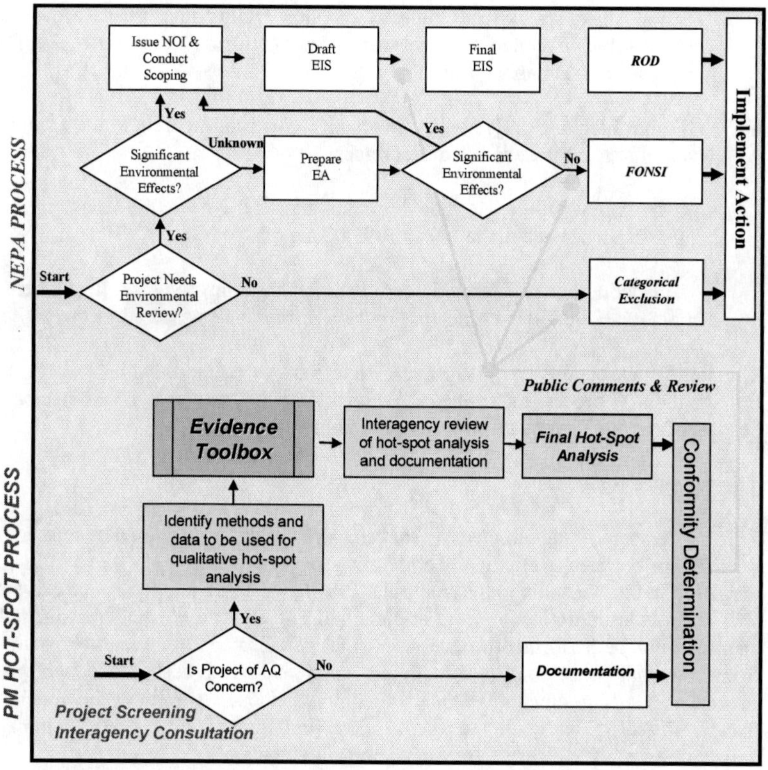

Figure 3: Relationship of PM Hot-Spot Process to the NEPA Process

This documentation should be agreed upon through the ICG process and should clearly support the conclusion that potentially new or worsened future violations either will or will not be created due to the project in combination with changes in background air quality concentrations. Should the qualitative analysis determine that the project will create new or worsened violations, mitigation measures may be needed to reduce project related emissions and any local air quality impacts. EPA's

March 2006 guidance provides the following examples of projects of air quality concern that would be covered by 40 CFR 93.123(b)(1)(i-iv):

- A project on a new highway or expressway that serves a significant volume of diesel truck traffic, such as facilities with greater than 125,000 annual average daily traffic (AADT) and 8% or more of such AADT is diesel truck traffic;

- New exit ramps and other highway facility improvements to connect a highway or expressway to a major freight, bus or intermodal terminal;

- Expansion of an existing highway or other facility that affects a congested intersection (operated at Level-of-Service D, E, or F) that has a significant increase in the number of diesel trucks;

- Similar highway projects that involve a significant increase in the number of diesel transit buses and/or diesel trucks;

- A major new bus or intermodal terminal that is considered to be a "regionally significant project" under 40 CFR 93.01; and,

- An existing bus or intermodal terminal that has a large vehicle fleet where the number of diesel buses increase by 50% or more, as measured by bus arrivals.

The following sections of this paper outline various methods which can be used through the ICG process to perform a PM qualitative hot-spot analysis when a project has been determined to be a "project of air quality concern."

The evidence toolbox

Existing research and modeling studies are inconclusive in determining the relationship between traffic and potential hot-spot concentrations; as a result, and as discussed in the preamble to the final rule, EPA has determined that quantitative hot-spot analyses cannot be completed at this time. Therefore, a PM hot-spot qualitative analysis must be performed to assess potential new or worsened violations resulting from the implementation of the project combined with background air quality concentrations. The qualitative analysis requires evidence to support the conclusion that a "project of air quality concern" either will or will not adversely impact a nonattainment area's efforts to reach its air quality goals.

The evidence toolbox outlines several methods relating to analytical computations, however, these computations are primarily used to parallel the results of other accepted scientific studies and aid in illustrating the project's contribution to existing $PM_{2.5}$ concentrations. Through the ICG process, individual evidence items, or a combination of several items, can be selected from the toolbox and used to support the conclusions of the analysis. This section outlines these analytical methods, collectively known as the evidence toolbox:

- Comparison to other monitor locations (impacts of relevant completed projects)
- Historic monitor reading trends
- Future forecast of monitor trends
- Emissions by source category
- Regional emissions and emission factor trends
- Impact on sensitive land use in the project vicinity
- Relative impact of projects on regional emissions quantities
- Research studies addressing dispersion of PM emissions

Comparison to Other Monitor Locations

Comparing to another location with similar characteristics is a method indicated in the guidance as a potential approach for demonstrating that a new project will meet statutory conformity requirements. It entails reviewing existing highway or transit facilities that were constructed in the past and built in locations similar to the proposed project location. A comparison of a monitoring location with a completed project with similar traffic characteristics and roadway influences and a monitor location in proximity to the proposed project could be conducted in order to determine if a project will create or worsen air quality violations. EPA's AirData website (http://www.epa.gov/air/data/index.html) can be used to obtain a listing of existing PM monitoring locations and the associated traffic volumes in the vicinity of those monitors.

In addition to the monitor comparison method suggested in the guidance, and along the same lines, an examination of recently completed projects can be used in combination with regional monitor readings to evaluate whether a project may create or worsen air quality conditions. There are significant issues to consider including the size of the project, the distance to the closest monitors, wind directions, impacts on traffic diversions, and changes to background concentrations to determine a direct relationship.

Historic Monitor Trends

As per the qualitative hot-spot analysis guidance, annual average monitor readings within the vicinity of the project should be included in the existing conditions section of the qualitative analysis. Examining the trend of monitor readings may also be useful for illustrating the impacts of recently completed projects or in identifying trends in background concentrations.

Future Monitor Trends

Given that the pollutants which lead to regional haze can originate from sources located across broad geographic areas, EPA has encouraged states to address visibility impairment from a regional perspective. There are currently five regional planning organizations (RPO) which address regional haze and related issues. These organizations have been tasked with evaluating technical information to better understand how their states and tribes impact national park and wilderness areas across the country. Furthermore, they are to undertake the development of regional strategies to reduce emissions of particulate matter and other pollutants leading to regional haze. As a result of the nature of these tasks, RPOs can be an excellent source of PM forecast emissions which may be useful in a qualitative analysis.

In addition to information available from the RPO's, the EPA has projected PM emissions trends as a result of the promulgation and implementation of the Clean Air Interstate Rule (CAIR), which covers SO2 and NOx emissions in the Eastern U.S.; the Clean Air Mercury Rule (CAMR), which covers mercury emissions nationwide; and the Clean Air Visibility Rule (CAVR), which requires certain units - depending on their visibility impacts - to install pollution controls in certain areas of the country. These results can be found on EPA's website: http://www.epa.gov/cleanair2004/ and could provide additional evidence which can be used to demonstrate whether PM emissions are or are not anticipated to trend down in the future.

Emissions by Source Category

Understanding the potential and relative contribution of transportation sources to total PM emissions may provide some evidence as to the potential impact of transportation sources on localized hot-spot concentrations. National data obtained from EPA's AirData "Emissions by Category Report" for 2001 is summarized in Figure 4 and illustrates that 2001 highway vehicle emissions represent approximately 2% of the total $PM_{2.5}$, while off-highway vehicle emissions represent about 4%. It is important to stress that national data can vary greatly from regional and local data and transportation officials should remain aware of this when performing PM hot-spot analyses. Emissions by source category reports are available on the county level and can be obtained from the state environmental agency.

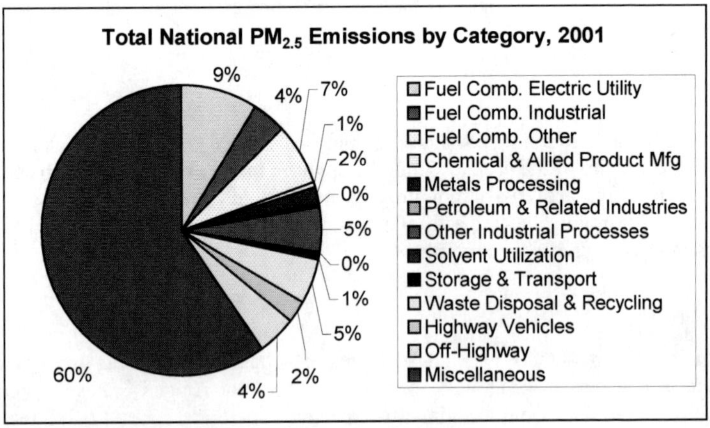

Figure 4: 2001 EPA AirData $PM_{2.5}$ Emissions by Category Report (Entire U.S.)
Regional Emissions and Emissions Factor Trends

Although there is little research providing a clear linkage between the emission quantities produced by regional MOBILE6.2 modeling and localized PM hot-spot concentrations, the results do indicate that transportation-related emission quantities will be reduced (by 50% or more between 2002-2020) despite growing regional vehicle miles traveled (VMT). This is primarily due to improvements in vehicle and fuel technologies and expected regional control strategies.

A further examination of emissions by vehicle type indicates the primary sources of highway and off-highway related $PM_{2.5}$ emissions are diesel vehicles. EPA's emission source by category data for 2001, available on the AirData website, can be used to determine what percentage of highway emissions are related to diesel vehicles. On the national level, this data indicates that approximately 67% of $PM_{2.5}$ highway source emissions are attributable to diesel vehicles, despite the fact that they comprise, on average, less than 10% of the vehicle fleet. This data substantiates the focus of the hot-spot analysis rule and guidance on diesel vehicle types. Again, it is important to note that national data can vary greatly from local data and transportation officials should remain aware of this when conducting a hot-spot analysis. Figure 5 provides an example from the Washington, DC – Maryland – Virginia $PM_{2.5}$ nonattainment area. This example illustrates the estimated impacts of future vehicle technologies and control strategies on $PM_{2.5}$ emissions per EPA's MOBILE6.2 emission model. Significant reductions in heavy-duty diesel emission factors are expected within the next 15 years, which is the primary reason for the 50% reduction in future year emissions produced in the conformity analysis. These projected reductions in diesel truck emission factors are expected to reduce the impact of highway vehicles on hot-spot concentrations in future years.

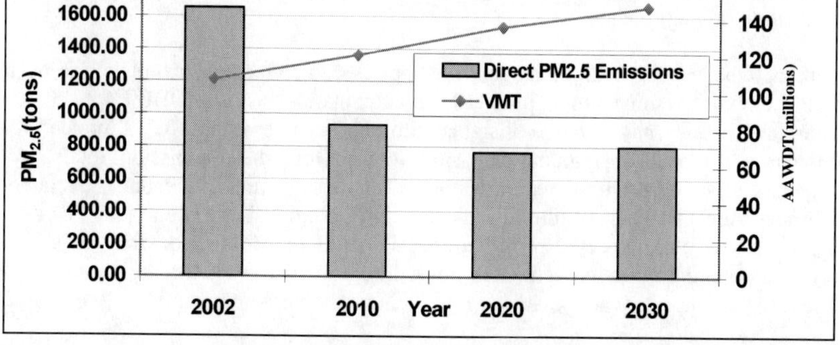

Figure 5: Comparison of Annual VMT and $PM_{2.5}$ Emissions

Impact on Sensitive Land Use in the Project Vicinity

Due to the uncertainty in evaluating and addressing whether a hot-spot concern could be created from the project, it may be useful to consider nearby landuse, including households, schools, hospitals, churches, etc., as a potential risk assessment. Several data sources including project specific studies or EPA's EnviroMapper data can be used to identify key landuse and distances from the project study area.

Although not specifically addressed in the PM hot-spot guidance, Appendix B of FHWA's Mobile Source Air Toxics (MSAT) Guidance indicates several factors that should be considered when crafting a qualitative analysis for MSATs (note that diesel particulate matter is one of the MSATs). One of these factors is the following:

"Projects that create new travel lanes, relocate lanes or relocate economic activity closer to homes, schools, businesses and other sensitive receptors may increase concentrations of MSATs at those locations relative to No Action."

Some research efforts have focused on determining the potential dispersion of highway-related PM emissions based on distance from the roadway (see "Research Studies Addressing Dispersion of PM Emissions" below). These studies can be referenced in combination with the land use locations to identify whether a potential area may experience health risks due to the project impacts.

Relative Impact of Projects on Regional Emissions Quantities

Another item discussed within the MSATs guidance, and potentially applicable to PM hot-spot analyses, as a factor to consider in a qualitative analysis is the net regional impact of the project. Projects that divert traffic volumes or facilitate new development may generate additional fine particulate matter emissions in the local project area; however, such activity may be attracted from elsewhere in the region. As a result, on a regional scale, there may be no net change in emissions or even an overall benefit. The above data may not eliminate the need for potential mitigation measures within the project vicinity but should certainly be considered in the evaluation of the project.

Although the MOBILE6.2 model does not apply speed correction factors with respect to PM emissions, some simple project-level computations using MOBILE6.2 vehicle emission factors may provide insights into the relative impact of a project as compared to local or regional emission totals. How these emission totals are dispersed to areas within the region is not well-understood and thus cannot be related to direct concentrations within the project vicinity. Regional emission factors can be applied directly to expected project impacts (delta) on VMT and idling delay to produce potential emission estimates (most likely for the project completion year). These delta impacts can be divided by the total emission quantities (e.g. from conformity analysis or other analyses) for roadway, sub-regional, county, or other aggregations.

Research Studies Addressing Dispersion of PM Emissions

There is limited information about fine particle contributions to ambient $PM_{2.5}$ levels as generated from roadway traffic. However, there has been some research conducted which can be used to assist in evaluating the potential $PM_{2.5}$ hot-spots that might occur near roadways, considering the contributions from other nearby roadways as well.

Understanding how PM concentrations dissipate from highways is an important point to consider in addressing potential impacts on nearby monitors and/or sensitive land uses. Available studies indicate that particulate matter can vary significantly at distances from the roadway. One such study (Zhu et al, 2002) provides an example of this relationship based on test studies of black carbon (which originates as ultra-fine or fine particles from primary sources during incomplete combustion of carbon-based fuels, for example, diesel engines, wood burning, and poorly maintained industrial and residential heating). These relationships indicate that black carbon (which is assumed as a surrogate for $PM_{2.5}$ emissions in this discussion) can decrease significantly, especially for heavy-volume freeways, at distances between 100-300 meters from the roadway travel lanes. Such relationships may be useful in evaluating potential project impacts on nearby land uses and monitor locations.

Conclusion

Since EPA has determined that quantitative hot-spot analyses cannot be completed at this time, qualitative analyses must be performed in order to ensure that projects located in PM nonattainment areas and deemed to be "projects of air quality concern" are properly evaluated for their potential to create new or worsened air quality violations.

This paper focuses on addressing the challenges of performing detailed qualitative PM hot-spot analyses through the development and utilization of various evaluation options, which can be combined to form an evidence toolbox for a nonattainment area. The evidence toolbox outlines several methods relating to analytical computations, however, these computations are primarily used to parallel the results of other accepted scientific studies and illustrate the relative proportion of project emissions to total regional emissions. The toolbox should be updated regularly with the most recent available local information, such as monitor readings and traffic data. When properly maintained and referenced, the evidence toolbox can be a valuable aid to transportation officials and ICGs tasked with demonstrating whether a transportation project located in a PM nonattainment area may have any adverse effect on future PM emissions.

References

Environmental Protection Agency (EPA). (2006). "$PM_{2.5}$ and PM_{10} Hot-spot Analyses in Project-Level Transportation Conformity Determinations for the New $PM_{2.5}$ and Existing PM_{10} National Ambient Air Quality Standards; Final Rule."

Environmental Protection Agency (EPA) and United States Department of Transportation (USDOT). (2006). "Transportation Conformity Guidance for Qualitative Hot-spot Analyses in $PM_{2.5}$ and PM_{10} Nonattainment and Maintenance Areas."

Federal Highway Administration (FHWA). (2006). "Interim Guidance on Air Toxic Analysis in NEPA Documents."

North Capital Region Transportation Planning Board, Metropolitan Washington Council of Governments (MWCOG). (2005). "Fine Particulates (PM2.5) Standards Air Quality Conformity Assessment, Supplement to the Air Quality Conformity Determination of the 2005 Constrained Long Range Plan and GY2006-2011 Transportation Improvement Program for the Washington Metropolitan Region."

Zhu, Y., Hinds, W. C., Kim, S., Shen, S., and Sioutas, C. (2002). "Study of Ultrafine Particles Near a Major Highway with Heavy-Duty Diesel Traffic." *Atmospheric Environment* 36.

Results of Implementing Aggressive PM Reduction on Non-road Construction Equipment at Two Lower Manhattan Project Sites

R. Raman[1]

[1]Civil Engineer, MTA Capital Construction Projects, Broadway, New York, N.Y. 10004; email: ramesh.raman@nyct.com

Abstract

The Metropolitan Transportation Authority, Capital Construction Company in New York is managing the construction of two projects in Lower Manhattan, funded via federal government post-9/11 recovery allocations. Lower Manhattan is one of the largest business districts in the nation, and the two projects – the South Ferry Terminal Station and the Fulton Street Transit Center – will improve transit access to, and help to revitalize, downtown Manhattan.

With dozens of large-scale projects in the planning and construction stages in Lower Manhattan, many having lengthy and overlapping construction periods, MTA CC has aggressively implemented Environmental Performance Commitments (EPCs), developed in conjunction with other agencies building in Lower Manhattan. Heightened community awareness of air quality issues in Lower Manhattan has made the clean diesel emissions program one of the most important of the EPCs to implement. The U.S. EPA is using MTA CC's clean diesel emissions program as a model of success to induce its implementation in other areas of the country.

The clean diesel emissions program includes use of Ultra Low Sulfur Diesel (ULSD) fuel, use of Tier 2 engines and diesel particulate filters (DPF) on all off-road construction engines of 50 horsepower (hp) or greater. MTA CC worked closely with retrofit vendors, equipment/engine dealers, rental agencies, and contractors to implement and manage this program. Between the two projects almost 150 pieces of equipment have been eligible for the retrofit technology and an approximate 86 percent compliance rate has been maintained. Given that these measures reduce particulate matter emissions by more than 90 percent, and that construction diesel equipment is a significant contributor to poor air quality, the program provides significant air quality benefits to Lower Manhattan.

This paper reviews MTA CC's clean diesel emissions program, describes the challenges faced to implement the program, and describes the results including compliance statistics and cost factors.

MTA CC's Lower Manhattan Transportation Recovery Projects

MTA CC is constructing a new South Ferry Terminal Station at a cost of approximately $489M, which consists of a structural box (approximately 1700' by 50') beneath Battery Park at the tip of Manhattan. Construction has included installing mini piles to support two active train tunnels, relocation of utilities, removal of over 180,000 cubic yards of rock and 110,000 cubic yards of soil (via drill-and-blast and mechanical excavation), pouring concrete and installation of station finishes. The completed station will consist of an upper level, housing NYCT personnel and material, and a lower level of two subway tracks with a single island platform.

At Fulton Street, a multi-level Transportation Center will be built to serve as a focal point for entry into the 12 subway lines that serve lower Manhattan, most of which date back to the early 1900s. Underground concourses will be constructed to connect the subway lines and link the Transit Center to the New Jersey PATH system and the World Trade Center site. Currently work is underway on the Dey Street concourse, to be completed by April 2008 at a cost of $171M. The total cost of the Fulton Street Transit Center project is approximately $888M.

Environmental Performance Commitments

In addition to MTA CCs projects, a number of other projects are under construction or planned for development in Lower Manhattan, including the permanent PATH Station, the WTC Transportation Hub, Route 9A Reconstruction, and the WTC Redevelopment and Memorial Center. Working with the Federal Transit Administration (FTA) of the Lower Manhattan Recovery Office, an advance set of mitigation measures called Environmental Performance Commitments (EPCs) were developed and are being implemented by the project sponsors -- MTA CC, the Port Authority of New York and New Jersey (PANYNJ), the New York State Department of Transportation (NYSDOT), and the Lower Manhattan Development Corporation (LMDC) -- to minimize cumulative environmental impacts during construction. The EPCs include requirements related to dust, noise, vibration, traffic control, and disruption to businesses and residents, in addition to clean diesel emissions controls, which is the subject of this paper.

Regulatory Diesel Emissions Controls

Under the Clean Air Act, National Ambient Air Quality Standards (NAAQS) have been established for particulate matter with an aerodynamic diameter of less than or equal to 2.5 micrometers and 10 micrometers ($PM_{2.5}$ and PM_{10} respectively) and sulfur dioxide (SO_2), pollutants that are of concern during construction since they result from the combustion of fossil fuels. New York City has been designated as a non-attainment area for both PM_{10} and $PM_{2.5}$, both of which can act as a substrate for the adsorption of other pollutants, often toxic and some likely carcinogenic compounds. $PM_{2.5}$ is of particular concern since the smaller fraction of the particle size range has the ability to reach the lower regions of the respiratory tract. The U.S. EPA estimates that construction equipment contributes about a third of annual mobile PM emissions in the U.S., the second largest source behind total highway at 40 percent. While there are around two million off-road construction vehicles in the U.S. compared to around 200 million cars and trucks, a typical 175-horsepower bulldozer emits as much particulate matter as 500 cars. The September 11, 2001 events heightened community concerns and awareness related to the potential health effects of particulate matter, so minimizing particulate matter diesel emissions became extremely important for the downtown construction projects.

The first tier of a four-tier progression of U.S. EPA emissions standards for new off-road diesel engines was phased in (by horsepower rating) from 1996 to 2000. Prior to the 1996 regulations, off-road engine emissions were unregulated (and much more polluting). Emission controls on these pre-1996 unregulated diesel engines remains voluntary. Subsequently, more stringent Tier 2 standards were introduced from 2001 to 2006 and yet more stringent Tier 3 standards phased in from 2006 to 2008. PM values for 175 horsepower Tier 2 and Tier 3 engines are 62 percent better than Tier 1. (Note that while the PM standard is the same for Tier 2 and Tier 3 engines, NO_x values for Tier 3 engines are about 38 percent better than Tier 2 for 175 horsepower engines).

The U.S. EPA issued a diesel sulfur rule affecting highway fuel in early 2001 and, in 2004, finalized the diesel sulfur rule for off-road fuel. U.S. EPA's highway rule involves a phase-in of ULSD fuel by requiring that, beginning in 2006, 80 percent of highway fuel produced in refineries be 15-ppm sulfur, and the remaining 20 percent at the 500-ppm standard (called Low Sulfur Diesel fuel). By 2010, all highway diesel fuel will be 15-ppm. The non-road rule utilizes a "two step" approach, phasing non-road fuel to 500 ppm initially, then to 15-ppm in 2010 in conjunction with the highway diesel requirements. Compliance with the rule contains some "hardship provisions" that allow some small refiners to continue producing off-spec diesel for several years, particularly off road fuel. Although all diesel fuel will be ULSD by 2010, the U.S. EPA does not yet require service stations and truck stops to sell ULSD fuel and its current availability is market driven.

The New York State Environmental Conservation Law was amended in February 2007 to include a new section 19-0323 that requires the use of ULSD and best available retrofit technology for both on- and off- road heavy duty diesel vehicles for all state projects.

Table 1: non-road diesel engine emission standards for pm (g/bhp-hr)

Model Year	Tier 1 PM	HP Range	Tier 2 PM	Model Year
2000	0.75	Less than 11	0.60	2005
2000	0.60	11 - 25	0.60	2005
1999	0.60	25 - 50	0.45	2004
1998	----	50 - 100	0.30	2004
1997	----	100 - 175	0.22	2003
1996	0.40	175 - 300	0.15	2003
1996	0.40	300 - 600	0.15	2001
1996	0.40	600 - 750	0.15	2002

Source: U.S. EPA

Clean Diesel Emission Control Program

The Clean Diesel Emission Control Program that has successfully been implemented by MTACC at the South Ferry Terminal Station and Fulton Street Transit Center projects includes the following:

1. Use of Tier 2-compliant equipment for all off-road engines greater than 50 hp;
2. Use of ULSD in off-road equipment with engine rating of 50 hp and above;
3. Use of DPF in off-road equipment to further reduce emissions unless technically not feasible, in which case a Diesel Oxidation Catalyst (DOC) is considered.

In addition to these measures, idling times are limited to no more than three minutes, and engines are located away from fresh air intakes whenever possible.

Tier 2 Benefits

Aside from the significantly better performance of Tier 2 engines, PM "loading" on the filters is less likely to occur with Tier 2 equipment as compared to older engines since Tier 2 engines generate less PM emissions to begin with. Also, poor maintenance tends to be more of a problem with the older equipment, which also causes elevated PM emissions that would contribute to "loading". With non-tier engines, the order of magnitude of PM generation is so much higher that retrofit manufacturers might decline to offer a passive DPF, because of the danger of PM overload in the filter.

The requirement of Tier 2 engines motivated contractors to replace aging equipment. Very quickly contractors and sub contractors became aware of engine tier requirements and started asking for Tier 2 engine equipment. Market preparedness by the rental agencies and equipment dealers helped respond to the need.

ULSD Benefits

Unlike on-road diesel engines, construction engines were not regulated until recently and therefore used diesel fuel of any sulfur level. ULSD offers up to 13 percent reduction in PM as well as significant reductions in sulfur and other pollutants. When ULSD is used with DPF the reductions are over 90 percent for PM. ULSD also reduces corrosion and wear of exhaust system components and engine parts and offers more flexibility in the selection of retrofit technology.

USLD can cause older engines to malfunction. Problems cited with older engines include:

1. Seal leak: USLD contains less aromatics that cause seals to shrink. Aged nitrile rubber seals tend to lose elasticity and are unable to adapt to lower aromatics in ULSD and can leak.
2. Lubricity: Naturally occurring lubricity agents in diesel fuel are removed during the process of reducing sulfur to 15 ppm, which can cause engine problems. However, to overcome the lubricity issue, ASTM adopted lubricity specification D975 for all diesel fuels and this standard went into effect from January 1, 2005
3. Fuel economy: processing ULSD reduces aromatics content and density resulting in up to one percent less energy per gallon. However, these reductions are negligible for fuel economy and horsepower in off-road engines.

DPF Benefits

On MTA CC projects, only "passive" DPFs were used, because "active" DPFs were not available at the time of program inception.[1] Passive DPFs filter more than 90 percent of the PM emissions from engines of all horsepower.

Challenges to Implementation

Challenge 1: Education

Educating the contractors and sub contractors about the diesel emission control requirements was found to be extremely important so that the best possible equipment (within the contractor's means) could be in operation at the earliest possible time. Almost two years prior to construction, MTA CC communicated with rental agencies, dealers, and manufacturers of the upcoming need for machines with Tier 2 engines in Lower Manhattan. MTA CC also made contractors aware of the new requirements

[1] The passive DPF uses the heat of the exhaust gas to burn-off PM from the filter, thus regenerating the filter capacity. Active systems, on the other hand, are not dependant on the exhaust gas temperature for filter regeneration but use an auxiliary means to generate heat (e.g. a diesel burner integrated in the exhaust system). Active DPFs are used on equipment such as cranes, because the application does not generate enough heat to burn off the PM. We employed DOCs on equipment such as cranes, since passive DPFs are not applicable and active DPFs were not available.

and shared a preliminary copy of the specifications with the industry. Commitments from United Rental Agency and equipment dealers were obtained that ensured Tier 2-engines would be included in their annual procurement for the most commonly anticipated equipment types.

Initially some contractors confused DPFs with DOCs. A DOC is a catalytic converter that promotes the oxidation of carbon monoxide, hydrocarbon and soluble organic fraction portion of particulate matter. It also oxidizes sulfur dioxide to sulfate. DPFs are preferable to DOCs because they are much more effective in reducing PM emissions. There was also confusion about verified technology, which is described below under *Challenge 3*.

Introducing these new specifications required MTA CC personnel to take on additional roles including: being available as a single-point help resource; offering to research and provide practical solutions to the contractor; being a coach; and following-up on issues. DPF suppliers were knowledgeable and helpful in providing information about their products. They were also willing to work with the contractors and agencies to ensure their products satisfied the requirements of the program.

Challenge 2: Resistance to Change Standard Practices

Some manufacturers and their dealers initially did not agree to allow DPF installation on their machines, citing the dangers of potential backpressure caused by PM "loading" despite a backpressure monitor in the cabs. The backpressure monitors serve as an early warning system to cab operators if an unsafe backpressure condition should arise. High backpressure or no backpressure during engine use could mean either that the filter was blocked or that it was ruptured, respectively. It should be noted that none of these conditions were experienced.

Engine warranties with respect to DPFs were initially a subject of great debate because of equipment owner's fear of damage to engine from high back pressure and the lack of endorsement to the use of DPF from the original equipment manufacturer (OEM). Generally equipment and engine manufacturers allow use of auxiliary devices, for example Caterpillar's warranty states that when other manufacturers auxiliary devices, accessories, and consumables are used on Caterpillar equipment, the Caterpillar warranty is not affected simply because of such use Finally, it should be noted that no engine failures have been reported.

Contractors were also resistant to using ULSD, claiming that they would need equipment specifically designated to the Lower Manhattan work that could not be used elsewhere due to a fear of cross contamination of ULSD with regular diesel with higher sulfur content. This fear proved to be unfounded since off road equipment dedicated to Lower Manhattan sites remained on site for a considerable period of time and were fueled on site. Irrespective of horsepower all diesel engines received only ULSD fuel. Additionally, even though MTA NYCT Clean Fuel buses program had already created a market for ULSD fuel, it was originally only available from a single supplier. It is now readily available in the New York metropolitan region.

Challenge 3: Availability of Quality DPFs

Retrofits for non-road engines are custom made. Unlike off-the-shelf parts, retrofit procurement has to be planned at least 4 to 6 weeks before construction equipment is deployed. Many retrofit suppliers were more interested with large volume production and not with customization for our off-road equipment. As a result, the supplier base for off-road retrofit is not as extensive as for on-road retrofits.

The U.S. EPA and California Air Resources Board (CARB) publish a list of verified commercially ready retrofit technologies. The U.S. EPA has a DPF verification program where products undergo a standardized test protocol; generally a 4-cycle ISO8178 test. The program eliminates the need for project sponsors to self-qualify vendors' products.

In the northeast, there was only one manufacturer of U.S. EPA verified products that was willing to serve the off- road market. In order to expand the available supplier list, VERT-certified suppliers (a European certification process) were included as acceptable. Information on approved manufacturers of DPFs is available on the U.S. EPA[2] and CARB[3] web sites

MTA CC took actions to ensure the availability of verified DPFs by making sure that retrofit manufacturers were aware of the future need in Lower Manhattan. In addition, MTA CC acted as a liaison between the retrofit filter manufacturers and the rental agencies and equipment dealers, provided contact information to contractors, kept tabs on orders placed, and followed through with the installation. These steps were necessary to ensure the timely implementation of the controls.

Challenge 4: DPF Installation

The first DPFs used on our projects were larger than the actual mufflers in the equipment and required some extra fittings, such as piping and clamps, for initial installation. Poorly installed DPFs resulted in leaky joints, which became evident within a few days of operation because a black ring would form on the outside of the filter. Dealers, rental companies and some contractors who used off-site maintenance shops for the installation were typically more successful. Today's custom designs allow for DPF installation within the confines of the engine compartment as a muffler replacement. This innovation makes installation safer, easier and more effective.

---

[2] EPA- http://www.epa.gov/etv/verification/verification-index.html

[3] ARB - http://www.arb.ca.gov/diesel/verdev/verdev.htm

Challenge 5: Enforcement and Monitoring Compliance.

Despite the proactive educational measures described above, a certain amount of flexibility had to be afforded to contractors at outset of each contract, particularly for smaller subcontractors. A request for a waiver from the Tier 2 requirement was possible, although not encouraged. MTA CC follows the following guidelines in making allowances for hardship claims made by the contractor:

Accept:	Tier 2 + dpf or Tier 1 (waiver submitted) + dpf
Consider:	Tier 2 + dpf (ordered from manufacturer/will be installed within a specified timeframe)
Avoid:	Tier 1 +dpf (ordered from manufacturer/will be installed within a specified timeframe)
Disallow:	Non-tier engines + dpf Non tier engines

Verifying the tier of the equipment brought on site was particularly problematic, since engine data plates are not standardized, do not state tier level directly, and in many instances the information on the engine data plates had to be researched for tier status. Additionally, access to engine data plates within the tight confines of the engine compartment was typically available only when the engine was running, which made recording data difficult.

Initially, tracking fuel consumption and quality was difficult because contractors purchased fuel from different distributors and different contractors on the same site had equipment models that were the same. It was difficult to match equipment to fuel distributor to contractor to gallons of ULSD. To resolve the difficulties with tracking compliance, a simple but effective solution was implemented, where the prime contractor negotiated a fuel contract with a single fuel distributor. The fuel distributor was asked to bar code each piece of equipment filled on site and provide an automated report that captured the date, time, quantity and quality of fuel supplied. An unexpected benefit to this solution was that the fuel reports served as a cross check to the machine inventory on site.

DPF verification was based on physical inspection at site. Once verified for compliance, equipment received a very visible numbered green sticker stating ‘Low Emission’ that served as an easy means to communicate to those unfamiliar with the details of the program.

Program Results/Conclusion

As of December 2006, data from Fulton Street and South Ferry projects indicate that there have been almost 185 pieces of equipment operated on the sites of which 147 pieces were eligible for the Clean Diesel Emissions Controls (i.e. 50 horsepower and above). Of the 147 pieces, about 86 percent were Tier 2 engines and 13 percent were Tier 1 engines. Of the 147, approximately 72 percent were Tier 2 DPF-retrofitted and eight percent were Tier 1 DPF-retrofitted. Another six percent was a combination of Tier 1 or Tier 2 engines with DOCs.

The DPFs varied in cost according to horsepower, ranging from about \$6,000 to \$9,000 for engines between 84 and 155 horsepower. For engines between 210 and 800 horsepower, the cost of a DPF ranged from about \$10,000 to \$33,000. These costs exclude the cost of installation.

ULSD fuel was used exclusively, with no negative implications to the contractor's equipment or equipment fleet. The cost difference at the rack between ULSD and the low sulfur diesel fuel alternative was as little as \$0.04 rising up to \$0.40, with an average value of \$0.17. Between South Ferry and Fulton Street Transit Center, as of December 2006, approximately 380, 000 gallons of ULSD fuel was consumed at a cost, resulting in an incremental cost of roughly \$65,000.

Positive feedback on the Clean Diesel Emissions Program was received from construction workers and, in particular, the equipment operators (who are around machines eight hours per day), who appreciated breathing cleaner air as a result of the program. Positive press articles were generated and community support for the projects was enhanced because of the program.

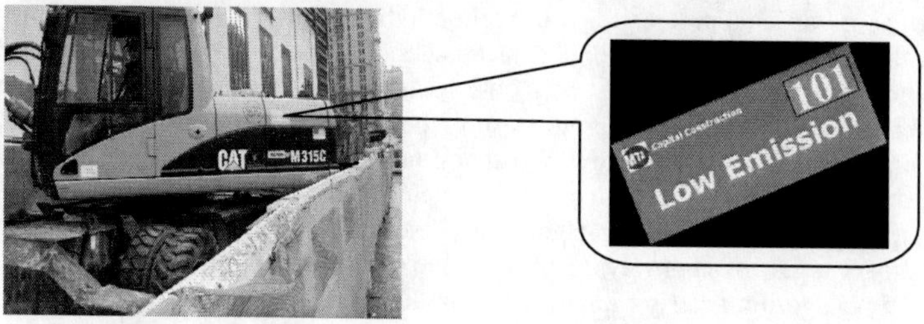

Figure 1. Excavator with Low-Emission Sticker

Figure 2. Excavator with DPF

Investigating the Impact of Ramp Metering from Simulation Results

S. Farahani [1] and B. Hashemloo [2]

[1] University of Waterloo, Department of Civil and Environmental Engineering, Transportation Group, N2L 3G1, Waterloo, 200 University Avenue West, PH (519)8884567;email: sfarahan@uwaterloo.ca
[2] University of Waterloo, Department of Civil and Environmental Engineering, Transportation Group, N2L 3G1, Waterloo,200 University Avenue West, PH(519)8884567;email: bhasheml@uwaterloo.ca

Abstract

As an efficient traffic control strategy to ameliorate freeway traffic congestion, ramp-metering has been successfully applied in the US. However, the applicability and effectiveness of a ramp-metering strategy are required to be investigated during the pre-implementation phase in order to ensure the success of the implementation. Traffic simulation models can provide a quick and cost-effective way to test and evaluate a ramp-metering algorithm prior to implementation on a freeway network.
In this report, one ramp-metering algorithm (Fixed-time ramp metering) has been evaluated over a part of Highway 85 in Waterloo, Ontario. Simulation results show that fixed time algorithm can not improve the condition of Highway 85; in some cases, the algorithm may even delay traffic and increased pollution.

Introduction

The use of traffic signals on-ramps at freeways is called ramp metering. They are installed to control the rate of vehicles that enter freeways from ramps, to keep it below the critical volume of a freeway in order to control the demand and moreover, prevent movement of vehicles in a group entering the freeway upstream of the signal to decrease the weaving phenomenon at the merge area.
Ramp metering is a well-known technique for freeways. In fact, various techniques of ramp control were used in the late 1950s and through 1960s in Chicago, Detroit, and Los Angeles. By the early 1990s Ramp metering systems expanded to nearly twenty metropolitan areas (Pearson, 2002).

The performance of a metering system depends on the metering rate (the specific entrance rate for vehicles from ramp to the freeway) and a control strategy. Modes of metering operation can be divided into three primary categories: fixed-time traffic responsive, local traffic responsive and coordinated traffic responsive. A fixed-time ramp-metering plan is based on historical traffic information and established on a time-of-day basis (Papageorgious, 2000). For local traffic responsive operation, the metering rate is based on prevailing traffic conditions in the vicinity of the ramp. Coordinated traffic responsive ramp metering operation seeks to optimize a multiple-ramp section of a highway, often with the control of flow through a bottleneck as the ultimate goal (Salem, 1995).

There are two ways to evaluate the performance of ramp metering systems: field operational tests and computer simulations. Although field tests provide the ultimate evaluation tool they are expensive, time consuming, and may be affected by uncontrolled events.

When a new project or changes to the current system need to be made, it is doubtful that full scale field tests can improve operations in a timely manner. Transportation projects, because of their high investment costs, require thorough investigation about benefits and applicability. Simulation can be used to evaluate new policies, operating procedures and decision rules without disrupting ongoing operations of the real system for maximum efficiency (Liu, 2001).

In this report, Integration as one kind of simulation model with its capabilities in estimating of travel time, fuel consumption, vehicle emissions and vehicle delay will be used for the investigation of changes after ramp metering action.

Integration as a microscopic model has significant random components and attempts to represent the dynamic nature of traffic phenomena by simulating individual vehicles, such as shockwaves, gap acceptance, and weaving (Aerde, 2002).

A fixed-time strategy, that is based on historical demands without use of real time measurements and on a simple static model is used for metering of this location's ramps.

In this suggested approach by Wattle-Worth (Papageorgious, 2000), a freeway with several on-ramps and off-ramps is subdivided into sections. Each section contains one on-ramp. So,

$$q_j = \sum_{i=1}^{j} \alpha_{ij} r_i \qquad \alpha_{ij} \in [0,1] \qquad (1)$$

where:
qj is the mainline flow of section j, ri is the on-ramp volume of section i and α_{ij} expresses The portion of vehicles that

enter the freeway in section i and do not exit the freeway upstream of section j. To avoid congesting there are two limitations:

$$q_j \le q_{capj} \qquad \forall j \tag{2}$$

$$r_{j,\min} \le r_j \le \min\{r_{j,\max}, d_j\} \tag{3}$$

where:
q_{capj} is the capacity of section j, d_j is the demand while $r_{j,max}$ is the ramp capacity at on-ramp j.

At the end, the objective of this approach is to maximize the number of served vehicles which is equivalent to minimize total travel time or maximize the total traveled distance which is shown on equation 4.

$$Max \sum_j \Delta_j q_j \tag{4}$$

where:
Δ_j is the length of section j ,q_j is the mainline flow of section j, and the objective function is to maximize the total travel distance in the whole network.
These formulations lead to linear-programming problem that may be solved by using Simplex method and broadly available computer codes.

Study Site Attributes

The Waterloo Region is located about one hour west of Toronto in Ontario. The region includes of the cities of Kitchener, Waterloo, and Cambridge plus four townships (Figure1).

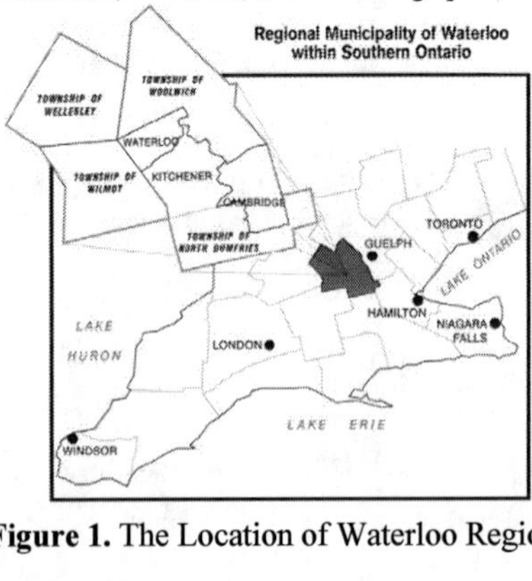

Figure 1. The Location of Waterloo Region

Overall population of this area is almost 460,000 persons according to 2005 census.

Highway 85 at Waterloo region is the main freeway which connects to Hwy 401 through Hwy, 7 and 8.
In recent years, there have been a number of major traffic congestions within the region of Waterloo due to traffic incidents on mentioned Highways (Waterloo Region report, 2005)
The selected section of Hwy 85 for modeling purposes in this paper is 4.11 kilometers long and contains 4 on-ramp, 4 off-ramps and 6 on and off loops in north and southbound direction (Figure2).

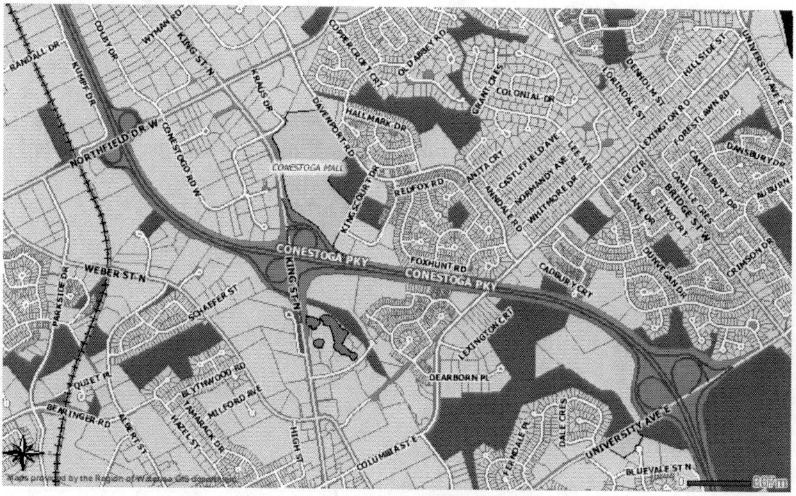

Figure 2. The geographic map of the selected section

Modeling of the Project in Integration

The project is modeled by 41 nodes or points, which 33 of them are defined as intermediate nodes. These nodes are connected by 140 links or arcs in the network and conduct the demand from 8 remaining hypothesis nodes at the heads of the network into system.

As a very first step in preparing data, the coordinates of each node was interpreted to understandable inputs for Intergrasion Software. Links that connect the upstream and downstream node in the network should have associated number of lanes, length, saturation flow rate, free-speed, speed-at-capacity, and jam density which obtained from the region of Waterloo municipality (Table.1).

Table1. One-way highway and street characteristics for links modeling

	Highway 85	University Avenue	King street	Northfield DR	On and Off ramps
Number of Lanes	2	2	2	2	2
Free speed (Km/h)	100	50	60	60	50
Speed at capacity	70	40	48	48	40
Base saturation flow rate (vph)	2000	1600	1600	1600	1100

On-ramps are modeled as two separate sequential links. The first link is coded such that it is controlled by a ramp metering traffic signal and can store any queues. The second link then allows for the acceleration by vehicles from the ramp meter stop line to the actual freeway-ramp merge area.

Three different conditions of origin-destination demands (Low, Average, and High) are modeled by considering the information from Regional Municipality of Waterloo.

Modeling Fixed Time Algorithm for the Study Site

In a Fixed-time strategy, the whole freeway should be segmented so that each section contains one on-ramp. For this study site, the Highway 85 is subdivided in to 3 parts containing one on-ramp in each part (Figure3).

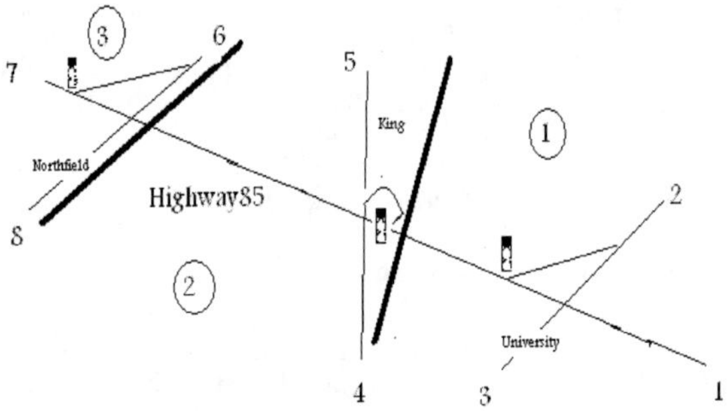

Figure 3. Diagram of subdivided of Highway 85 to three sections

Considering three conditions of traffic demand, there are three linear programming problems that call for branch-and-bound methods (Simplex Solution) for results.

The sample formulation of this linear model for average demand is shown in below.

$$\text{Max } (2.63q_1 + 1.52q_2 + 0.65q_3) \qquad (5)$$

$$q_1 : 2000$$
$$q_2 \leq 2000 \qquad (6)$$
$$q_3 \leq 2000$$

$$150 : r_1 : \min\{1100,810\}$$
$$150 : r_2 : \min\{1100,640\} \qquad (7)$$
$$150 : r_3 : \min\{1100,430\}$$

$$q_1=0.34r_1$$
$$q_2=0.47r_1+0.18r_2 \qquad (8)$$
$$q_3=0.65r_1+0.41r_2+0.32r_3$$

Estimating Cycle Time and Split Phases

Once the linear model has been solved for travel demand, the cycle time of signals can be estimated based on-ramp volume. The duration of cycle time and green phase is found with:

$$Cycletime = \frac{3600}{r} \qquad (9)$$

$$g = \frac{r}{r_{sat}}(Cycletime) \qquad (10)$$

where:
g is the green phase duration, r_{sat} is the ramp capacity flow and r is ramp Volume. The cycle time and split phases (Considering signals with two sections of Red and Green) for three on-ramps in three conditions of O-D demands are shown in Table.2.

Table2. Cycle time and split phases (Seconds) for signal on-ramps

		Low Demand	Average Demand	High Demand
Ramp on part# 1	Cycle Time	8.5	6	5.92
	Red Phase	5.26	2.8	2.64
	Green Phase	3.24	3.2	3.27
Ramp on part #2	Cycle Time	10	5.90	5.62
	Red Phase	6.73	2.6	2.36
	Green Phase	3.27	3.3	3.26
Ramp on part #3	Cycle Time	12.85	8.4	8.27
	Red Phase	9.57	5.1	5
	Green Phase	3.27	3.3	3.27

In reality, it is recognized that ramp metering should not be specified for volumes below 180 veh/h on ramp, because of possible malfunctions, enforcement issues and equity concerns for users who travel from arterials. Even in this case waiting time of 6 to 9 seconds at red signal of ramps in the low demand condition on Highway 85 is not justified, because more people admire to select highway which results congestion in highway.

For solving this problem of fixed-time strategy, there should be switch on or off of ramp in the different time periods of day. For example it can be operated during the morning and afternoon peak hour period regarding the historical demands.

Integration Features

Several simulation runs with different random seeding were conducted for different time periods under three demands conditions.

The first 60 seconds of each simulation run were treated as "warm-up" periods by the software itself and the data of this period are thrown out. Following, the brief description of some effectiveness measurements by Integration is provided. Vehicles are modeled into five classes: light and heavy duty cars, light and heavy duty truck, in addition to buses.

Estimation of Link Travel Time

Integration determines the link travel time for any given vehicle by providing that vehicle with a time card upon its entry to any link. This time card is retrieved when the vehicle leaves the link. So, the differences between these times provide a direct measurement of the link travel time.

Estimation of Vehicle Delay

This model estimates vehicle delay every deci-second as the difference in travel time between travel at the vehicle's instantaneous speed and travel at free-speed.

Estimation of Vehicle Fuel Consumption

Integration computes the speed of vehicles each deci-second. This capability permits the estimation of fuel consumption rate for a vehicle each second on the basis of its current speed and acceleration (Rakha et al, 2000; Ahn et al, 2001). The model in Integration was developed using data that are collected on a chasis dynamometer at the Oak Ridge National Labs (ORNL). This model use instantaneous speed and acceleration level as independent variables and estimation are executed every second for every vehicle in the network.

Estimation of Vehicle Emissions

The key urban transport emissions that affect air quality for all classes of vehicles are CO(Carbon Monoxide), Nitrogen Oxides (NOx), Ozone(O_3), particulate matter (PM_{10}), Hydro Carbon(HC), and Sulfur Dioxides (Potter, 2003; Ahn, 1998).

Emission estimation also operates on a second-by-second basis for six main categories of urban pollutions as it is shown in Table 3

Comparison of Simulation Results

In continue the comparison of several measurements is provided for without control and control case for the average and low demand which are more accurate cases. In the high traffic demand, there is a spillover of traffic from ramps to arterials, and Integration has not this ability to change the signal timing during the running period in order to decrease the red phase on signals.

Table3. Results of simulation running for Average and Low demand

	Average Traffic Demand		Low Traffic Demand	
	No-Control	Ramp-Control	No-Control	Ramp-Control
Vehicle trips –Number of trips	3505	3505	1850	1850
Vehicle Km-Total traveling distance	6963.19	6963.39	3665.84	3665.76
Vehicle Stops – Number of stops	3791.34	4098.26	1437.85	1772..95
Total traveling Delay (Seconds)	329317.53	208626.88	11481.19	14335.38
Stopped Delay	11182.98	10484.09	213.70	553.40
Accel /decel Delay	318134.53	198142.72	11267.49	13781.98
Fuel Consumption (L)	686.09	589.45	264.30	298.80
HC(g)	198.71	192.81	96.53	101.53
CO(g)	5500.20	5405.92	2662.72	2856.72
NOx(g)	801.17	791.05	423.89	460.89
CO_2(g)	1438221.75	1353316.75	605018.06	655801.09
Total Link Travel Time (Veh-min)	10818.07	8806.33	3001.67	3678.78
Average Network Speed (Km/h)	38.75	47.61	73.53	72.40

Inspecting the Table3, shows that during the installation of ramp metering total vehicle stops has been increased because of additional stops at ramp signals, however change of speed (acceleration and deceleration) in the merge area without signals on ramps has been resulted more traveling delay for average traffic demand. In addition, the results show an increment for fuel consumption and total green house emission components that adapts to the researches by Rakha et al (2002) and Ahn(1998); vehicle fuel consumption is more sensitive to running speed condition than it was to vehicle stops. The objective function which is to minimize total travel time has been observed for average demand with provided higher network speed.

For low traffic demand since the vehicles have to stop at ramp's signals when there is not any congested condition and needs to be stopped , unnecessary stopped, accel, and decel delay has been experimented which results more fuel consumption and vehicle emission components.

Conclusion

The simulation results show that the effectiveness of ramp control has been influenced severely by traffic demand condition. Average demand with providing necessary stops of vehicles at ramps' signals prevent acceleration and deceleration of vehicles at the merge area of highway and ramp, thus the benefits such as reduced total travel time, delay, fuel consumption and GHG emissions are observed for this scenario.

Low and high traffic demand respectively with needless stops and spillover of traffic from ramps to arterial causes deficiencies that diminish the benefits of ramp metering. As a result, fixed time strategy with the setting based on historical rather than real-time data may be a crude simplification because:

- Demands are not constant, even within a time-of-day.
- Demands may vary at different days, e.g. due to special events.
- Demands change in the long term leading to "aging" of the optimized settings.
- Incidents and farther disturbances may perturb traffic conditions in a non predictable way.

Traffic-responsive coordinated strategies which determined the signal's phases at each second based on the current condition of traffic demand are potentially more efficient even with respect to their cost.

References

Aerde. M.V.,and Assoc. (2002). "Integration Release2.30 for windows users guide, Volume I, II."

Ahn. K.,and Rakha. H. (2001). 'Estimating Vehicle Fuel Consumption and Emissions based on Instantaneous Speed and Acceleration Levels." *Journal of Transportation Engineering.*

Liu. H.X., and Recker. W. (2001). "Development of a simulation laboratory for evaluating ramp metering algorithm." *Intelligent transportation system conference.*

Papageorgious. M., and Kotsialos. M. (2000). "Freeway ramp metering: an overview." *IEEE intelligent transportation system conference.*

Pearson. R., and Black. J. (1999). "Intelligent Transportation system, ramp metering." http:// Calccit.org /its decision/

Potter. S. (2003). "Transport Energy and Emissions, Urban Public Transport." *Journal of Public Transportation.*

Rakha. H., and Ding. Y. (2003). "Impact of Stops on Vehicle Fuel Consumption and Emissions." *Journal of Transportation Engineering.*

Salem. H., and Papageorgious. M. (1995). "Ramp metering impact on urban corridor traffic: field results." *Transportation Research Record,* PartA,Volume29.

A Microscopic Approach to Modeling Air Quality Impacts of HOV Lane Conversion

K. Boriboonsomsin[1] and M. Barth[2]

[1,2]College of Engineering - Center for Environmental Research and Technology (CE-CERT), University of California, Riverside, 1084 Columbia Ave, Riverside, CA 92507; PH (951) 781-5791; FAX (951) 781-5744; email: kanok@cert.ucr.edu, barth@cert.ucr.edu

Abstract

There has been an increasing trend in converting some of the under-utilized high-occupancy vehicle lanes to mixed-flow lanes. This paper presents a methodology to model and evaluate the air quality impacts of such conversion using an integrated modeling tool that uniquely combine a microscopic traffic simulation tool with a modal emissions model. A freeway stretch in Southern California was used as a case study to demonstrate the deployment of the integrated tool. The results for this particular cast study show that the lane conversion will not benefit air quality if the induced demand is more than 5% of the current demand.

Introduction

High-occupancy vehicle (HOV) lanes have been implemented on many freeway systems as a measure to improve mobility, trip time reliability, and air quality (NCHRP, 1998). With regards to air quality, current federal policies encourage construction of HOV lanes and restrict funding for mixed-flow (MF) lanes in air quality non-attainment areas (FHWA, 2002). However, there has been criticism that some HOV lanes are not effectively utilized and should be converted to either MF lanes or high-occupancy toll (HOT) lanes (California LAO, 2000). Examples of HOV-to-MF lane conversion include those that occurred on I-80 and I-287 in New Jersey in 1998 (Turnbull and DeJohn, 2000) while HOV-to-HOT lane conversion has been underway nationwide in the past few years (FHWA, 2006).

Significant changes to HOV lanes, such as conversion to MF lanes, may be subject to review by the relevant state and federal agencies. Federal approval is also required if such a conversion is located in an air quality non-attainment or maintenance area. In particular, the conformity determination of Transportation Plan or Transportation

Improvement Program must be re-evaluated (FHWA, 2002). The traditional method of determining transportation conformity involves the use of a regional travel demand model and emission factors from conventional emission models (e.g. MOBILE or EMFAC). This method has limitation in the sense that it is not sensitive to operational effects such as the differences in traffic dynamics between HOV and MF lanes. This paper presents the deployment of an integrated microscopic modeling tool that uniquely combines the PARAMICS traffic simulation software with the Comprehensive Modal Emissions Model (CMEM) to assess the air quality impact of HOV-to-MF lane conversion.

PARAMICS is high-fidelity traffic simulation software that is in wide use. One of its key advantages is that it has an open architecture for integrating plug-in modules to perform specific simulation functions through Application Programming Interfaces. The CMEM plug-in for PARAMICS was developed and tested extensively at the University of California, Riverside Center for Environmental Research and Technology (Barth et al., 2001). Effort has been made to keep it up-to-date and compatible with new versions of PARAMICS as they were released. Because PARAMICS simulates the movement of each individual vehicle, it is capable of producing second-by-second vehicle trajectories (location, speed, and acceleration) that are necessary inputs for CMEM. During the simulation, the CMEM plug-in gathers driving trajectories of all vehicles being simulated and calculates the corresponding vehicle emissions.

Methodology

A 12-mile stretch of SR-91 East in Riverside County, California, was used as a case study. It was chosen because of the availability of reliable loop sensor data along the section. Its median lane is a limited access HOV lane with seven ingress/egress locations (see Figure 1). The freeway stretch and associated travel demand were coded in PARAMICS. Traffic in the model network was simulated and calibrated to the weekday morning peak hour traffic in April 2006. The existing traffic condition during this period was primarily uncongested in the HOV lane and mildly congested in the MF lanes. Approximately, the HOV lane was in level of service (LOS) B or C for about 80% of the time. The MF lanes were in LOS C or D for about 80% of the time.

Both MF and HOV lanes were calibrated. A majority of the calibration effort was placed on adjusting demand tables and HOV lane choice. The simulated volumes were checked against the volumes collected by the Freeway Performance Measurement System (PeMS, 2006). The simulated travel times were compared with the travel times calculated from probe vehicle runs. The calibration continued until all criteria set by the California Department of Transportation (Dowling et al., 2002) were met. For instance, the final *GEH* statistics of total link flows for MF and HOV lanes were 2.4 and 2.8, respectively; both below the set criterion of being less than 4. For more details regarding model calibration, please see (Boriboonsomsin and Barth, 2006). After being well calibrated, the model network was used to simulate traffic and estimate associated vehicle emissions. Further, another model network with the median lane being converted to an MF lane was created and vehicle emissions from this network were also estimated.

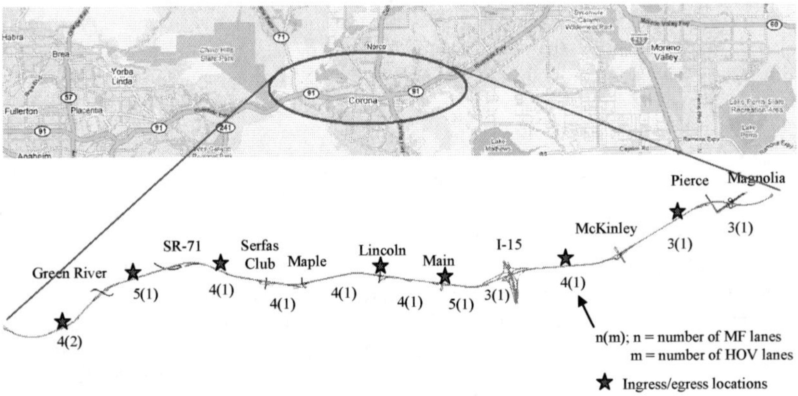

Figure 1. Study site and network characteristics [credit for the map: http://maps.google.com].

In addition to the existing traffic condition, sensitivity analyses were conducted to investigate how the two networks would perform if the traffic condition changed. Two key variables were studied—overall vehicle demand and the percentage of HOVs in the traffic mix (%HOV). The increase in vehicle demand will affect the operational performance of the network as there will be more vehicles using the limited capacity of the network. As the demand approaches the network capacity, congestion may occur and the pollutant emissions will be likely to increase. The %HOV is important in the sense that it determines how many vehicles are eligible to use the HOV lane. This value is also concerned with whether the split between eligible and ineligible vehicles is balanced with the split between the roadway capacity or not. For example, for a freeway section with three MF lanes and one HOV lane, the HOV lane accounts for 25% of the total capacity. If the %HOV is 10%, the HOV lane will be under-utilized and the remaining 90% of the traffic will be forced to use the 75% capacity of the freeway. Nevertheless, the impact may not be significant if the overall vehicle demand is well below the total capacity. In essence, both variables and the interaction between them are important factors in determining how a freeway with HOV lane would operate. Two demand levels in addition to the current demand level were 5% and 10% growth. Two %HOV values in addition to the existing 25% value were 18% and 32%. In total, they formed up nine scenarios, which were simulated in both networks.

Results

For a fair comparison, only emissions from mainline links were reported here. For each scenario, four to six simulation runs were made using different seed numbers to ensure statistical significance of the average value at 5% alpha level. The average values and standard deviations of the estimated emissions of all scenarios are summarized in Table 1. According to the table, it is observed that under the same traffic condition the median MF network produces less emission mass than the median HOV network for every pollutant. This is mainly due to the slightly better and smoother flow of traffic in the median MF network as the vehicles (both HOVs and non-HOVs) in this network have no restriction in their lane choice. The largest emission differences are for carbon monoxide (CO), followed by hydrocarbon (HC). The median MF network produces about 20-25% less CO and 15-20% less HC than the median HOV network. For nitrogen oxides (NO_x) and carbon dioxide (CO_2), the differences are comparatively lower. The median MF network produces less NO_x and CO_2 in the order of 10% or less.

Table 1. Estimated emissions.

(a) Median HOV Network

Scenario	Demand Growth	%HOV	CO (kg)		HC (kg)		NO_x (kg)		CO_2 (kg)	
			Avg.	Std.	Avg.	Std.	Avg.	Std.	Avg.	Std.
1	0%	18%	3,432	62	52	1.4	194	6.5	43,957	1,098
2	0%	25%	3,320	76	50	1.3	191	5.0	42,592	763
3	0%	32%	3,332	31	51	0.6	187	4.4	41,719	626
4	5%	18%	3,856	40	57	0.5	207	5.5	47,638	881
5	5%	25%	3,747	27	56	1.4	206	4.4	46,596	717
6	5%	32%	3,665	77	55	1.2	201	4.5	45,213	927
7	10%	18%	4,469	122	65	0.8	229	5.7	53,956	980
8	10%	25%	4,190	92	61	1.0	217	5.7	50,255	931
9	10%	32%	4,248	99	62	1.4	222	1.7	50,620	818

(b) Median MF Network

Scenario	Demand Growth	%HOV	CO (kg)		HC (kg)		NO_x (kg)		CO_2 (kg)	
			Avg.	Std.	Avg.	Std.	Avg.	Std.	Avg.	Std.
1	0%	18%	2,593	29	43	0.1	180	3.9	40,100	429
2	0%	25%	2,528	28	42	0.7	175	6.1	38,875	481
3	0%	32%	2,577	44	43	1.0	174	3.3	38,774	204
4	5%	18%	2,874	53	47	1.1	187	6.9	42,628	1,088
5	5%	25%	2,859	46	47	1.1	191	5.6	42,680	582
6	5%	32%	2,829	51	46	0.9	184	8.7	41,615	1,073
7	10%	18%	3,228	19	52	0.8	210	6.9	47,704	857
8	10%	25%	3,301	66	53	0.7	206	1.2	47,205	578
9	10%	32%	3,251	50	52	1.1	208	5.7	46,800	843

Discussion

The presented results show that under the same overall vehicle demand and %HOV in the traffic mix the median MF network produces less total emissions mass. However, a change in transportation supply (e.g. add a new lane) will usually result in a shift in system equilibrium, due in part to induced demand. This has been well discussed in the literature, for instance, (Johnston and Ceerla, 1996) and (Rodier and Johnston, 1997). In the case of HOV-to-MF lane conversion, increased capacity for non-HOVs after the conversion has taken place may cause an increase in overall vehicle demand on the freeway network. For example, vehicles from other routes may change their travel route to use this freeway. Also, some carpoolers and transit riders may no longer do so and turn to driving alone. Consequently, the %HOV will also change in the decreasing direction. Therefore, the evaluation of emissions impact of HOV lane conversion should not be done on the basis of the same vehicle demand and %HOV.

It is of interest to determine how much additional vehicle demand the freeway can take before the lane conversion will have adverse impact on air quality. Based on the sensitivity analysis results in Table 1, the estimated emissions of the median MF network are normalized by the estimated emissions of the median HOV network for the *existing condition* (Scenario 2). If the normalized values are smaller than 1, then the lane conversion is considered to have positive impact on air quality because it reduces emissions. If the normalized values are greater than 1, then the lane conversion is considered to have negative impact because it increases emissions. The normalized values for each pollutant are plotted and shown in Figure 2 for two levels of %HOV (only two are shown for brevity). In this figure, the % growth in demand at which the normalized value is equal to 1 is the "even emissions point". This means the lane conversion which induces more than that percentage of vehicle demand will result in higher emissions. According to Figure 2, findings are:

- The even emission point is pollutant-dependent. CO_2 is the pollutant with the lowest even emission point while CO is the pollutant with the highest even emission point.
- The range of the even emission point for the HOV-to-MF lane conversion of this particular case study is 5-11%.
- The minimum even emission points are 5%, 5%, and 6% for %HOV of 18%, 25%, and 32%, respectively.

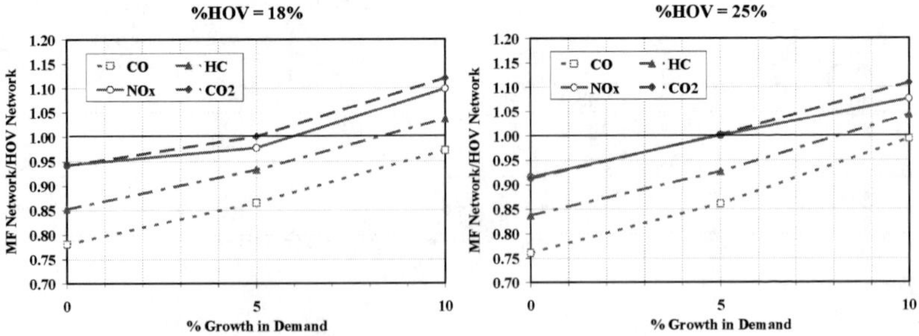

Figure 2. Emissions impact of HOV-to-MF lane conversion.

Conclusion

As discussed earlier, the HOV-to-MF lane conversion is likely to decrease %HOV in the traffic mix. Therefore, the lane conversion in this case study would not benefit air quality if the additional vehicle demand is to be more than 5% of the current demand. This information can be used in concert with the estimated vehicle demand from a regional travel demand model to formulate the most effective freeway lane use policy in both short term and long term. For example, the lane conversion could be done temporarily (e.g. 2-3 years) when the immediate induced demand would not exceed 5%. After that, it could be terminated when the induced demand and the natural growth in demand together would become greater than 5% of the current demand.

References

Barth, M., Malcolm, C., and Scora, G. (2001). "Integrating Comprehensive Modal Emissions Model into ATIMS transportation modeling frameworks." *California PATH Research Report UCB-ITS-PRR-2001-19*, August, 48 pp.

Boriboonsomsin, K. and Barth, M. (2006). "Evaluating and modeling the effectiveness of HOV lanes at improving air quality." *Final report to the California Department of Transportation,* December, 122 pp.

California Legislative Analyst's Office (LAO). (2000). "HOV lanes in California: Are they achieving their goals?" *Legislative report*, Sacramento, CA, July.

Dowling, R., Holland, J., and Huang, A. (2002). "Guidelines for applying traffic microsimulation modeling software." *Final report to California Department of Transportation,* September, 107 pp.

Federal Highway Administration (FHWA). (2002). "Federal-aid highway program guidance on high occupancy vehicle (HOV) lanes." <http://www.fhwa.dot.gov/operations/hovguide01.htm> (Feb. 7, 2007).

Federal Highway Administration (FHWA). (2006). "Converting high-occupancy vehicle (HOV) lanes to high-occupancy toll (HOT) lanes." <http://www.ops.fhwa.dot.gov/tolling_pricing/value_pricing/projtypes/hovhot lanes.htm> (Feb. 7, 2007).

Freeway Performance Measurement System (PeMS). (2006). <http://pems.eecs.berkeley.edu/> (Sep. 30, 2006).

Johnston, R. A. and Ceerla R. (1996). "The effects of new high-occupancy vehicle lanes on travel and emissions." *Transportation Research Part A,* 30(1), 35-50.

National Cooperative Highway Research Program (NCHRP). (1998). "HOV systems manual." *Report 414, Transportation Research Board*, Washington, D.C..

Rodier, C. J. and Johnston, R. A. (1997). "Travel, emissions, and welfare effects of travel demand management measures." *Transportation Research Record* 1598, 18-24.

Turnbull, K. F. and DeJohn, T. (2000). "New Jersey I-80 and I-287 HOV lane case study." *Report no. FHWA-OP-00-018, Federal Highway Administration*, Washington, D.C.

Study of Influence of Lane Restrictions on Vehicular Emissions under Heterogeneous Traffic Flow

S. P Anusha[1], R. Sivanandan[2], and S. K. Senthilraj[3]

[1]Former Research Scholar, Department of Civil Engineering, Indian Institute of Technology Madras, Chennai-600 036, India; email: anushanair1@yahoo.co.in
[2]Professor, Department of Civil Engineering, Indian Institute of Technology Madras, Chennai-600 036, India; Ph: +91-44-22574275; email: rsiva@iitm.ac.in
[3]Project Officer, Department of Civil Engineering, Indian Institute of Technology Madras, Chennai-600 036, India; email: senthilrajsk@yahoo.co.in

Abstract

This paper presents the findings of a study of motorized two-wheeler emissions under lane restricted and lane-less flow conditions in heterogeneous traffic. Lane restriction is defined as the movement of particular categories of vehicles on specified lanes (dedicated lanes), whereas lane-less flow movements are conditions when vehicles are free to change lanes. This study was conducted on selected mid-block road sections of different lengths in Chennai city, India. In the process of test runs, tailpipe emissions (CO, HC and NO) were measured using a portable gas analyzer at short intervals of time. These data were then synthesized to evaluate the total emissions from the vehicles for each test run. Instantaneous speeds of the vehicles were measured using an optical sensor fixed to the wheel. Consumption of fuel by the test vehicles were also measured using a petrol reading apparatus. Based on the analysis of field data collected specifically for this study, it is concluded that lane restricted flow generally produced reduced levels of tailpipe emissions compared to lane-less conditions. The impact of various factors such as, number of lane changes, average speed and fuel consumed on emissions were also studied. In order to understand the relationships between emissions and various traffic factors, regression models were tested and built using SPSS software. Highly heterogeneous traffic conditions and the lane changing behavior of drivers in traffic conditions prevailing in urban areas in India offered unique research opportunities in this study.

Keywords: Vehicular emissions, Lane restricted and Lane-less Movements, Heterogeneous Traffic Flow

Introduction

In this age of rapid industrialization, one of the biggest problems facing the society is the issue of clean air. Among the various sources contributing to air pollution, the automobile industry has emerged as the largest source of urban air pollution in developing countries, such as India. India is currently under a phase of rapid growth. Increase in urban population has resulted in unplanned urban development, higher demands for transport, energy and other infrastructure. Vehicles are the major source of air pollution in urban areas. Statistics indicate that during the year 2001 in India, the annual CO emission loads (in Thousand Metric Tones (TMT)) from vehicles in six metro cities were 293 in Delhi, 109 in Mumbai, 45 in Kolkata, 88 in Chennai, 118 in Bangalore and 129 in Hyderabad (Sengupta, 2004).

The emission of pollutants from vehicles are greatly influenced by acceleration, deceleration, idling and cruising. Vehicular accelerations and decelerations produce large amounts of emissions. The number of accelerations and decelerations carried out by the vehicles during travel are a function of characterstics of traffic flow. In unsteady traffic, vehicles experience numerous deceleration events without coming to a full stop or even sharply decelerating. These decelerations, combined with the accelerations, lead to high levels of emissions under stop-and-go or other variable speed conditions.

Traffic control measures have been found to have effect in reducing such vehicular pollution. These measures are strategies that reduce vehicle use or change traffic flow or congestion conditions to reduce vehicular emissions. Some of these measures include ensuring smooth flow of traffic without stoppages, creating dedicated bus lanes, encouraging alternate work hours, introducing driver training programme to reduce fuel consumption, etc. These control measures have to be studied in order to evaluate the benefits on their actual implementation.

Review of literature was conducted to ascertain the extent of work carried out in areas related to this study. Selected works are reviewed below. Cernuschi et al. (1995) reported on the analysis of emissions from light-duty vehicles conducted using a modal approach, which describes emissions for every single mode typical of traffic regimes (idle, constant speed, acceleration, deceleration). Frey et al. (2000) evaluated the effect of traffic signal coordination and timing on real-world on-road emissions. A key insight from this work is that emissions during idling are generally low compared to emissions during acceleration. The investigators reported that emissions are substantially influenced by traffic signalization, primarily because of the accelerations that occur after a red light, rather than because of deceleration or idling. Prakash and Allapat (1998) stressed on the point that vehicles during idling produce more pollutants. A steady speed of 25 to 45 km/h has been reported to be causing minimum pollution. Rouphail et al. (2000) studied the feasibility of using a portable instrument to collect real-world on-road tailpipe emissions data for CO, NO and HC. The investigators reported that measured emission rates (on a gram per second basis) were found to be highest during the acceleration driving mode from zero to approximately 40 mph. Another key finding is that measured emissions tend to increase with traffic congestion, since there are more acceleration events. Sivacoumar and Thanasekaran (2000) studied the vehicular pollution load in Chennai city in India by conducting a detailed survey of vehicular pollution. The investigators developed a highway pollution simulation model based on traffic density of vehicles along with respective emission factors Teng et al. (2001) developed the ONROAD vehicle exhaust emission model for estimating CO and HC emissions. This model establishes relationships between the on-road vehicles exhaust emission rates and a vehicle's instantaneous speed profile. The researchers conducted regression analysis by selecting CO and HC emissions as dependent variable and instantaneous speed, acceleration or deceleration rate, ambient temperature and humidity as independent variables. Varhelyi (2002) conducted a case study on the effects of small roundabouts on emissions and fuel consumption and reported that when a roundabout replaced a signalized junction, CO emissions decreased by 29%, NO_X emissions by 21% and fuel consumption by 28%. Yu and Qiao (2004) studied the relationships between the emissions and model variables such as speed, acceleration and engine RPM. The investigators reported that, car emissions are relatively higher when driving not only around intersections, but also on arterial streets, where decelerations/ accelerations are normally needed.

The above review has highlighted the importance of tackling vehicular emissions. It has also presented a cross section of studies addressing the problem. Under heterogeneous traffic conditions in India, only a few studies have been conducted to monitor tail-pipe emissions from vehicles. In urban roads in India, the carriageway space is used by different types of vehicles by occupying available spaces, without lane discipline. The objective of this research was to study reduction in vehicular tail-pipe emissions under lane restricted conditions vis-à-vis lane-less flow conditions, for two-wheeler vehicles of different age models, and to study the impact of various factors such as, number of lane changes, average speed and fuel consumed on emissions.

Lane restriction is defined as the movement of particular category of vehicles on a specified lane assigned to them. Figure 1 represents a line drawing of lane restricted (1a) and lane-less (1b) traffic flow conditions in a road stretch of the case study. The median lane is dedicated for use by two-wheelers and three-wheelers, the central lane for light motor vehicles and the left lane for buses and trucks (In Indian roads, the traffic moves on the left side of the road). It is believed that vehicles following a lane while driving results in lower emission rates. Lane-less flow movements are conditions when vehicles have the freedom to move by changing lanes. This movement of changing lanes causes the vehicle to accelerate and decelerate according to the traffic flow situations, thereby emitting higher emissions from the exhaust.

Methodology

The crux of the methodology of this study relied on field measurements of tailpipe emissions of selected vehicles while they traversed a mid-block section under varying conditions. An overview of the methodology is shown in Figure 2. The data required for the study included pollutant concentrations, instantaneous speeds and fuel consumption during each test run. A gas analyzer with exhaust probe attached to the tail pipe of the vehicle was used to measure the pollutant concentrations and an optical sensor with speed counter and a data logger assembly measured the instantaneous speeds. These were measured for the selected vehicles (two-wheelers of different ages). These test vehicles were run on different study stretches of varying distances, with the instruments fitted. Each pass was taken as one test run. Many runs were made for each of the vehicles in each of the cases of lane restricted and lane-less movement. Data was collected in the form of pollutant concentrations for emissions and wheel speed for instantaneous speed. The pollutant concentrations were converted into total emissions for each run using appropriate equations. Thus, the total pollutants and speed data for test runs for different conditions were obtained. From this, the percentage reduction in emissions for lane restricted conditions vis-à-vis lane-less condition were evaluated. Further, the interest here was also to examine the variations in emissions and fuel consumption vis-à-vis speeds, and variations in emissions with fuel consumption. Also, regression models were developed to test and build relationships between vehicular emissions and factors such as, type and age of vehicle, fuel consumed, and number of lane changes.

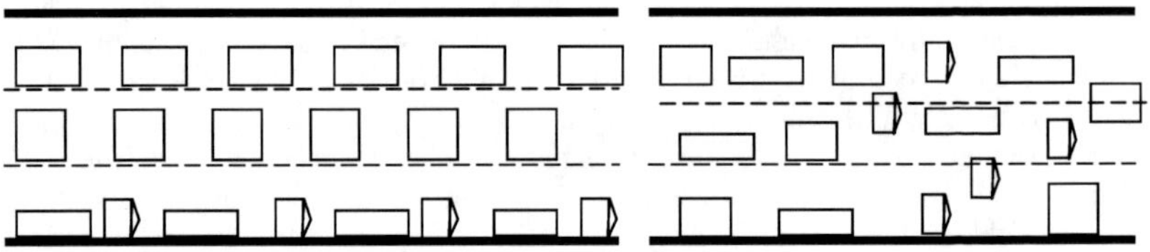

1a) Lane Restricted Flow **1b)** Lane-less Flow

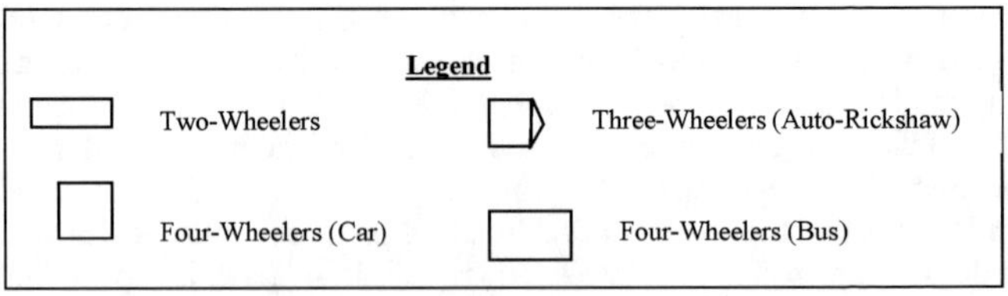

Figure 1. Traffic Flow Conditions in a Case Study Road Stretch

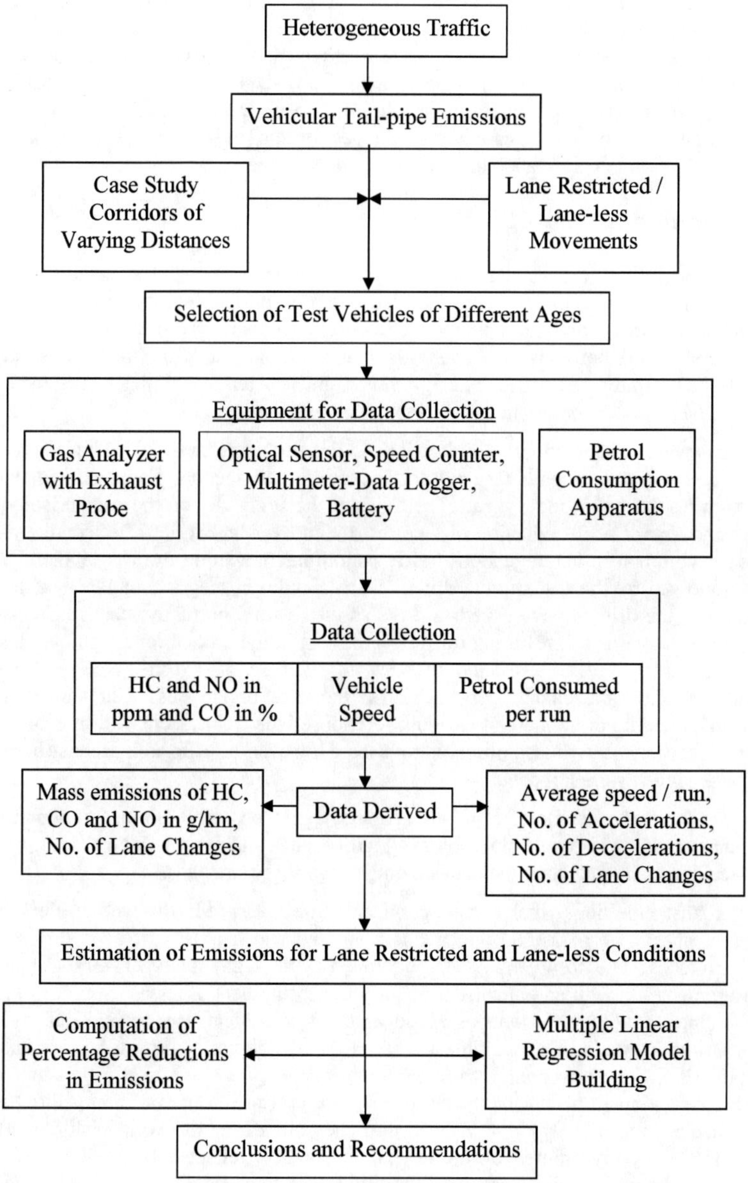

Figure 2. Overview of Methodology

Since the focus of the study was on evaluating the influence of lane restrictions, mid-block sections were chosen as the study stretches, where both lane restricted and lane-less flows could be possible. Also, these sections had to have multiple lanes to analyze the effect of number of lane changes carried out by the test vehicles on vehicular emissions. Four sections, one in Anna Salai and three in GST road in Chennai city were chosen as case study stretches for this study. These sections varied from 1 to 2.5 km in length and had three lanes in each direction.

Data Collection and Analysis

Three two-wheelers (Make: Hero-Honda, Model: Splendour) were selected for the test runs. These test vehicles were identified based on the availability and the age of the vehicles (old, medium and new vehicles). Vehicles registered prior to 1997 were considered as old, between 1997 and 2002 as medium age, and 2003 or later as new vehicles. The study was carried out by running the selected test vehicles on selected road stretches under conditions of lane restriction and lane-less flow.

Test runs were conducted in both directions in the study stretches, by taking U-turns at the intersections at both the ends of the mid-block section. Each pass in one direction was taken as one test run. The focus of the study was on the comparison of the relative merit of introduction of lane restriction over lane-less traffic conditions, which is common in heterogeneous traffic conditions prevailing in urban India. The rules followed by the test vehicle drivers were simple: in the case of lane-less test conditions, the drivers were asked to drive as they would normally, changing lanes when necessary due to prevailing traffic conditions; in the case of lane-restricted test runs, the drivers were instructed to stick to their lanes, irrespective of traffic conditions. The lane changing behavior (in the case of lane-less tests) are quite realistic, since the drivers are at freedom to choose lanes, consistent with prevailing traffic conditions. It was also observed to be so. However, rules for the drivers (based on the behavior) were not developed.

Table 1 shows the details of test runs and study stretches. The number of runs for various stretches was constrained by availability and condition of test vehicles, and practical issues in collecting the data in some of the study stretches.

The exhaust emissions, fuel consumption and average speeds of two-wheelers of different ages were measured under lane restricted and lane-less flow on different road stretches. The test vehicle was equipped with a gas analyzer for pollutant measurement and with an optical sensor for the measurement of speed. Simultaneous measurement of the instantaneous speed of the test vehicle was carried out at 5-second time intervals, in synchronization with that of the gas analyser. An assembly of optical sensor, speed counter and a multimeter-datalogger was specifically adopted for the measurement of instantaneous speeds. The optical sensor was clamped to the two-wheeler in such a way that it faced the inner side of the tire with a white mark painted. The IR ray emitted by the optical sensor gets reflected back whenever it encounters the white paint marked on the tire, which happens once in every rotation of the tire, while running. The reflected ray sends signal to the speed counter which displays the RPM of the tire, which in turn was recorded by data logger. The recorded

RPM of the vehicle tire was converted into speed through suitable calculations. By knowing the speed of test vehicles at the beginning and at the end of 5-second intervals, the corresponding accelerations were calculated.

Table 1. Number of Test Runs Conducted on Study Stretches

Vehicle Model	Study Stretch	Year of Manufacture	Number of Test Runs		Total Test Runs
			Lane Restricted Conditions	Lane-less Conditions	
Hero-Honda Splendor	1	1995	20	20	40
		2005	17	27	44
	2	1995	15	15	30
		2000	19	20	39
		2005	15	15	30
	3	1995	10	10	20
		2000	9	11	20
		2005	10	8	18
	4	1995	14	14	28
		2005	14	10	24
	Total		**143**	**150**	**293**

The exhaust probe of gas analyzer was fixed to the tailpipe of the test vehicle and readings were recorded automatically. A Kane Automotive Gas Analyzer was used in the emission tests. This equipment has been designed to be used on gasoline, petrol, LPG or CNG powered engines. Microprocessor based exhaust gas analyzer has been designed for continuous measurement of gases like CO, CO_2, and HC based on Non Dispersive Infrared (NDIR) principle. Oxygen and NO are measured using automotive electrochemical sensors. It is a portable data measurement device for vehicle emissions. The analyzer can store 250 sets of full test results in its internal memory. Test results can be recorded manually by the press of a key or automatically timed. Data stored in the Analyzer can also be uploaded, diagnosed graphically and / or converted into spreadsheets for service records. The equipment has a length of 22 cm, width of 12 cm and depth of 5.5 cm. The weight of the equipment is around 1 kg. It has clip handle to secure to exhaust and it has a 4m long hose.

A petrol reading apparatus similar to a laboratory burette with graduated markings was connected by tube to the carburetor of the vehicle. This connection was temporarily done on the vehicles to note the initial and final level of petrol to determine the fuel consumed per run by the vehicle. The pollutants of interest for the case study were CO, HC and NO. CO was recorded in percentage (%) and HC and

NO in parts per million (ppm) at five second intervals. For each of the test runs, the total of each pollutant in grams were evaluated. The method adopted to estimate emissions is shown below.

Total mass of exhaust gas (g) = Mass of fuel + Mass of air
Mass of fuel = Density of fuel (petrol) × Volume of fuel consumed in c.c
Mass of air = Air-Fuel (A/F) ratio × Mass of fuel
A/F ratio for two-wheeler vehicles is taken as 15.

Vehicular emission data for each vehicle was collected for both lane restricted and lane-less flow conditions. The total emission during each run was found out from the following equation.

Emission (g/s) = $A \times B \times C / (D \times 10^6)$
where,
A = Average emission concentration (ppm)
B = Molecular weight (g) of the component being measured (HC/CO/NO)
C = Total mass of exhaust gas (g)
D = Molecular weight (g) of exhaust gas (assumed as that of N_2)
The molecular weight of Nitrogen is 28g.

Generally, it can be expected that vehicle emissions under lane-less traffic flow condition will be more than those under lane restricted flow condition. This reduction can be expressed as a percentage, as follows:

Percentage reduction in pollutant = $((P_{LL} - P_{LR})/ P_{LL}) \times 100$
where,
P_{LL} = Total pollutant (g/run) under lane less conditions
P_{LR} = Total pollutant (g/run) under lane restricted conditions

Speeds of the vehicles were displayed by the speed counter and were recorded simultaneously by the data logger.

Results and Discussions

The gas analyzer recorded the pollutant concentrations at 5-second time intervals. The data set from each of the test runs were analyzed separately to determine the average pollutant concentrations during each test run in lane restricted and lane-less test runs. The observed pollutant concentrations were checked with reasonable upper limits for each of the pollutants for two-wheeler. Those test runs which produced pollutants beyond these limits were not included for the analysis purpose.
Considering lane restricted and lane-less conditions together, it can be seen from Table 2 that HC values ranged from 0.301 to 0.528 g/km, CO from 2.487 to 6.096 g/km and NO from 0.018 to 0.091 g/km. From the average values of emissions for lane-less and lane restricted conditions, it is seen that that the lane restricted test runs produce less emissions compared to lane-less test runs.

Table 3 shows the percentage reductions in emissions achieved by lane restriction for each of the vehicles. From the above table, it can be seen that for the test vehicles, the reductions in HC, CO and NO range from 12 to 24%, 2 to 27% and 10 to 17%, respectively. From this, it is evident that the lane restriction as a traffic control measure helps in reducing the emissions. Statistical hypothesis test also revealed that the emissions were higher for lane-less conditions as compared to lane restricted conditions.

Table 2. Average Emissions of HC, CO and NO under Lane Restricted and Lane-less Conditions

Vehicle Type	Year of Manufacture	Emissions					
		Lane Restricted Conditions (g/km)			Lane-less Conditions (g/km)		
		HC	CO	NO	HC	CO	NO
Two-Wheeler	1995	0.301	5.988	0.018	0.342	6.096	0.020
	2000	0.399	4.573	0.082	0.528	5.453	0.091
	2005	0.380	2.487	0.058	0.487	3.416	0.070

Table 3. Percentage Reduction in Emissions for Lane Restricted Conditions*

Vehicle Type	Year of Manufacture	Percentage Reductions		
		HC	CO	NO
Two-Wheeler	1995	12.2	1.8	11.2
	2000	24.3	16.1	9.5
	2005	21.9	27.2	17.4

*Compared to lane-less conditions

Impact of Various Factors on Emissions

The impact of various factors such as, average speed, fuel consumed, number of accelerations, etc. on emissions (HC, CO, NO) were explored by plotting graphs between suitable parameters, as shown in Figure 3 below.

From the graphs, it is seen that at low and high ranges of speeds, the emissions were higher, indicating a range of optimal speeds in between them corresponding to least emissions. The speeds for minimum emissions were found to be in the range of 58 –

62 km/h. Similarly, it is seen that optimal speed corresponding to least fuel consumption was 53 km/h.

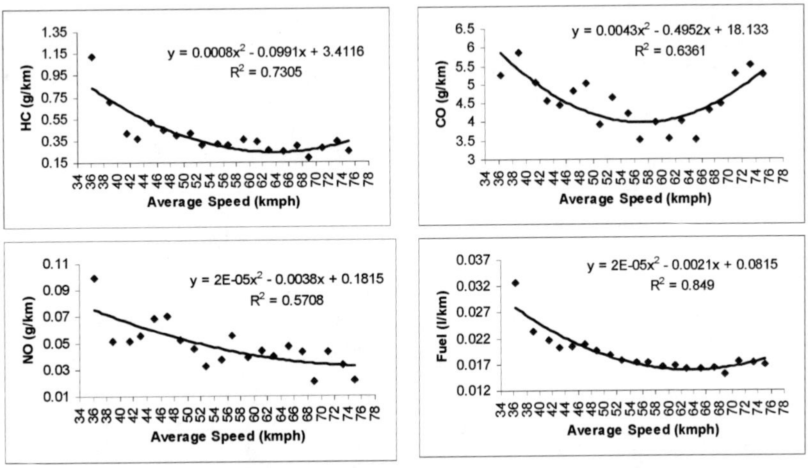

Figure 3. Variations in Emissions (g/km) and Fuel Consumption (l/km) with Average Speed (kmph) for Two-Wheelers

The influence of number of accelerations on emissions was also studied. From the analysis of accelerations by the test vehicles, it was seen that the maximum frequency of accelerations occurred in the range of 0.1 to 0.4 m/s^2 in most cases. The urban traffic in the study stretches limited the test vehicles from carrying out aggressive accelerations. Even though it is generally expected that accelerations accompany lane changes, from the study it was found that in urban traffic as prevailing in India, it need not be so. The drivers of the test vehicles maneuvered through the spaces available between the other vehicles, resulting in slightly aggressive driving compared to that of the lane restricted test runs. It was observed lane shifts were carried out by merely crossing the lanes rather than by exerting sharp accelerations.

Regression Model

In order to understand the relationships between emissions (HC, CO and NO) and various traffic factors, regression models were tested and built. In particular, the interest was on evaluating the number of lane changes as a factor in these models. Multiple Linear Regression analysis using SPSS software was carried out to develop these models. The independent variables tested included fuel consumed (l/km), number of lane changes per kilometer, number of accelerations per kilometer and speed. Dummy variables were employed to represent the types and ages of the vehicles. *Enter* and *Stepwise* methods of regression analysis were performed. Table 4 shows the SPSS output for best regression models using *stepwise* method. Entering variables and their coefficients are shown.

Table 4. Regression Models for HC, CO and NO Emissions for Two-wheelers (*Stepwise* Method) (SPSS Output)

Emissions (Dependent Variable)	Variables Included	Coefficients	Std. Error	R Square	F	t	Sig.
HC	*Constant*	-0.313	0.055	0.524	63.885	-5.692	.000
	fuel	29.610	2.617			11.314	.000
	#ln_ch	0.02231	0.005			4.058	.000
	new	0.195	0.037			5.234	.000
	med	0.160	0.043			3.711	.000
CO	*Constant*	-0.287	0.309	0.800	231.619	-.931	.353
	fuel	312.870	14.694			21.292	.000
	new	-2.785	0.209			-13.307	.000
	med	-1.626	0.242			-6.729	.000
	#ln_ch	0.167	0.031			5.398	.000
NO	*Constant*	-0.03643	0.006	0.631	98.991	-5.931	.000
	#ln_ch	0.002426	0.001			3.952	.000
	med	0.06028	0.005			12.541	.000
	new	0.04710	0.004			11.316	.000
	fuel	2.654	0.292			9.082	.000

The parameter of interest, namely, the number of lane changes, which was identified to analyze the effect of lane restriction on vehicular emissions, was found to be significant in the models. This establishes the fact that the number of lane changes carried out by the vehicles in heterogeneous traffic is an important contributor to emissions. In addition to this, fuel consumed is also an important predictor of emissions. Parameters regarding to vehicle maintenance were not considered directly; however, the age of the vehicle which is related to maintenance were considered in building the regression models. As for age of the vehicle, the signs of the coefficients attached to these variables are not as expected in all cases. For example, an old age vehicle will be predicted to emit lower emissions than a new one. While this may appear to be counter-intuitive, the models cannot be faulted. This is because of the fact that maintenance of the vehicles are an important factor in emissions; this was confirmed, for example, in one case where the older vehicle emitted lesser emissions than a newer one, since it was better maintained. This aspect is mentioned in the research work by Wasburn et al. (2001), where it was cited that the emission level variance between poorly maintained and properly maintained vehicles within the same model year can be greater than emission level variance between properly maintained vehicles significantly different in age. Thus, these models can be taken as representative of the given set of data.

Conclusions

The present study evaluated the tail-pipe emissions of selected test vehicles in lane restricted and lane-less traffic conditions, thereby assessing the relative merit of lane restriction as a control measure. Extensive field test runs were conducted on two-wheeler vehicles on four road stretches in Chennai city, India. On-road emissions from the test vehicles were measured for these runs. Based on the analysis of field data collected specifically for this study, the following conclusions are made:

1. Lane changes carried out by the vehicles in heterogeneous traffic were determined to be an important contributor to emissions.
2. Lane-restricted flow generally produced reduced levels of tail-pipe emissions of HC, CO and NO. The data collected in this study indicated reductions of up to 24% for HC, 27% for CO and 17% for NO, considering average values of emissions per km for two-wheeler vehicles.
3. Statistical hypothesis tests were conducted to infer whether lane restricted runs produced lesser emissions than those for lane-less cases. This was found to be true.
4. The driving speeds for minimum emissions were found to be in the range of 58-62 km/h. Similarly, the driving speed for minimum fuel consumption were found to be around 53 km/h for two-wheeler vehicles. However, it must be noted that these values are for limited number of test vehicles and under conditions adopted in this study. Also, not all speed ranges could be tested due to the nature of urban traffic conditions.
5. Multiple linear regression models developed in this study have brought out the contributors to emissions in the light of lane restriction control (vis-a vis lane-less flow) in heterogeneous traffic conditions.

The above conclusions may support the case for considering lane restrictions on certain types of roads as a control measure for reducing overall vehicular emissions. However, these conclusions are based on limited tests on selected two-wheeler vehicles on certain road stretches. Further tests on more vehicles of different types and ages on different case studies will help confirm and generalize the conclusions.

Acknowledgements

This research study was made possible through funding support provided by Indian Institute of Technology Madras under the Inter-disciplinary Research Program.

References

Cernuschi, S., Giugliano, M., Cemina, A., and Giovannini, I. (1995). "Modal analysis of vehicle emission factors." *Science of the Total Environment,* 169, 175-183.

Frey, H.C., Rouphail, N., Unal, A., and Colyar, J. (2000). "Emissions and traffic control: An empirical approach." *Proc., 9th CRC On-Road Vehicle Emissions Workshop*, San Diego, California.

Prakash, S. P. V., and Allapat, B. J. (1998). "Control strategies for automobile pollution." *Indian Journal of Environmental Protection,* 19, 185-192.

Rouphail, M., Frey, H. C., Colyar, J. D., and Unal, A. (2000). "Vehicle emissions and traffic measures: exploratory analysis of field observations at signalized arterials." *Proc., 80th Annual Meeting of Transportation Research Board,* Washington.

Sengupta, B. (2004). "Experience of Air pollution control in last two decades in India." *International Conference on Better Air Quality,* Agra, India, Dec.

Sivacoumar, R., and Thanasekaran, K. (2000). "Emission factors and emission estimation for Indian vehicles- a case study." *Journal of Institution of Engineers*, 81, 28-31.

Teng, H., Grace, Y., Byun, J. H., Nakanishi, Y., and James, E. (2005). "Modeling vehicle emissions in hot-stabilized conditions using a simultaneous equation model." *Journal of Transportation Engineering,* 131, 762-770.

Varhelyi, A. (2002). "The effects of small roundabouts on emissions and fuel consumption: a case study." *Transportation Research*, Part D, 7, 65-71.

Washburn, S., Seet, J., and Mannering, F. (2001). "Statistical modeling of vehicle emissions from inspection/maintenance testing data: an exploratory analysis." *Transportation Research*, Part D, 6, 21-36.

Yu, L., and Qiao, F. (2004). "Collection and evaluation of on-road vehicle emission and activity data." <http://www.harc.edu/harc/Projects/AirQuality/Projects>, (November 16, 2005).

The Emissions Score, a Composite Measure for Ranking Transportation Control Measures and Similar Projects in Terms of Emissions Benefits

R. J. d'Abadie[1] and R. G. Kaiser[2]

[1]Michael Baker Jr., Inc., 1304 Concourse Drive, Suite 200, Linthicum, MD 21090-1014; PH (410) 689-3400; FAX (410) 689-3401; email: RDABDIE@MBAKERCORP.COM
[2]Michael Baker Jr., Inc., 1304 Concourse Drive, Suite 200, Linthicum, MD 21090-1014; PH (410) 689-3400, FAX (410) 689-3401; email: RKAISER@MBAKERCORP.COM

Abstract

Transportation Control Measures (TCMs) are often a vital component of the overall plan adopted by transportation agencies to achieve their air quality goals. However, in an environment of limited funding only the most effective projects can be implemented, and agencies require metrics to help determine which TCMs to pursue. The comparison of different TCMs is complicated by the fact that projects which address one pollutant may not impact another, and while Ozone is analyzed on a daily basis, $PM_{2.5}$ and its precursors are evaluated using an annual comparison. There is a need to simplify the analysis and selection of projects by providing a metric or composite measure that summarizes total emissions benefits in a clear and concise way.

This paper focuses on a metric developed for the New Jersey Department of Transportation in support of their recent analysis of Reasonably Available Control Measures (RACMs.) The Emissions Score was developed as a way to collapse both annual and daily estimates of emission benefits for multiple pollutants into a single, consistent value. This metric considers the annual and daily emissions targets for each pollutant, the estimated current inventory for each pollutant and a series of weighting factors that emphasize the pollutants of most concern to the region. The Emissions Score was used to rank the proposed projects in terms of overall impact. Ultimately this rating, combined with other factors such as cost effectiveness and political feasibility, was used to articulate the effectiveness of the TCM projects proposed in the study.

Introduction

Air quality issues associated with criteria air pollutants including Volatile Organic Compounds (VOCs), Nitrous Oxides (NO_X) and Particulate Matter (PM) have become an increasing concern for regulators and transportation planning agencies. Research has shown that these pollutants are harmful at levels lower than previously thought and as such the regulations governing them have become increasingly more stringent. Ozone and its precursors (VOCs and NO_X) remain an issue for regions striving to meet their air quality goals. Particulate matter, specifically fine particulates with a diameter of less than 2.5 micrometers ($PM_{2.5}$), have also become a special concern. At the same time, all agencies are faced with limited resources and as such must strive to identify the most efficient use of available funding. This issue is addressed directly in the Safe, Accountable, Flexible, Efficient Transportation Equity Act: A Legacy for Users, or SAFETEA-LU, specifically in regard to the allocation of funds from the Congestion Mitigation and Air Quality (CMAQ) program:

"States and MPOs will give priority in distributing funds for projects and programs to diesel retrofits and other cost-effective emission reduction activities, and cost-effective congestion mitigation activities that provide air quality benefits." (USEPA, 2006)

The difficulty in allocating funds efficiently is that the comparison of projects is complicated by a number of factors. Projects may not address all pollutants, so while one project may primarily address $PM_{2.5}$ another may impact only ozone precursors. The health risks of the various pollutants are also thought to vary greatly between one another. In research by Delucchi and McCubbin, the health costs associated with one ton of VOC, NO_X and $PM_{2.5}$ emissions were shown to be in an approximate ratio of 1:16:152 respectively (McCubbin and Delucchi, 1999). At the same time the overall inventory of emissions varies greatly between pollutants. Tables 1a and 1b summarize the 2002 periodic emissions inventory contained in the New Jersey State Implementation Plan (SIP). While it is estimated that $PM_{2.5}$ has roughly 10 times the health costs of an equal amount of NO_X, the overall inventory of $PM_{2.5}$ is $1/88^{th}$ of that of NO_X.

Additional complications in ranking projects based on their emissions benefits include:

- Precursors to ozone are analyzed for a summer day, while $PM_{2.5}$ emissions are estimated on an annual value basis. While Carbon Monoxide (CO) emissions were not considered in this analysis, if they were to be added a third dimension would be involved as CO is analyzed for a winter day.
- In New Jersey, the boundaries of the non-attainment areas for ozone and $PM_{2.5}$ are different. Table 2 summarizes which counties belong to the various non-attainment areas. Because of these differences a project that helps an area achieve its ozone goals may not fall within either $PM_{2.5}$ non-attainment

area. Also, the boundaries for the previous 1-hour standards for Ozone are different than those of the new 8-hour standard, further complicating any comparisons.

- In the air quality regulations it is the precursors to ozone that are considered, while $PM_{2.5}$ emissions are analyzed both directly and for some of its precursor pollutants. At the same time NO_X is not only a precursor to ozone but to $PM_{2.5}$ as well. Since not all precursor pollutants will become either ozone or $PM_{2.5}$ they should be considered to have less of an overall impact.

While agencies would like to rank mitigation projects in terms of their overall impact, the difficulty becomes finding a common ground on which to base the comparison. In addition to the issues highlighted above, an agency may wish to emphasize one pollutant over another if, for example, a region is in attainment for $PM_{2.5}$ but not for ozone. Any metric developed must allow for a degree in versatility in order to tailor it to the needs of end users.

The Analysis of Reasonable Available Control Measures for New Jersey

The federal Environmental Protection Agency's (EPA's) interpretation of the Clean Air Act requires that regions work to expedite the attainment of the National Ambient Air Quality Standards (NAAQS) if at all possible. Specifically, the guidance on this subject states the following:

"Sections 172(a)(2)(A) and 181(a) of the act require ozone nonattainment areas for to attain the ozone NAAQS as expeditiously as practicable and provide outer-limit dates for attainment based on an area's classification. Furthermore, section 172(c)(1), provides for "the implementation of all reasonably available control measures as expeditiously as practicable." (USEPA, 1999)

In response to this requirement, the New Jersey Department of Environmental Protection (NJDEP) requested that NJDOT develop a list of Reasonable Available Control Measures (RACM) for mobile source emissions and analyze the potential emissions benefits of each. The intention was to demonstrate that all reasonable measures were being taken that might expedite the attainment date. While this analysis was initiated in regards to $PM_{2.5}$, impacts on ozone were also included in the overall evaluation. A total of 26 project types and overall project categories were identified and analyzed as part of this effort, including both on-road and off-road projects where NJDOT would have some influence. Initially the projects were ranked separately for each pollutant at the request of NJDEP and NJDOT. Tables 3 and 4 lists the projects, their estimated impacts by pollutant and provide the initial rankings. Unfortunately, a great deal of variation was seen in the rankings both between the different pollutants and the analysis timeframes (summer day for ozone verses annual reductions for $PM_{2.5}$) and a single ranking could not readily be developed from these individual values.

Development of the Emissions Score

The creation of a value to help in ranking projects was done largely by applying existing guidance and research into a simple linear equation. From the issues described above, three primary factors were considered in the development of the metric that eventually became the Emissions Score:

1) The absolute value of the emission reductions for each pollutant had to be factored such that the values for all of the pollutants could be compared despite having significantly different ranges. Concurrently, the different time periods for which each pollutant as well as the varying geographic areas covered by each non-attainment area had to be considered

2) The relative impact of each pollutant on health, or the relative severity of each pollutant needed to be incorporated

3) While VOC and NO_X are regulated emissions they are only precursors and as such may not go on to form either Ozone or $PM_{2.5}$, the pollutants of most concern. $PM_{2.5}$ on the other hand is also considered directly and therefore should carry additional weight. Any metric developed needed to reflect this difference.

To address the first issue, the total emission reductions for each individual pollutant was divided by the total emissions reported in the 2002 conformity periodic emissions inventory provided by NJDEP. The purpose of this was to reduce the absolute emission values to a relative share of overall emissions for each pollutant, time period, and non-attainment area. The resulting value eliminates the fact that the absolute magnitude of total emissions vary greatly for each pollutant.

To address the fact that the relative health impacts of each pollutant vary significantly, findings by Robert F. Westcott used to compare the cost effectiveness of diesel retrofits versus current CMAQ projects were adopted (Westcott, 2005). In this research it is argued that relative health costs of an equal amount of VOC, NO_X, and $PM_{2.5}$ emissions are in a ratio of approximately 1:4:9.45. These ratios were used considered as potential coefficients in the formulation of the Emissions Score.

Finally, in the May 2006 NJDEP State Implementation Plan (SIP), a NO_X to VOC substitution ratio was discussed, specifically in regard to the McGuire Air Force Base conformity budget. It was argued that while NO_X emissions are expected to go up due to operations at the base, this is will be offset by a significant decrease in VOC emissions. NJDEP, using EPA guidance (USEPA, 1993), determined that a substitution rate of 1.04 tons of VOC to 1.0 tons of NO_X was an acceptable ratio. For the purposes of ranking projects by emissions, it was felt that this was a very conservative value, in particular since most of the RACM projects showed a marked decrease in VOC emissions while NO_X emissions generally decreased by a much smaller percentage. As this ratio was developed solely for ozone, it was thought not to be applicable in regard to NO_X as a $PM_{2.5}$ precursor.

No guidance could be found in regard to evaluating NO_X as a precursor to the formation of $PM_{2.5}$. The amount of NO_X that goes on to form $PM_{2.5}$ is likely only a small percentage of the overall inventory. However, given the fact that $PM_{2.5}$ has significantly higher health impacts relative to ozone, it was decided to weigh the annual NO_X value by a factor of two to emphasize the potential impacts as a precursor.

Combining the above factors into a single formulation, the Emission Score was calculated as follows:

$$\text{Emission Score} = \frac{1.04(VOC_R)}{VOC_{Inventory}} + \frac{(NOx_{R\ Daily})}{NOx_{Daily\ Inventory}} + \frac{2(NOx_{R\ Annual})}{NOx_{Annual\ Inventory}} + \frac{9.45(PM_{2.5\ R})}{PM_{2.5\ Inventory}}$$

Where:

VOC_R	= Summer Day VOC Emission Reductions Estimated for the project
$VOC_{Inventory}$	= Total Summer Day VOC Emissions from the 2002 Periodic Emissions Inventory for Ozone for all nonattainment counties
$NOx_{R\ Daily}$	= Summer Day NOx Emission Reductions Estimated for the project
$NOx_{Daily\ Inventory}$	= Total Summer Day NOx Emissions from the 2002 Periodic Emissions Inventory for Ozone for all nonattainment counties
$NOx_{R\ Annual}$	= Annual NOx Emission Reductions Estimated for the project
$NOx_{Annual\ Inventory}$	= Total Annual NOx Emissions from the 2002 Periodic Emissions Inventory for $PM_{2.5}$ for all nonattainment counties
$PM_{2.5\ R}$	= Annual $PM_{2.5}$ Direct Emission Reductions Estimated for the project
$PM_{2.5\ Inventory}$	= Total Annual $PM_{2.5}$ Direct Emissions from the 2002 Periodic Emissions Inventory for $PM_{2.5}$ for all nonattainment counties

The emissions score was calculated for all RACM projects and the results are shown in Table 5. Overall, the rankings based on the Emissions Score favored diesel retrofits as well as projects that resulted in significant decreases in overall Vehicle

Miles of Travel (VMT). Operational improvements such as signal improvements and Intelligent Transportation Systems (ITS) did not rank as highly as they did solely from an ozone perspective. This reflects the fact that the Emissions Score places a higher priority on $PM_{2.5}$, and projects that enhance operational conditions by increasing average speed, reducing unstable traffic flow and minimizing idling tend to have a smaller impact on $PM_{2.5}$. Note that the emissions benefits for the NJDOT RACM analysis relied on emission rates developed using the EPA software MOBILE 6.2. The calculation of $PM_{2.5}$ emission rates in MOBILE 6.2 is not sensitive to changes in operating conditions (such as speed) although research has shown this not to be the case. Future analysis using the forthcoming MOVES software package may yield different results.

The Emissions Score is intended as a helpful indicator in ranking projects. It is not intended to be the sole measure used in such an evaluation. Cost effectiveness, political feasibility, local priorities and other factors must be considered to ensure that selected projects are not only effective but can also be realistically implemented.

Future Enhancements

The Emissions Score as it is currently calculated reflects the data available at the time of the NJDOT RACM analysis. Additional data expected in the upcoming year will help to further refine this value. Soon to be released work such as the upcoming conformity determinations including application of the new 8-hour standard for ozone, as well as new studies reviewing the heath impacts of $PM_{2.5}$ could be used to refine the coefficients used in the Emission Score. While SO_2 was not included as part of the Emissions Score it could be added at a future time if this precursor is determined to be an issue. CO was also not included in the Emissions Score as New Jersey is in attainment for this pollutant, however it too could be added in the future as needed and as data permits. The current version of the Emissions Score is intended for statewide evaluations, but it could also be applied to MPOs or to specific nonattainment areas simply by substituting appropriate values for the emissions inventories. As the Emissions Score is calculated for more projects, a range of typical values can be determined and normalizing the metric between 0 and 100 will become a possibility. Project costs were deliberately excluded in the formulation of the Emissions Score, however it would be possible to develop a cost/benefit ration using this metric. Also, the formula underlying the Emissions Score is relatively straight forward and other pollutants, in particular the various Air Toxics, could be included in the overall score at a later date.

References

McCubbin, D., and Delucchi, M. (1999). "The Health Costs of Motor-Vehicle-Related Air Pollution." *Journal of Transport Economics and Policy*, Vol. 33, Part 3, pp. 253-286.

New Jersey Department of Environmental Protection. (2006). "*State Implementation* Plan (SIP) Revisions for the Attainment of Maintenance of 8-Hour Carbon Monoxide National Air Quality Standard, 1-Hour Ozone National Air Quality Standard, and Fine Particulate Matter National Air Quality Standard; and the 2002 Periodic Emissions Inventory."

South Coast Air Quality Management District. (2000). "Multiple Air Toxics Exposure Study (MATES-II), Final Report."

Transportation Research Board of the National Resources Council. (2002). "The Congestion Mitigation and Air Quality Improvement Program: Assessing 10 Years of Experience."

USEPA, Office of Air Quality Planning and Standards. (1993). "NOx Substitution Guidance."

USEPA. (1999). "Guidance on the Reasonably Available Control Measures (RACM) Requirement and Attainment Demonstration Submissions for Ozone Nonattainment Areas."

USEPA. (2006). *SAFETEA-LU Fact Sheets on Highway Provisions*, http://www.fhwa.dot.gov/safetealu/factsheets.htm

Westcott R. F. (2005). "Cleaning the Air: Comparing the Cost Effectiveness of Diesel Retrofits vs. Current CMAQ Projects." Emissions Control Technology Association.

Tables 1a and 1b: 2002 Periodic Emissions Inventory by Non-Attainment Area (From NJDEP)

Ozone Non-Attainment Area	Emissions (Tons/Summer Day)	
	VOC	NO_X
NY-NJ-CT	179.96	298.17
PA-NJ-MD-DE	93.14	162.86

$PM_{2.5}$ Non-Attainment Area	Emissions (Tons/Year)	
	NO_X	$PM_{2.5}$
NY-NJ-CT	110,142	1,243
PA-NJ-DE	27,857	387

Table 2: Non-Attainment Areas within New Jersey by County (1-Hour Ozone Standard)

County/MPO		Ozone Non-Attainment Area		$PM_{2.5}$ Non-Attainment Area	
County	MPO	NY-NJ-CT	PA-NJ-MD-DE	NY-NJ-CT	PA-NJ-DE
Atlantic	SJTPO		□		
Bergen	NJTPA	□		□	
Burlington	DVRPC		□		□
Camden	DVRPC		□		□
Cape May	SJTPO		□		
Cumberland	SJTPO		□		
Essex	NJTPA	□		□	
Gloucester	DVRPC		□		□
Hudson	NJTPA	□		□	
Hunterdon	NJTPA	□			
Mercer	DVRPC		□	□	
Middlesex	NJTPA	□		□	
Monmouth	NJTPA	□		□	
Morris	NJTPA	□		□	
Ocean	NJTPA		□		
Passaic	NJTPA	□		□	
Salem	SJTPO		□		
Somerset	NJTPA	□		□	
Sussex	NJTPA	□			
Union	NJTPA	□		□	
Warren	NJTPA	□			

Table 3: Estimated 2009 Emissions Reductions for New Jersey RACM Projects

Control Measures	Estimated Emissions Reductions				
	VOC Tons/Day*	NO_X Tons/Day*	NO_X Tons/Year	$PM_{2.5}$ Tons/Year	SO_2 Tons/Year
Limit use of Recreational Watercraft	-2.8068	-0.6222	-12.0910	-1.2079	-0.5355
Limit the Use of Lawn and Garden Equipment	-0.0794	-0.0107	-1.5429	-0.0765	-0.0072
Retrofit Construction Equipment	-0.3696	0.0000	0.0000	-25.6413	0.0000
Retrofit Switch Yard Locomotives	0.0000	-0.3423	-92.4122	-3.5566	0.0000
Construction Equipment Idling Restrictions	-0.1937	-2.1739	-481.4903	-7.9833	0.0000
Fuel tax increase	-3.0634	-2.8938	-1020.867	-117.8744	-4.3151
Implement Pay-as-you-drive vehicle insurance.	-0.3829	-0.3617	-127.6084	-14.7268	-0.5394
Truck idling restrictions	-0.3322	-3.2297	-505.9726	-16.7991	-0.9669
Impact of Various Transit Projects	-0.1209	-0.1211	-30.2900	-0.9436	-0.5920
Effect of No Fare Increase	-0.0717	-0.0765	-18.0880	-0.5070	-0.3100
Adoption of Smart Growth Land Use Policies	-0.46	-0.79	-162.93	-3.59	-1.28

TCMs - Programs and ordinances to facilitate non-automobile travel provision and utilization of mass transit,	-0.3088	-0.2589	-59.0	-1.8	-1.1
Clean Fleets Replacements	-0.0335	-0.0574	-9.1990	-0.1880	-0.1180
High emitter vehicle detection (dirty screening)	-2.998	-2.463	-793.141	-6.147	-3.635
Electric Vehicles at Transit Stations	-0.0002	-0.0002	-0.0398	-0.0011	-0.0007

Table 3 (Continued): Estimated 2009 Emissions Reductions for New Jersey RACM Projects

Control Measures	Estimated Emissions Reductions				
	VOC Tons/Day *	NO_X Tons/Day *	NO_X Tons/Year	$PM_{2.5}$ Tons/Year	SO_2 Tons/Year
School Bus Replacements	-0.0080	-0.2700	-67.2885	-3.2474	0.0000
IdleAire Installations	-0.0064	-0.0620	-4.6954	-0.1559	-0.0090
Transit Bus Replacements	-0.0186	-0.4205	-234.6982	-10.5797	-0.0084
Heavy Duty Diesel Engine Replacements	-0.1971	-8.2252	-2326.28	-138.06	-0.1750
School Bus Retrofit	-0.0281	0.0000	0.0000	-9.3418	0.0000
Improved Signal Coordination	-0.0207	-0.0100	-2.0562	-0.3299	-0.0123
Commercial Vehicle Information Systems and Networks (CVISN).	-0.0465	-0.9954	-207.6093	-5.3915	-0.0585
Implementation of Express E-Z Pass	-0.0240	-0.0045	-0.8800	-0.0003	0.0000

Toll Collection					
Incident Management/Service Patrols	-0.0967	-0.0438	-10.1738	-1.7299	-0.0633
Speed Limit Adherence	0.36	-4.87	-1502.29	0.00	0.00
Statewide Expansion of Bicycle Facilities	-0.0066	-0.0018	-0.42	-0.02	-0.01

Table 4: Ranking of RACM Projects by Estimated Emission Reductions

Control Measures	Project Ranking By Total Individual Emissions Reductions				
	VOC Tons/Day*	NO_X Tons/Day*	NO_X Tons/Year	$PM_{2.5}$ Tons/Year	SO_2 Tons/Year
Limit use of Recreational Watercraft	3	9	14	16	7
Limit the Use of Lawn and Garden Equipment	13	20	19	20	16
Retrofit Construction Equipment	6	25	23	3	18
Retrofit Switch Yard Locomotives	25	12	11	12	21
Construction Equipment Idling Restrictions	10	6	6	8	20
Fuel tax increase	1	4	3	2	1
Implement Pay-as-you-drive vehicle insurance.	5	11	10	5	6
Truck idling restrictions	7	3	5	4	5
Impact of Various Transit Projects	11	15	25	25	25
Effect of No Fare Increase	14	16	26	26	26
Adoption of Smart Growth Land Use Policies	4	8	9	11	3
TCMs - Programs and ordinances to facilitate non-automobile travel provision and utilization	8	14	13	14	4

of mass transit,					
Clean Fleets Replacements	16	18	16	18	9
High emitter vehicle detection (dirty screening)	2	5	4	9	2
Electric Vehicles at Transit Stations	24	24	22	22	17

Table 4 (Continued): Ranking of RACM Projects by Estimated Emission Reductions

Control Measures	Project Ranking By Total Individual Emissions Reductions				
	VOC Tons/Day *	NO_X Tons/Day *	NO_X Tons/Year	$PM_{2.5}$ Tons/Year	SO_2 Tons/Year
School Bus Replacements	21	13	12	13	22
IdleAire Installations	23	17	17	19	14
Transit Bus Replacements	20	10	7	6	15
Heavy Duty Diesel Engine Replacements	9	1	1	1	8
School Bus Retrofit	17	26	24	7	19
Improved Signal Coordination	19	21	18	17	12
Commercial Vehicle Information Systems and Networks (CVISN).	15	7	8	10	11
Implementation of Express E-Z Pass Toll Collection	18	22	20	23	23
Incident Management/Service Patrols	12	19	15	15	10
Speed Limit Adherence	26	2	2	24	24
Statewide Expansion of Bicycle Facilities	22	23	21	21	13

Table 5: Emissions Score and Project Rankings for the NJDOT RACM Analysis

Control Measure	Total Emissions Benefit Ranking	
	Emissions Score (x 100)	Rank
Limit use of Recreational Watercraft	-4.6509	14
Limit the Use of Lawn and Garden Equipment	-0.1498	23
Retrofit Construction Equipment	-25.9867	3
Retrofit Switch Yard Locomotives	-5.1930	13
Construction Equipment Idling Restrictions	-10.0600	9
Fuel tax increase	-103.3637	2
Implement Pay-as-you-drive vehicle insurance.	-12.9022	8
Truck idling restrictions	-20.7523	4
Impact of Various Transit Projects	-0.8905	18
Effect of No Fare Increase	-0.5215	20
Adoption of Smart Growth Land Use Policies	-3.8297	15
TCMs - Programs and ordinances to facilitate non-automobile travel provision and utilization of mass transit	-1.8894	16
Clean Fleets Replacements	-0.2966	22
High emitter vehicle detection (dirty screening)	-12.3399	7
Electric Vehicles at Transit Stations	-0.0011	26
School Bus Replacements	-5.2329	12
IdleAire Installations	-0.5501	19
Transit Bus Replacements	-13.6895	6
Heavy Duty Diesel Engine Replacements	-189.2436	1
School Bus Retrofit	-13.8858	5
Improved Signal Coordination	-0.3731	21
Commercial Vehicle Information Systems and Networks (CVISN).	-7.2968	10
Implementation of Express E-Z Pass Toll Collection	-0.0172	25
Incident Management/Service Patrols	-1.4638	17
Speed Limit Adherence	-6.1454	11
Statewide Expansion of Bicycle Facilities	-0.0246	24

The Influence of a Noise Barrier and Vegetation on Air Quality near a Roadway

G. E. Bowker,[1] R. Baldauf,[2,3] V. Isakov,[4] A. Khlystov,[5] W. Petersen,[4] E. Thoma,[2] and C. Bailey[3]

[1] U.S. Environmental Protection Agency, Office of Research and Development, National Exposure Research Laboratory, Atmospheric Modeling Division
[2] U.S. Environmental Protection Agency, Office of Research and Development, National Risk Management Research Laboratory
[3] U.S. Environmental Protection Agency, Office of Air and Radiation, Office of Transportation and Air Quality
[4] NOAA/Atmospheric Sciences Modeling Division (In Partnership with the U. S. Environmental Protection Agency)
[5] Department of Civil and Environmental Engineering, Pratt School of Engineering, Duke University.

Abstract

A growing number of epidemiological studies conducted throughout the world have identified an increase in occurrence of adverse health effects for populations residing, working, or attending school near major roadways. In addition, several air quality studies have identified increased concentration levels of certain pollutants near high traffic volume roads. The U.S. Environmental Protection Agency (EPA) has begun a research program investigating the relationship of traffic activity, environmental conditions, and near road air quality. As part of this program, the EPA is investigating the influence of noise barriers, vegetation, and other roadside structures on air pollutant concentrations near the road.

This presentation integrates results from an air quality modeling assessment and air quality monitoring measurements to identify how noise barriers and vegetation near roads may impact local air quality. Air quality measurements were collected at sites with and without noise barriers and vegetation along a stretch of limited access highway in Raleigh, North Carolina, USA during the summer of 2006. This study allowed an assessment of the potential influence of these structures on near-road air quality. These structures influence pollutant transport and dispersion in the near-field (<300 m). Preliminary results suggest that, under some meteorological conditions, noise barriers and vegetation may reduce air pollutant concentration levels downwind of the barrier. These results may provide useful information in assessing the role of roadside structures on near road air quality for future land use decisions.

Introduction

Recently, a large number of studies have found associations between public health and proximity to major roadways (Baldauf, 2007). One aspect integral to understanding these effects is adequate prediction of pollutant concentration levels in the near-road environment. Since extensive spatial and temporal measurements of pollutant concentration levels are usually not a financially-viable alternative, numerical models (e.g., CALINE, Benson, 1992) are often used. These air quality models must be capable of predicting the complex transport and dispersion of pollutants in the near-road environment. The transport and dispersion of pollutant from roadways is governed by many factors, including the predominant wind direction, wind speed, stability of the atmosphere, the number and composition of the traffic fleet, and the geometry of the roadway area. This work focuses on the last of these factors, the effects of local obstacles, such as noise barriers, buildings, and vegetation on the initial transport and dispersion of pollutants from the roadway. These obstacles may restrict pollutant transport, or may increase initial dispersion through mixing (Tan et al., 1977; Madders and Lawrence, 1985; Swamy and Lokesh, 1993; Kingston et al. 1988; Hölscher et al., 1993).

We chose to explore the effects of noise barriers on pollutant transport and dispersion when the predominant wind was roughly perpendicular to the roadway using an empirically based high-resolution diagnostic wind-field and dispersion model, called Quick Urban & Industrial Complex (QUIC; version 4.3; Los Alamos National Laboratory/University of Utah). With the geometry of the buildings, input meteorology, and source characteristics, the QUIC model produces time-average solutions for wind velocity and concentration (Pardyjak and Brown, 2001, 2002, 2003; Williams et al., 2004; Bowker et al., 2006; Gowardhan et al. 2006a; 2006b; Pol et al., 2006). Using the QUIC model, we examined three basic scenarios: (1) a "base case" where the terrain was completely open (e.g., no noise barrier, buildings, or vegetation); (2) a situation similar to the base case except a single isolated noise barrier was present ("barrier only case"); and (3), a highly-resolved situation with a noise barrier, vegetation, and buildings (representing the "field study site" location in Raleigh, North Carolina USA). The model simulation results for the last case were compared with the dispersion patterns of ultrafine particles (UFP; particulate matter with an aerodynamic diameter less than 100 nanometers) measured using a mobile monitor. Since the mobile monitoring was not done directly on the highway, we chose to focus our analysis on the effects of noise barriers and vegetation on pollutant concentration patterns downwind of the roadway, rather than directly on the roadway.

Methods

The three model domains (base, noise barrier only, and study site) were constructed based on the dimensions of the "study site" domain, a 0.35 km^2 area (about 700 m along the roadway, by 500 m away from the road, and 50 m in height) in Raleigh, North Carolina as shown in Figure 1. The three domains were modeled within QUIC at 2 m resolution in each dimension. The "base" case domain was flat with no barriers, buildings, or vegetation. The "noise barrier only" case was similar to the base case except a barrier was present parallel to the roadway. The location and length of the barrier (it only extended for about half the length of the domain) reflected the configuration at the field site. The "study site domain" contains all three situations. A suburban neighborhood is present in the region on the lee side of the noise barrier. This area is characterized by one- and two-story residential houses, with many tall trees interspersed among the houses. A noise barrier (the red line in Figure 1) separates this area from the roadway. The second part of the study domain is open, with scattered isolated stands of trees, and no noise barrier. A wide (but short) building used as an adult educational facility occupies the central open area (marked with a B in Figure 1). Upwind of the study site (e.g. to the south of the roadway) is a relatively uniform stretch of scattered trees and residential and office buildings (typically one or two stories tall) typical of a suburban area extending for several kilometers.

Figure 1. Plan view of the study site. The white arrow shows the wind direction. The green (teal) line shows the route of the mobile van. The red line shows the location of the noise barrier.

Within QUIC, the noise barrier, buildings, and stands of trees were modeled as solid, non-porous, rectangular blocks, with heights approximated from photographs taken at the site. Bowker et al. (2006) used the QUIC model to explore airflow patterns within stands of vegetation (mesquite bushes), while modeling the bushes as solid blocks. A comparison between QUIC predictions and field measurements of wind velocity suggested that, while the wind within the vegetation is unnaturally predicted to be non existant, the model appeared to accurately predict the major flow features around the vegetation. This includes the locations of wake zones, recirculation cavities, and areas experiencing channeling of flow.

For the QUIC simulations of the Raleigh site, the wind direction chosen was 6 degrees from perpendicular to the roadway (coming from the Southwest at an angle of 214 degrees relative to North, shown with a white arrow in Figure 1), matching the average meteorological conditions measured at the study site during the several-hour time-period air quality measurements were collected. The incoming boundary layer was modeled as logarithmic (roughness length of 0.05 m), characterized by a wind speed of 2 m/s at a reference height of 7 m.

For each of the three domains, the roadway was simulated using two line sources, one to represent each of the two traffic directions. These line sources extended along the length of the domain and were 12 and 32 m from the barrier, respectively. The pollutant from the vehicle emissions was simulated using approximately 3 million neutrally buoyant "particles". Particles were randomly released along the line source during each model time step. During the simulation, the concentration predictions were recorded after the situation reached equilibrium (when the average concentrations no longer changed with time).

Measurements of UFP taken at high spatial and temporal resolution at the study site were compared with the results of the QUIC "study site" simulation. . The measurements were made at 10 Hz using a Differential Mobility Analyzer – Condensation Particle Counter (DMA-CPC) placed in a van which was slowly driven through the study site. UFP concentrations for both 20 and 75 nm particles were recorded as functions of location using a Global Positioning System (GPS) and time. The driving route is shown in Figure 1 and included the area directly behind the noise barrier, areas within the residential neighborhood protected by the noise barrier, and the open area next to the roadway. Approximately 2 hours of measurements were made (10-minutes driving time per route) during which time, the wind was essentially constant and perpendicular to the roadway. Using the spatial information from the GPS and the measurement time, we determined the location of each UFP concentration measurement, then averaged all the measurements collected at each point (within a 20 m by 20 m area), and developed a spatial map of pollutant concentration (Khlystov et al., 2006). This technique assumes that the traffic emissions are roughly constant for the course of each driving route, and across the study time period. In order to directly compare the QUIC simulations with the UFP measurements, the concentration values were normalized using the median concentration value in the open area parallel to the roadway (Fig. 2, the red area along the roadway at an East-West distance from 402 to 550 m).

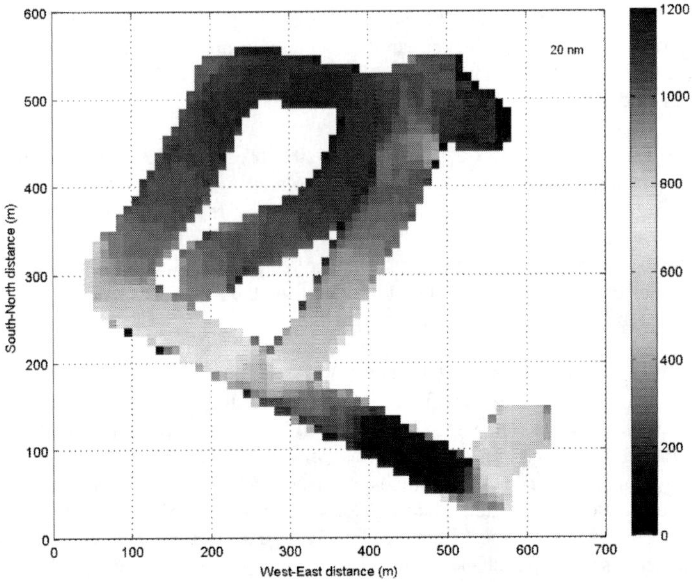

Figure 2- Spatial map of 2-hour average concentrations of UFP from mobile measurements.

Results

The QUIC simulations suggest the noise barrier, vegetation, and buildings influence the roadside airflow and pollutant dispersion patterns. As shown in Figure 3, the highest concentrations in the open terrain case are seen directly off the road with decreasing concentrations with increasing distance from the road. When a noise barrier is in place, pollutants are transported over the barrier, creating a zone of low concentration behind the barrier. The pollutants reach the ground (plume reattachment) at a North-South distance of about 70 m (about 55 m downwind of the barrier). The location of the plume reattachment point as well as concentrations is affected by the size of the recirculation cavity in the lee of the barrier. Previous studies suggest that this cavity can extend between 3 and 12 wall heights downwind, is well-mixed so pollutant concentrations are relatively constant and usually lower compared with the roadside values (between 0 and 80 percent) (Nokes and Benson, 1984; Paul-Carpenter and Barboza, 1988; Hölscher et al., 1993; Swamy and Lokesh, 1993). Our results suggested that the size of the recirculation cavity was at the upper end of previous study results (modeled as 10 barrier heights), and the amount of vertical mixing and concentrations in the cavity were less than observed in previous studies. These results require further exploration. When a noise barrier and vegetation are present near the road (as was the case for the field site case, Figure 3c), enhanced vertical mixing of the pollutants occurs. This mixing increases concentrations directly behind the barrier (relative to the barrier-only case), decreases ground-level concentrations further from the barrier, and eliminates the plume reattachment point. For both scenarios with noise barriers, elevated pollutant concentrations occur on the roadway.

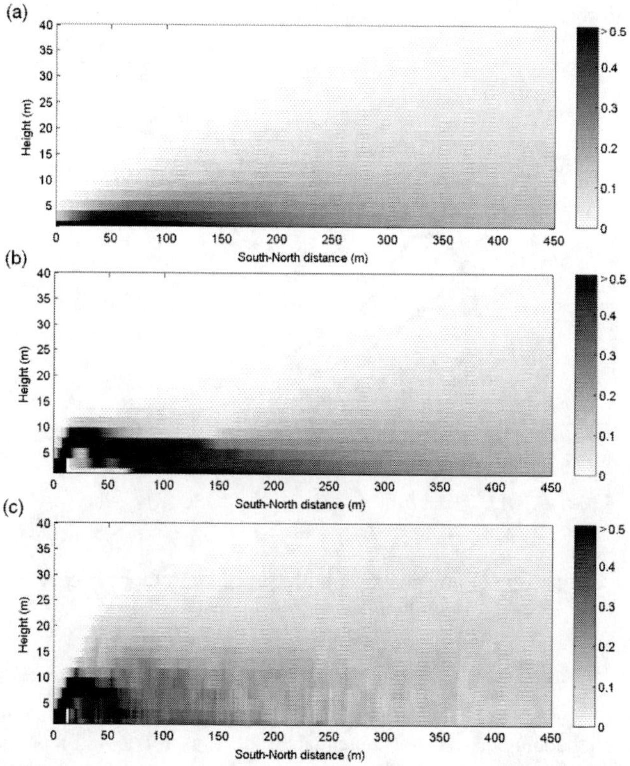

Figure 3- Vertical cross-sections of normalized modeled pollutant concentrations downwind of a highway with (a) no barrier, (b) barrier only, and, (c) the study site, a barrier with vegetation.

Figure 4 shows a comparison of air pollutant concentrations at breathing height level as a function of downwind distance for all three modeling scenarios. Additionally, Figure 4 shows the measured concentration of 20 nm UFP as a function of downwind distance for two situations at the field site (the open area and the residential area in the lee of the noise barrier). The highest concentrations were found near the roadway (modeled base case and measured UFP concentration clear situation) when the area was open (no barriers, vegetation, or buildings). In all cases, the pollutant concentrations decrease with distance from the roadway. The relative change in UFP concentration with distance appears to be similar to the modeling results for both the open area and the vegetated area behind the noise barrier situations (Figure 4). In the lee of the noise barrier, the measured UFP concentrations were about 60% of concentrations in the open area. This compares well with the modeled concentration in the lee of the barrier for the "study site", which is about 50% of the concentration in the open area.

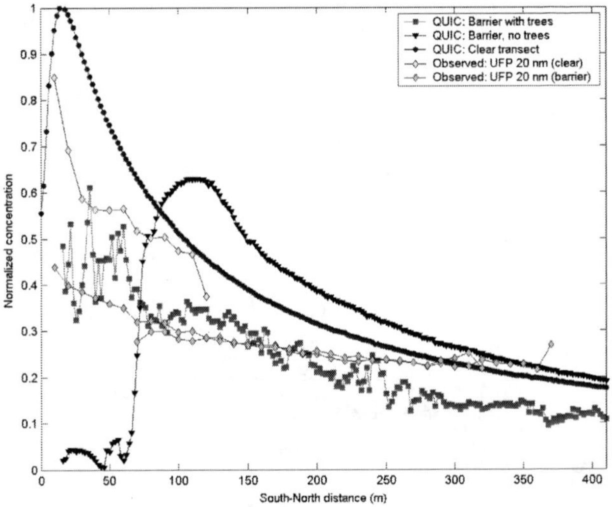

Figure 4-Comparison of downwind air pollutant concentrations from QUIC model with mobile van measurements.

Summary

The results suggest that the QUIC model, when combined with other numerical modeling methodologies, has the potential to be a useful tool for predicting air pollutant concentrations near roadways. The model was capable of analyzing the effects of complex terrain and structures adjacent to the road. The model also predicted similar decay rates as measured by a mobile monitoring vehicle under simple and complex conditions. In some instances, fine-scale modeling (using a model such as QUIC) could be integrated with other regulatory emission and dispersion models for future near road assessments.

The results of this work also suggest that noise barriers and vegetation adjacent to highways may mitigate air quality concentrations near the road, diluting concentrations through increased mechanical mixing. However, the effect of noise barriers alone may not always result in a decrease in pollutant concentrations at varying distances from the road. Study results suggest that the barrier lifts the plume over the wall, and the plume returns to ground at a distance from the road. This leads to very low concentrations within the cavity of the noise barrier but higher concentrations further from the barrier than if the barrier were not present. When vegetation is adjacent to the noise barrier, the pollutant concentrations in the lee of the structure are higher than with a barrier only, but lower than if no barrier were present. At further distances, modeling suggested that the presence of a noise barrier and vegetation resulted in lower concentrations than a barrier only or no barrier at all. Finally, the presence of a barrier led to increased pollutant concentrations on the highway. The integration of fine-scale meteorological modeling tools (e.g., QUIC) with emissions modeling may provide useful information in assessing the need and impact of roadside structures on near road air quality.

Disclaimer

The research presented here was performed under the Memorandum of Understanding between the U.S. Environmental Protection Agency (EPA) and the U.S. Department of Commerce's National Oceanic and Atmospheric Administration (NOAA) and under agreement number DW13921548. This work constitutes a contribution to the NOAA Air Quality Program. Although it has been reviewed by EPA and NOAA and approved for publication, it does not necessarily reflect their policies or views.

References

Baldauf, R., Black, K., Martinez V., Gaber M., Thoma E., and Costa D. (2007). "Traffic Emission Impacts on Air Quality Near Large Roadways." *TRB Transportation Planning and Air Quality Conference*, Orlando, FL, July 9-11.

Bowker, G. E., Gillette, D. A., Bergametti, G., and Marticorena, B. (2006). "Modeling flow patterns in a small vegetated area in the northern Chihuahuan Desert using QUIC (Quick Urban & Industrial Complex)." *Environmental Fluid Mechanics 6,* 359–384, DOI 10.1007/s10652-005-6021-8.

Benson, P. E. (1992). "A review of the development and application of the CALINE3 and 4 Models." *Atmospheric Environment* 26, 379-390.

Gowardhan, A. A., Brown, M. J., Decroix, D. S., Pardyjak, E., and Nelson, M. A. (2006). "Evaluation of the QUIC Pressure Solver using wind-tunnel data from single and multi-building experiments." *Sixth Symposium on the Urban Environment / 14th Joint Conference on the Applications of Air Pollution Meteorology with the Air and Waste Management Assocation.* Atlanta. pp. J8.2. 12.

Gowardhan, A. A., Brown, M. J., Williams, M. D., and Pardyjak, E. (2006). "Evaluation of the QUIC Urban Dispersion Model using the Salt Lake City URBAN (2000) Tracer Experiment Data – IOP 10." *Sixth Symposium on the Urban Environment / 14th Joint Conference on the Applications of Air Pollution Meteorology with the Air and Waste Management Assocation.* Atlanta. pp. J6.3. 13.

Hölscher, N., Höffer, R., Niemann, H-J., Brilon, W., and Romberg, E. (1993). "Wind tunnel experiments on micro-scale dispersion of exhausts from motorways." *The Science of the Total Environment* 134, 71-79.

Kingston, L., Leharne, S., McPhee, E. (1988). "A survey of vehicular lead deposition in a woodland ecosystem." *Water, Air, & Soil Pollution* 38, 239-250.

Khlystov, A., Ma, Y. (2006). "An on-line instrument for mobile measurements of the spatial variability of hexavalent and trivalent chromium in urban air." *Atmospheric Environment* 40, 8088-8093

Lidman, J. K. (1985). "Effect of a Noise Wall on Snow Accumulation and Air Quality." *Transportation Research Record* 1033, 79-88.

Madders, M., and Lawrence, M. (1985). "The contribution made by vegetation buffer zones to improved air quality in urban areas." *In D.O. Hall, N. Myers, and N.S. Margaris (eds.) Economics of ecosystems management.* Dr. W. Junk Publishers, Dordrecht. Netherlands. ISBN 90-6193-505-9.

Nokes, W. A., and Benson, P. E. (1984). "Carbon monoxide concentrations adjacent to sound barriers." Office of Transportation Laboratory, California Department of Transportation. *Report FHWA/CA/TL-84/04.*

Pardyjak, E., and Brown, M. (2001). "Evaluation of a fast-response urban wind model - comparison to single-building wind-tunnel data," Int. Soc. Environ. Hydraulics, Tempe, AZ. pp LA-UR-01-4028, 6.

Pardyjak, E., and Brown, M. (2002). "Fast response modeling of a two building urban street canyon", *4th AMS Symposium on the Urban Environment,* Norfolk, VA. pp. LA-UR-02-1217.

Pardyjak, E., and Brown, M. (2003). "QUIC-URB v1.1 Theory and User's Guide." *Los Alamos National Laboratory.*

Pardyjak, E. R., Brown, M. J., and Bagal, N.L. (2004). "Improved velocity deficit parameterizations for a fast response urban wind model." *84th AMS Conference: Modeling Urban Land Surfaces and Buildings: Part 2.* Seattle, WA. pp. 7.4 5.

Paul-Carpenter, S., Barboza, M. J. (1988). "Effects of highway noise barriers on carbon monoxide levels." *81st Annual meeting of the Air Pollution Control Association (APCA),* Dallas, TX.

Pol, S. U., Bagal, N. L., Singh, B., Brown, M. J., and Pardyjak, E. (2006). "Implementation of a new rooftop recirculation parameterization into the QUIC fast response urban wind model." *Sixth Symposium on the Urban Environment / AMS Forum: Managing our Physical and Natural Resources: Successes and Challenges.* Atlanta. Pp. JP1.2 19.

Swamy, K.T., and Lokesh, K.S., (1993). "Lead dispersion studies along highways." *Indian Journal of Environmental Health* 35:33, 205-209.

Tan, K.T., Lepp, N.W. (1977). "Roadside vegetation: an efficient barrier to the lateral spread of atmospheric lead? " *Arboricultural Journal* 3(2), 79-85.

Wang, S.C., Flagan, R.C. (1990). "Scanning electrical mobility spectrometer." *Aerosol Science and Technology* 35, 718–727.

Williams, M., Brown, M., Boswell, D., Singh, B., and Pardyjak, E. (2004). "Testing of the QUIC-PLUME model with wind-tunnel measurements for a high-rise building," *Fifth Symposium on the Urban Environment, Vancouver, B.C., Canada,* pp. LA-UR-04-4296, 10.

Thriving with Neighborhood Electric Vehicles

T. Cosgrove,[1] J. E. Pedri,[2] and R. O. Watkins[3] G. Capik,[3] L. Rubio,[3] S. Ainsworth,[3]

[1]Councilmember, City of Lincoln, 640 Fifth Street Lincoln, CA-95648; PH (916) 645-4070
[2]P.E., Director of Public Works, City of Lincoln, 640 Fifth Street Lincoln, CA-95648; PH (916) 645-8547
[3]MHM Engineers and Surveyors, 1082 Sunrise Avenue, Suite 100, Roseville, CA-95661; PH (916) 783-4100

Setting

The City of Lincoln (City) is located approximately 30 miles northeast of Sacramento, California in Placer County. From its founding in 1859 until the mid 1990s, the City was affectionately known as a sleepy little town dependent upon surrounding farmers and the Gladding McBean clay products manufacturing plant. In the 1990s, City leaders had knowledge of the issues described in *Edge City, Life on the New Frontier,* by Joel Garreau. This book describes the explosive population growth of towns which lie on the edge of metropolitan areas. The City's population slowly crept up to 8,500 by 1997, and once the expansion of the greater Sacramento area reached Lincoln, the City needed to accommodate the population influx and support new jobs. The City's population is expected to reach 40,000 in 2007.

The City council ramped-up facilities and services to accommodate this rapid growth, and the success of their efforts is evident today. Examples of new improvements include: expansion of the school system, law enforcement, fire protection, water transmission and supply, new environmentally friendly wastewater treatment and reclamation facilities, drainage systems, parks, roads, bridges, bike trails, roadways, historical preservation areas, and transportation system improvements.

In 2002, as a member of the Sacramento Area Council of Governments (SACOG), Lincoln began working with the six-county, twenty-two-city region to develop the *Blueprint* – a voluntary fifty-year regional land use plan. In an effort to more closely link transportation planning to local land use decisions, the Blueprint has been incorporated into Lincoln's recent General Plan update.

The City is also working with County officials on the Placer County Conservation Plan (PCCP). Implementation of the PCCP entails land preservation and land restoration to help mitigate development impacts of endangered species and wetlands over the next fifty years. If early estimates hold firm, approximately 57,000 acres of land will be preserved for the purpose of mitigating impacts associated with urban/suburban development.

The City has received recognition and awards for their civic programs:

- The 2006 All-America City© Award was given by the National Civic League (NCL). It is the nation's longest running and most prestigious civic recognition program. It recognizes communities whose citizens work together to identify and tackle community-wide challenges and achieve uncommon results.
- 2005 Grand Prize League of California Cities, Helen Putnam Award for Lincoln's Wastewater Treatment Reclamation Facility.
- 2006 League of California Cities, Helen Putnam Award for Excellence in the Public Works, Infrastructure and Transportation category for their Neighborhood Electric Vehicle (NEV) Transportation Plan to foster the use of NEVs.

Introduction

This paper is about fostering and accommodating the use of NEVs in the City of Lincoln. NEVs are small, electric-powered personal vehicles, and are an ideal transportation alternative for local trips. The NEV Transportation Plan (Plan) was developed in response to requests from a group of citizens, and assembled to implement the community's vision to offer residents safe NEV access to downtown Lincoln and other commercial areas. The City shared this vision with owners of developable land and existing and proposed business owners in Lincoln. The Plan was approved by the City Council on August 8, 2006. This paper will summarize legal constraints and opportunities, energy and cost considerations, air quality benefits, design considerations, and NEV Transportation Planning in the City of Lincoln. Extraordinary benefits have been found in planning for the use of NEVs throughout the City and beyond.

Legality

With the advent of the golf cart and the runabouts used by park maintenance crews, a new class of federally-recognized vehicles emerged. While they may look like a golf cart to the casual observer, NEVs are actually motor vehicles that can be driven on public streets with certain restrictions which include: a driver's license, Vehicle Identification Number (VIN), registration, insurance, and adherence to vehicle safety standards. In 1994, the Federal Department of Transportation defined the street-legal Low Speed Vehicle (LSV) in the Code of Federal Regulations. NEVs are a federally-recognized sub-class of LSV. NEVs are limited to 25 miles per hour (mph) by federal requirements, and may be driven on streets with speed zones of 35 mph or less.

The City's first efforts to develop a transportation plan were aimed towards golf carts. Like several other cities, Lincoln adopted a Golf Cart Transportation Plan within a large active adult community, Sun City Lincoln Hills (SCLH). (The Golf Cart Plan description was prepared by Fehr & Peers Transportation Consultants, June 2006.) In

Lincoln, golf carts can be inspected by police, and issued a permit, provided they are *modified* to include safety equipment such as modified brakes, headlamps, taillamps, reflectors and stop lamps for nighttime operation. Modified golf carts reach a speed of 20 mph – five mph faster than normal golf carts. According to a recent Transportation Synthesis Report (CTC & Associates LLC, 2006), "In 1998, the National Highway Traffic Safety Administration [NHTSA] officially recognized NEVs as a form of transportation. Since then, 37 states have passed legislation allowing these vehicles to be driven on roads with posted speed limits of 35 miles per hour or lower."

The State of California has permitted registration of modified golf carts as NEVs in accordance with the NHTSA, Federal Motor Vehicle Safety Standards, Title 49 Code of Federal Regulations Section 571.500. **Table 1** below clarifies some of the distinctions between modified golf carts, NEVs and other similar vehicles.

Table 1. Vehicle Designation

Designation	Weight (lbs.)	Estimated Number in City (2006)	Top Speed (mph)	Registration Requirement	Access Area
Conventional Golf Cart	< 1,300	Not used on Streets	15	None	golf courses
Modified Golf Cart	< 1,300	800	20	City Permit	Striped paths/side streets within golf-cart plan area
Modified Golf Cart Registered by the State as NEV	< 1,300	200	20-25	License Plate	Access throughout City within NEV Plan
Manufactured Neighborhood Electric Vehicle "NEV"	≤ 1,800	400	25	License Plate	Access throughout City within NEV Plan
Low-Speed Vehicle "LSV" Gasoline Powered	≤ 1,800	None	20-25	License Plate	Access throughout City within NEV Plan

Legislation

Through the efforts of the City's Department of Public Works and City Council, a State of California Assembly Bill (AB) 2353 was drafted to provide the City with flexibility in planning for NEV use within the city limits. It was approved by the California state legislature in 2004 and became law on January 1, 2005. This new law enabled both the City of Lincoln and the adjoining City of Rocklin, to develop NEV Transportation Plans based on their specific needs.

The NEV is classified as an Advanced Vehicle in the 2005 Federal Energy Policy Act, Title VII, Part 2. Section 721. This Congressional Act established a 'grant pilot program' for the proliferation of alternative fueled vehicles.

Benefits

The benefits from expanding NEV use include, but are not limited to: energy savings, improved air quality, greater mobility for impaired drivers, cost savings, community cohesion, and support of local businesses. This Plan will provide an alternative approach to transportation planning that will save energy and reduce emissions.

Energy Benefits

In his State of the Union Addresses (2006 and 2007), President George W. Bush discussed the need to end the nation's dependence on imported oil. President Bush also expressed a desire to see a twenty-percent reduction in the nation's gasoline consumption over the next ten years. The increased usage of NEVs will support fuel-saving goals established by both the federal government and the State of California.

According to the July 1, 2002 report to the California Energy Commission titled *Demonstration of Neighborhood Electric Vehicles (NEVs)*, NEVs achieve an 'energy equivalent' of 150 mpg, as compared to 27 mpg for a standard, gasoline-powered vehicle. This 2002 report explains the methodology used to calculate the energy equivalency. Researchers utilized a California Air Resource Board (ARB) energy meter equipped with LCD kWh counters.

The energy required to operate an NEV is less than one-fifth when compared to a conventional automobile. The City has approximately 1,400 NEVs and golf carts, each driving approximately 1000 miles per year according to a survey of NEV owners (conducted by the City in 2003). NEV users in Lincoln save a combined 42,500 gallons of gasoline each year.

According to a recent press release from Global Electric Motorcars GEM© (2007), electric vehicles have reached significant milestones related to energy benefits. With more than 33,000 GEM vehicles on the road, 7.5 million gallons of gasoline were saved as of 2006.

In the past, electric power plants would generate energy using mostly non-renewable resources. In recent years, major electric companies such as Pacific Gas & Electric (PG&E) are increasingly generating energy from renewable resources. The argument that the generation of electric power negates the clean benefits of electric vehicles is disputable. PG&E currently supplies 12% of its energy from renewable sources such as solar, wind, biomass, geothermal, and hydroelectric. According to PG&E (News Release, March 13, 2007), more than 50% of the electricity comes from generating resources that emit no or low carbon dioxide, the primary contributor to global warming.

Air Quality Benefits

NEV trips made possible by the development of this plan produce a variety of air emission benefits to Lincoln and its citizens, and to the region's five-county air basin. Significant air quality improvements result from the use of small electric motors that emit no pollutants into the atmosphere. The City is located within the Sacramento Federal Non-attainment Area, a region federally designated as "severe non-attainment" of federal air quality standards for ozone NEVs provide real, quantifiable emission benefits for local and regional air attainment strategies. NEVs produce no tailpipe or evaporative emissions that contribute to air pollution and global warming.

Emission control systems on a gas-powered vehicle take time to rise to operating temperature, especially in winter. A report to the California Energy Commission dated July 1, 2002, prepared by Arthur D. Little, Inc., stated:

> "It is well documented that cold-start emissions have significant impact on air quality. Due to cold-start fuel enrichment, subsequent quenching of hydrocarbons in a cold engine, and the delayed attainment of proper operating temperatures of the catalytic converter, between 60 and 80% of the toxic air emissions from automobiles occur during the cold-start period."

The good news is that NEVs do not contribute to the pollution caused by cold-starts. The facts listed below were collected from a survey conducted by Global Electric Motorcars (2005):

- For NEV owners who also drive conventional motor vehicles, NEVs replace the use of cars and light trucks approximately two-thirds of the time.
- NEV owners use their NEVs every day.
- NEV owners make short trips. More than 75% of trips are three miles or less.
- On the average, two cold-starts per day are eliminated. 516 grams of (NMOG and NOx) pollution are eliminated each year just from the cold-starts of one vehicle.

A survey of NEV owners was conducted by the City of Lincoln in 2003. Applying GEM's cold-start data to NEV use in the City of Lincoln, at projected buildout of 5000 NEVs, 2.5 tons of cold-start pollution per year will be eliminated.

Total emission benefits from the City of Lincoln's NEV program are shown in **Figure 1**. NEVs eliminate NOx, CO, ROG and other toxic emissions that otherwise result from internal combustion-powered vehicles. The following assumptions were used to prepare the model with the urban emissions software, URBEMIS 2002, mobile source emissions estimation program:

- 5,000 NEVs at program buildout
- 2008 is the modeling year
- Each NEV will travel 1,000 miles/year

- NOx is the primary target; emission reductions annualized from summer conditions
- Only vehicle emissions were calculated (no industrial or construction emissions)
- Trip characteristics derived as 2.78 miles/each for 1,000 mile/year
- Trips calculated as home to work
- 95% light duty passenger car and 5% light duty truck ratio assumed

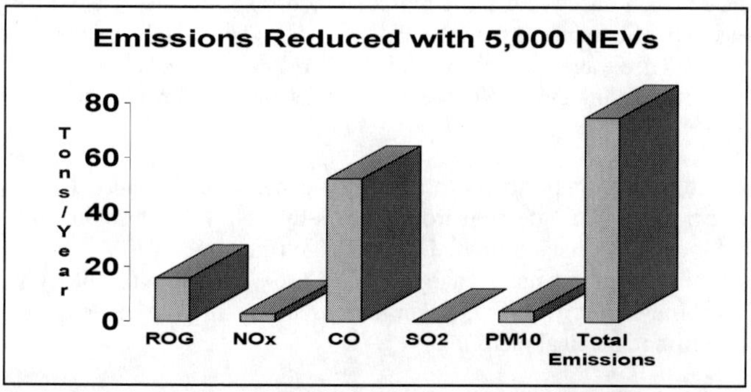

Figure 1. Emission Benefits

Cost Benefits

The initial cost to purchase and the operating costs of an NEV are substantially less than those of a conventional automobile. In fact, an NEV operates for about 20% of the cost of an automobile. Since NEVs cost less than automobiles to purchase, insure, and operate, the City's NEV Plan will enhance the quality of life for fixed-income and low-income residents. The MSRP for a GEM vehicle starts at $6,795. Used NEVs are available at reduced costs. Programs are in process and under development to support the use of NEVs through State and Federal incentives, grants and rebate programs.

To demonstrate the direct cost savings of electricity versus gasoline, the online GEM "Affordability Cost Calculator" was used (http://www.gemcar.com/affordability/). The Affordability Cost Calculator makes the conversion between electricity and gas to compare the energy efficiency and cost benefits of each. Assuming the rate for a unit of electricity is $0.10 per kilowatt hour (kwh), and driving an average of 20 miles per week, with the price of gasoline at $3.10 per gallon, the following results were calculated: compared to a vehicle that consumes 27 mpg, the cost to run an electric vehicle per year is $20.80, and the cost to run a gas powered vehicle per year is $119.41. Using an electric vehicle will save $98.61. This is assuming all of the charging will be done at the resident's home, and not using public charging station facilities. It's important to note, the higher gasoline prices soar, the higher the comparative savings.

Design Elements

The NEV Transportation Plan represents "inside-out transportation planning"; or planning from the user's perspective. NEVs are a supported "Smart Growth" option. According to the Smart Growth America organization, "Smart Growth transportation provides choice and convenience, and is coordinated with the way the community is growing." This Plan follows Smart Growth principals such as citizen-driven planning that coordinates development, transportation, revitalization of older areas, and preservation of open space and the environment. It also includes plans to provide better access to public transportation through pedestrian and bike lane improvements, strengthen and encourage growth in this existing area, encourage citizen and stakeholder participation in development decisions, and achieve a more cohesive mixed-land use.

NEV travel provides an opportunity to develop a cohesive community. NEVs travel at lower speeds and invite attention from passers-by. NEV use also supports local businesses. Since NEVs have a limited travel range (the typical NEV is designed to travel thirty miles on one battery charge.), NEV users will be more likely to shop locally. On thoroughfares within the Plan area, roadway striping extends to nearby shopping centers and medical facilities.

NEVs provide mobility and safety for people who cannot drive a high-speed automobile, including aging and disabled drivers. The slower speed at which an NEV travels is a contributing factor to the increased safety. Many senior drivers prefer to avoid high-speed traffic, and a designated NEV lane provides safe passage throughout the City. According to the City of Lincoln's Police Chief Brian Vizzusi, there have been no accidents reported involving NEVs.

A new program from the California Department of Motor Vehicles is just getting underway in the City. This program will allow mobility for aging drivers who can no longer qualify for a standard Class C California Driver's license. The DMV staff have tested and issued a new form of restricted driver's licenses for within the City.

A major factor for the success of the City's NEV program was the contribution from the local users. Residents with a common interest in NEVs assembled for the first time in October 2002. The group of residents grew, rapidly forming into an official Sun City Lincoln Hills LSV Group. The stated purpose of the group is "to educate, inform and socialize in our common interest of operating and maintaining a low speed vehicle conducive to operating within the Sun City Lincoln Hills community and neighboring areas." Group members can access news, a legislative watch section, current activities, calendar of events, and vehicle maintenance tips on their official website (www.lincolnhillslsv.com). Currently, there are over 100 members in the LSV group. The group attends many community events such as parades, festivals, the farmers market, and community meetings.

In a 2002 study conducted by the Green Car Institute for the Otay Ranch community, "Given a choice of travel modes for short trips, participants chose a NEV over their private cars 90% of the time. Study results also showed that of the trips taken in

NEVs, some 53% were for purposes defined as business or delivery, meaning trips of necessity. Some 33% of the trips taken were classified as leisure, while 14% were designated as other." In the City of Lincoln, the percent of trips used for leisure are much higher. The City is planning for public transit and commuter services that will soon accommodate and encourage the use of NEVs. For example, the City is designing a 500 parking space transit terminal near a new freeway interchange with provisions for commuter bus and future rail service to Sacramento and the San Francisco Bay Area. This facility will provide NEV parking spaces, charging stations, and signage for NEV use.

Early on, researchers recognized that new infrastructure standards and state codes would be needed to accommodate the NEV. According to the 1994 paper *Roadway Infrastructure for Neighborhood Electric Vehicles*, "Because the NEV is a new class of vehicle, NEV infrastructure will ultimately require new design guidelines set forth by AASHTO and FHWA." [AASHTO: American Association of State Highway and Transportation Officials; FHWA: Federal Highway Administration]. The City, with assistance of state transportation officials and the adjacent City of Rocklin, is working through the process envisioned by the 1994 researchers.

A major design goal of the City was to provide new standards and retrofit old standards to allow NEV users access to every part of the City. City plans were drafted to provide infrastructure improvements to allow for the safe, smooth flow of NEVs with pedestrians, bicycles, and other motor vehicles. NEV drivers will be able to make cross-town trips once the Plan is complete and the striping is in place. NEV routes within the NEV Transportation Plan area are shown in Figure 2.

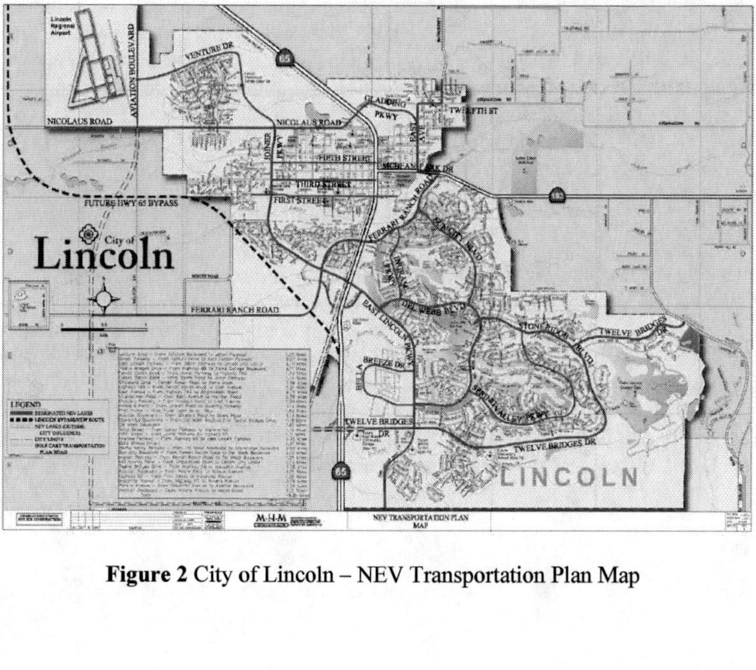

Figure 2 City of Lincoln – NEV Transportation Plan Map

Minimum street standards reflect similar striping design widths for NEV and/or golf cart travel in a typical golf resort community such as Sun City Lincoln Hills.

- Class I NEV routes provide a completely separate right-of-way for the exclusive use of NEVs, pedestrians, and bikes with cross-flow minimized.
- Class II NEV routes are designated as a separate, single-striped lane adjacent to traffic on streets 35 mph and greater.
- Class III NEV routes provide for shared use with automobile traffic on streets with a posted speed limit of 35 mph or less.

For NEVs to be viable, neighborhood services should be within a 5-10 mile radius of a populated neighborhood. The draft Lincoln General Plan update, scheduled for adoption this year, proposes seven villages, each with a build-out population of about 12,000. Each village layout was made with NEV travel in mind.

Challenges

One design challenge that has not been completely resolved is the left turn movement on roadways with speeds greater than 35 mph. Left turns are no problem in speed zones of 35 mph or less because at those speeds NEVs mingle well with other traffic. When striped NEV lanes are added to the shoulder of a thoroughfare, the NEV driver is faced with options similar to a bicyclist: dash across traffic into the left-turn pocket, or turn right on a side street and make a U turn. Engineers are currently working on recommendations and solutions with NEV user safety in mind.

Other challenges faced by developers during the initial stages of the City's program development were related to charging stations and parking stalls in commercial centers. Many questions have developed, including: How many parking spaces should be allocated for NEVs? Should these spaces be treated like handicapped parking and placed near an entrance? Should charging stations be incorporated into the parking lot lighting standards? Who should pay for the power service to the charging stations? These questions are being proactively negotiated in new development agreements. Land developers are anxious to accommodate the NEV customer and are providing parking spaces and charging stations as part of the new commercial sites.

Balancing the needs of bicyclists and pedestrians was another important consideration during the NEV planning process. Organized bicyclists have struggled for years to get adequate shoulders and roadside striping. Some bicyclists are willing to use the new NEV lanes but are reluctant to see a bike lane converted to a wider shared bike/NEV lane. Safety records show no issues with sharing a lane. All NEV striped lanes will have sufficient width to allow sharing with bicycles. Shared lanes are proven safe and cost effective. Adding single-lane striping instead of multiple- lane striping will also reduce costs and future maintenance.

Funding

The City has received Federal Congestion Mitigation and Air Quality (CMAQ) funding for the NEV Transportation Plan efforts. Federal funding was sought from the Sacramento Area Council of Governments (SACOG) and resulted in allocations totaling over $800,000. The City's Plan supports the goals of SACOG's Metropolitan Transportation Plan and Blueprint Project. To receive CMAQ funding, the City successfully demonstrated the cost-effectiveness of the funding dollars. The calculations were based upon several factors such as trip length, traffic volumes, emission reduction factors, auto trip reduction, and capital recovery factor (CRF). Placer County Air Quality Control Board (PCAPCD) funding assistance was also awarded for the City's NEV efforts.

Implementation

With Caltrans in attendance, the City conducted a public workshop on August 30, 2005 to discuss NEV issues. Signage, striping, lane spacing, and NEV lane designation priorities were discussed. The staff led the group through a consensus-building process. Caltrans took the resulting proposed standards to the California Traffic Control Devices Committee (CTCDC) for approval. On August 8, 2006 the City Council unanimously approved the NEV Transportation Plan which incorporated the CTCDC approved standards. The first NEV lane striping was incorporated, by change order, into an existing construction contract to build a railroad overcrossing. This one and a half mile thoroughfare project was almost complete when the extra signing and striping was added at a cost of $70,000. NEV lanes will be included in future roadway contracts in accordance with the adopted plan. The two-way bridge over Auburn Ravine will be sixteen feet wide. This structure will be signed and striped for Class I shared lanes for NEVs, bikes, and pedestrians.

The enabling AB 2353 legislation requires that the Cities of Lincoln and Rocklin, in consultation with state officials, submit an NEV report to the state legislature. This report is due January 1, 2008. In that report the City expects to take the view that this effort has been a tremendous success. From the law enforcement, infrastructure, and community point of view, this program needs to be available to all cities. Additional work on infrastructure and development standards is planned. These plans will be formalized and incorporated into statewide standards and codes. These two cities have been able to succeed because of the strong encouragement and cooperation of state transportation officials and the Placer County Air Pollution Control District. The benefits for air quality, elderly mobility, energy savings, and community lifestyle are clear. The City has contacted the League of California Cities with the intent to develop a statewide focus group to foster the use of NEVs.

Summary / Conclusion

NEV travel is rapidly increasing in the City of Lincoln. What was once considered a sleepy community has opened itself up to a tremendous opportunity for Smart Growth. This pilot NEV Transportation Plan will be an example to all cities in California. With the expected population growth, aging citizens, and future residential development, the timing is ideal. Many positive impacts will make themselves present, such as the ability for residents to easily travel to local services, medical facilities, and commercial centers as the City's NEV-planned infrastructure development is fully implemented. Neighboring cities are working together to promote a network of safe routes for NEV travel.

The NEV is a viable transportation alternative. The City plans to educate residents young and old about the benefits of driving NEVs. Families can use an NEV as a second vehicle for intracity travel and short trips, and high school/college students can use them to commute. The number of older Americans is expected to double over the next twenty-five years. Hopefully in the near future, the "restricted" driver's license program for the elderly will be codified and supported by all DMV offices. Within the text of AB 2353, Chapter 7 Neighborhood Electric Vehicle Transportation Plan, it states:

> "It is the intent of the Legislature, in enacting this chapter, to encourage discussions between the Legislature, the Department of Motor Vehicles, and the California Highway Patrol regarding the adoption of a new classification for licensing motorists who use neighborhood electric vehicles."

In accordance with the intent stated in AB 2353, it is recommended that the State of California expand the current restricted license program and establish an NEV driver's license program. In addition, it is recommended that an incentive program be established to encourage drivers with physical challenges to exchange their automobile driver's license for an NEV driver's license. This program should be aimed at those who no longer feel comfortable driving in high-speed traffic. One retired highway patrolmen suggested a one to two-year automatic extension incentive when exchanging licenses. The need for mobility in an aging population makes this suggestion worth pursuing.

The Cities of Lincoln and Rocklin are working toward nationwide uniformity of NEV Transportation Plan design elements. The trend of transportation using electric vehicles is catching on quickly. It is a tremendously desirable mode of transportation because it is good for the environment, economical, safe and fun. Transportation planning by local governments, in cooperation with the citizens in many communities, is proliferating throughout the nation. In the future, the standardization of signs, pavement markings, vehicle codes and design standards will assist with an efficient transition and modernization of roadways and infrastructure to allow for safe NEV travel.

References:

Arthur D. Little, Inc., (2002), "Demonstration of Neighborhood Electric Vehicles (NEVs)." California Energy Commission P600-02-020F. <http://www.energy.ca.gov/reports/2002-08-28_600-02-020F.PDF> (February 16, 2007).

Bein, R., (1991), "City of Palm Desert Golf Cart Transportation Program." William Frost & Associates, Palm Desert, California.

CTC & Associates LLC., (2006), "State Statutes on Neighborhood Electric Vehicles." *WisDOT Research & Communication Services Section. Transportation Synthesis Report:* <http://www.dot.wisconsin.gov/library/research/docs/tsrs/tsrelecvehicle.pdf> (February 16, 2007).

Department of Transportation, (1998), "National Highway Traffic Safety Administration, 49 CFR Part 571, [Docket No. NHTSA 98-3949], RIN 2127-AG58, Federal Motor Vehicle Safety Standards, Final Rule." <http://www.nhtsa.dot.gov/cars/rules/rulings/lsv/lsv.html#N_1_> (February 16, 2007).

Garreau, J. (1991), "Edge City, Life on the New Frontier." Doubleday, New York.

Stein, A.G., Kurani, K.S., and Sperling, D., (1994), "Roadway Infrastructure for Neighborhood Electric Vehicles." *National Research Council, Transportation Research Record No. 1444.* National Academy Press, Washington, DC.

GEM©, Global Electric Motorcars, A DaimlerChrysler Company, (2007), "Affordability - Cost Calculator." <http://www.gemcar.com/> (February 2, 2007).

GEM© Press Release, (2007), "A very big year – GEM passes 150 millionth mile marker in 2006." <http://www.gemcar.com/> (May 3, 2007).

Green Car Institute, (2002), "Study Proves Neighborhood Electric Vehicle's Value in Planned Communities." < http://www.greencars.org/newsrelease_nov20.html> (May 4, 2007).

Leslie, T. A., (2004), "State of California Assembly Bill 2353. Chapter 422. 'Neighborhood Electric Vehicles." Sacramento, CA.

Lincoln Hills LSV Club, (2007). <http://www.lincolnhillslsv.com/Introduction.htm> (February 2, 2007).

Oswald, L.J., (2006), "Impact of Neighborhood EV (NEV) Use in California on Air Quality." <http://www.arb.ca.gov/msprog/zevprog/symposium/presentations/oswald.pdf> (May 1, 2007).

PG&E News Department, (2007), "Pacific Gas and Electric Company Seeks to Increase Renewable Energy Resources for Its Customers." <http://www.pge.com/news> (May 1, 2007).

Sacramento Area Council of Governments, (2007), "Join the Project. See the Future." <http://www.sacregionblueprint.org/sacregionblueprint/> (February 19, 2007).

Smart Growth America, (2007), "Transportation." <http://www.smartgrowthamerica.org/transportation.html#smarttrans> (February 16, 2007).

The Impact of Biodiesel on Emissions from School Buses

M. Farzaneh, [1] J. Zietsman, [2] D. Perkinson, [2] and D. Spillane[2]

[1]Texas Transportation Institute, 3135 TAMU, College Station, TX 77483-3135, PH (979) 845-5932; email: mfarzaneh@tamu.edu
[2]Texas Transportation Institute, 3135 Tamu, College Station, TX 77483-3135

Abstract

This study investigated the impact of biodiesel (B20: 20 percent biodiesel, 80 percent conventional diesel) on oxides of nitrogen (NOx), hydrocarbons (HC), carbon monoxide (CO), and carbon dioxide (CO_2) emissions from diesel school buses. Two drive cycles were developed based on the rural and urban drive cycle data collected using global positioning system (GPS) technology. Five buses were selected according to the current model year mix in Texas and were driven following the developed drive cycles for three different fuels (TxLED, B20 market blend, and B20 all soy). A state-of-the-art portable emission measurement system (PEMS) unit was used to measure the pollutant emissions along with ambient weather condition, GPS readings, and vehicle engine data. The data were aggregated to represent the current Texas school bus fleet mix and rural versus urban miles driven. The results of statistical analysis showed that using B20 had no significant effect on the level of NOx and CO_2 emissions by the school buses. In contrary, it was found that the biodiesel resulted in a significant decrease of HC and increase of CO emissions.

Introduction

There are approximately 450,000 school buses in the United States traveling 4.3 billion miles a year to carry about 23.5 students (Hinch et al. 2002). More than 85% of these school buses are powered by diesel fuel (EPA 2006). Diesel engines are known to emit all of the criteria pollutants regulated by the National Ambient Air Quality Standards (NAAQS) established by the Clean Air Act including oxides of nitrogen (NOx), carbon monoxide (CO), carbon dioxide (CO2), hydrocarbons (HC), and particulate matter (PM). The use of alternative fuels and the addition of engine retrofits are two potential ways to reduce these pollutants. New standards set by EPA will go into effect in 2007 that will put more stringent emission standards in place for new heavy duty diesel engines including engines used for diesel school buses. However, replacing the existing fleet with these new cleaner buses will take a long time. For example, about 67% of the existing diesel school buses were manufactured between 1990 and 2002, and about 33% (129,000) of the diesel school buses are pre-1990 buses which are the heaviest polluters of the existing fleet (EPA 2006).

Emission reduction strategies for school buses are of a special importance since children are among the most vulnerable group to the effects of diesel emissions which can cause respiratory disease and aggravate long-term conditions such as asthma. Biodiesel is one of the promising means to lower the emissions from diesel powered vehicles. Biodiesel refers to a diesel-equivalent fuel derived from biological sources such as vegetable or animal oils, and can be readily used in diesel powered engines with little or no modification. Blends of 20% biodiesel with 80% petroleum diesel (B20) are considered the most common biodiesel blend for transportation section in the United States. Biodiesel is the only alternative fuel to have fully completed the health effects testing requirements (Tier I and Tier II) of the 1990 Clean Air Act Amendments. In Recent years several experimental studies have investigated the impact of biodiesel on the pollutants from diesel powered engines. These studies are not unanimous in their conclusions, especially in terms of changes in NOx emissions due to using biodiesel. Some studies suggest that the use of biodiesel may produce increases in NOx emissions concurrent with reductions in other pollutants, promoting the need for additional research and studies.

In 1995, the National Biodiesel Board and the National Renewable Energy Laboratory together with the University of Missouri Agricultural Engineering Department collected real-world data for inclusion in the Alternative Fuels Data Center (Schumacher et al. 1995). Qualitative and quantitative biodiesel fueling performance and operational data were collected from 10 urban mass transit buses at the Bi-State Development Agency in St. Louis, Missouri. Five buses were fueled on B20 blend and five buses were fueled on petroleum based low-sulfur diesel fuel (LSD). It was observed that the engines running on B20 produced lower levels of CO, HC, and PM. Slightly higher levels of NOx were noted, however, the increase was not different from the emissions that were recorded for the petroleum diesel control buses.

In October 2002, in connection with the increasing interest in the use of biodiesel, the EPA published the draft Technical Report, *A Comprehensive Analysis of Biodiesel Impacts on Exhaust Emissions* (EPA 2002), after conducting a comprehensive analysis of the emissions impacts of biodiesel using publicly available data. The investigation made use of statistical regression analysis to correlate the concentration of biodiesel in conventional diesel fuel with changes in regulated and unregulated pollutants. The researchers concluded that on average emissions decreased on B20 versus petroleum diesel by 10% for PM, 11% for CO, and 21% for HC, but increased by 2% for NOx. Furthermore, the researchers at EPA were unable to identify an unambiguous difference in exhaust CO_2 emissions between biodiesel and conventional diesel. More importantly, the authors noted, the CO_2 benefits commonly attributed to biodiesel are the result of the renewability of the biodiesel itself, not the comparative exhaust CO_2 emissions. Many state and environmental agencies have adopted EPA's position that use of B20 increases NOx emissions. However, recent research shows that B20 may decrease NOx emissions depending on the application.

The New Jersey Department of Transportation sponsored a research study at Rowan University to develop strategies for reducing diesel emissions from mobile sources such as school buses and Class 8 trucks. Hearne (2003) provided the results of this investigation performed to quantify the emission reduction capabilities of various alternative fuels, such as biodiesel, Ultra Low Sulfur Diesel (ULSD), and a blend of the two, when used to fuel school buses representative of those in use in the state of New Jersey. Hearne tested B20 (80% No. 2 petroleum diesel, 20% biodiesel), ULSD and B20/ULSD (80% ULSD, 20% biodiesel) and compared them to No. 2 diesel fuel in three different school buses. Test results showed that the B20 mixture alone has the ability to provide emissions reductions and potential to combine with other emissions reduction technologies for further reductions. He concluded that B20 and ULSD reduced CO and PM emissions by an average of 30% to 40%. These percentages increased to 70% for CO and 50% for PM when a B20/USLD blend was used. CO_2 emissions were not affected by the alternative fuels tested; however biodiesel provides CO_2 benefits because it is a renewable fuel. He also observed that NOx emissions were only slightly affected by the alternative fuels tested.

Hosatte and Lagacé (2003) tested biodiesel blends of B5 to B20 on 155 transit buses in downtown Montreal, Canada as part of the one year BIOBUS project. The project was a joint effort by the Canadian Renewable Fuels Association, the Fédération des producteurs de cultures commerciales du Québec (FPCCQ), Rothsay/Laurenco and the Société de transport de Montréal (STM). The study concluded that regardless of concentration or source, using biodiesel does not increase NOx emissions, and can even reduce them. It also substantially lowers the mass of PM.

A 2005 study conducted by North Carolina State University demonstrated a decrease in emissions of NO (a precursor to NOx) using soy-based B20 (Frey and Kim 2005). Researchers conducted a pilot study to demonstrate the use of B20 biodiesel fuel on approximately 1,000 vehicles in selected areas of North Carolina. Real-world, in-use, on-road emissions of selected heavy-duty diesel vehicles, including those fueled with B20 biodiesel and petroleum diesel, were measured during normal duty cycles using a PEMS. The results showed that on average fuel use and CO_2 emission rates were approximately the same for the two fuels, but average emission rates of NO, CO, HC, and PM decreased by 10, 11, 22, and 10 percent, respectively, for B20 biodiesel versus petroleum diesel.

Portable Emission Measurement System (PEMS) is an emerging technology that is capable of measuring mobile source emissions from on-road vehicles while operating under real driving conditions. This unique characteristic of PEMS enables the researchers to study the impacts of different strategies and emission control technologies under real-world conditions. In this study a state-of-the-art SEMTECH-DS PEMS unit manufactured by Sensors, Inc. was used to measure different tail pipe emissions from school buses.

This paper presents the result of an experimental study aimed at evaluating the effect of using biodiesel (B20) on NOx, CO, CO2, and HC. Five school buses from different model year groups were instrumented and tested at the Riverside Campus of Texas A&M University. The school buses were tested using a set of representative drive cycles which were developed using GPS data. The paper consists of four sections in addition to this Introduction. The next section explains the on-board emission measurement technology. The third section describes the experimental design of the study. The test results are presented and discussed in the fourth section and the fifth section contains the conclusions of the study.

On-Board Emission Measurement Device

On-board emission measuring is becoming the standard practice in mobile source emission measurement. Mobile testing is capable of providing accurate measurements while a vehicle is operated in real-world driving conditions. This means that the measured emissions are taking into account parameters such as pavement condition, wind resistance, realistic acceleration and deceleration maneuvers, etc. EPA plans to switch to mobile testing by 2010 for their emissions testing on heavy duty trucks (Kilcarr 2003). According to Karl Simon, deputy director of EPA's Office of Transportation and Air Quality, mobile testing is a way to ensure that all fleets are held to the same standard, and that there will be "no advantage gained from operating older engines (Kilcarr 2003)."

The Portable Emission Measurement System (PEMS) unit used in this study was the state-of-the-art SEMTECH-DS manufactured by SENSORS Inc. The SEMTECH-DS unit includes a set of gas analyzers, an engine diagnostic scanner, a GPS, an exhaust flow meter, and embedded software. Figure 1 shows the SEMTECH-DS system and the exhaust flow meter installed on a school bus.

Gas analyzers measured the concentrations of NOx (nitric oxide [NO] and nitrogen dioxide [NO2]), hydrocarbons (HC), carbon monoxide (CO), carbon dioxide (CO2), and oxygen (O2) in the vehicle exhaust. The engine scanner was connected to the vehicle engine control module (ECM) via a vehicle interface (VI) and provided speed, engine speed (RPM), torque, and fuel flow data. SEMTECH-DS used the Garmin International, Inc. GPS receiver model GPS 16 HVS to track the route, elevation, and ground speed of the vehicle under test on a second-by-second basis. The unit used the SEMTECH EFM electronic exhaust flow meter to measure the vehicle exhaust flow. This exhaust mass flow information and pollutants' concentrations then were used by The SEMTECH-DS and the post-processor application software to calculate exhaust mass emissions for all measured exhaust gases. The instrument's embedded software controls the connection to external computers via a wireless or Ethernet connection to provide the real-time control of the instrument. A Panasonic Toughbook laptop was used to connect to the SEMTECH-DS and to control the unit.

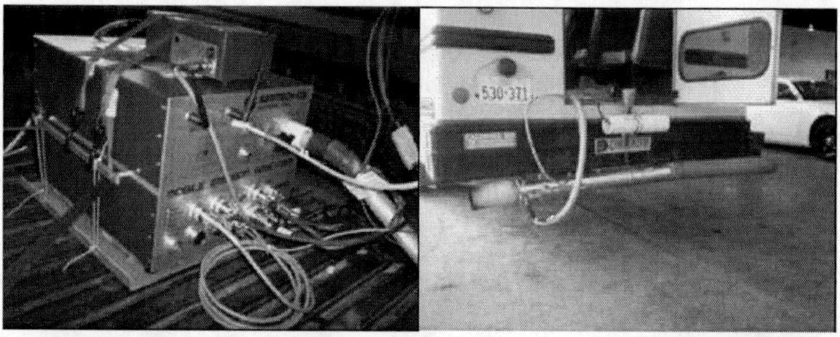

Figure 1. Components of the SEMTECH-DS unit installed on a school bus.

Experimental Design

The experimental design of the on-road data collection involved selecting the test vehicles, the test site and date, test fuels, driving cycles, and laying out the test protocol. This section provides a detailed description on these elements.

Test Vehicles, Dates, and Fuels

The Caldwell, Texas, Independent School District (CISD) provided five school buses as test vehicles for the study. The Caldwell independent school district was selected since it covers both urban and rural areas. Each bus belonged to a different model year group, which was adopted from TTI's report on *School Bus Emissions Reduction Program*, prepared for the Texas Department of Transportation (TxDOT) (Zietsman 2004). Each group represented a different emissions control and engine technology based on EPA's rule making since 1984. The composed average fleet emissions were calculated using the percentages of Texas' school bus in each group. Table 1 shows information about the school buses used in this study along with their share of the Texas school bus fleet.

All the test vehicles were type C school buses (conventional bus on a medium-duty truck chassis, with up to 31,000 lb gross vehicle weight) and were equipped with in-line, six-cylinder International engines (Series D466 and D466E). Each bus was loaded with 56 50-lb. sand bags (2,800 lbs) to replicate an average loading situation of approximately 30 children.

The data collection part of the study took place at TTI's test track located at Riverside Campus of Texas A&M University, Bryan, Texas. The Riverside campus is a 2,000-acre former Air Force base which is used for research and training purposes. The available test tracks consisted of a roadway network surrounding former barracks and other base buildings plus the former runways (longest straightaway 7,500 feet) which were used for testing in this study. The testing phase was conducted Monday, June 17 to Friday, June 28, 2006. These testing dates were typical summer days in central Texas. The conditions were mostly humid and sunny with temperatures in the mid to upper 90 degrees Fahrenheit.

Table 1: Age distribution of school buses in Texas.

Group	Percent of Texas Fleet	CISD Bus Numbers	Model Year
1978-1989	24%	Bus 3	1987
1990-1993	17%	Bus 21	1990
1994-1998	28%	Bus 20	1997
1999-2000	14%	Bus 30	2000
2001-2006	17%	Bus 6	2004

The three fuel blends examined in this study included:

- Base fuel (Fuel 1): Texas Low-Emissions Diesel (TxLED) Ultra Low Sulfur Diesel (ULSD) provided by Valero's Three Rivers Refinery, TX;
- B20 market blend (Fuel 2): 80 percent base fuel, 20 percent market blend biodiesel obtained from Austin Bio Fuels, LLC, TX; and
- B20 all soy (Fuel 3): 80 percent base fuel, 20 percent biodiesel made from soy and obtained from Austin Bio Fuels, LLC, TX.

The biodiesel fuels were obtained in the form of B100 (100 percent biodiesel) and were mixed (splash-blended) in appropriate proportions at the test site.

Drive Cycles

Researchers developed two drive cycles, one for rural regions and one for urban areas. The drive cycles were developed based on three criteria – each drive cycle should reflect typical school bus driving cycles, each cycle must be easy to follow, and each cycle must fit in the available test track.

Prior to the test, four Caldwell ISD school buses were equipped with GPS units for a complete school day to track morning and afternoon bus routes. The first two bus routes covered mainly rural areas, while the other two bus routes serviced urban areas. Researchers analyzed the speed profiles of all the buses and each cycle was separated into different components corresponding to different driving conditions. Each component was further examined and the average time spent in cruise and idle was calculated as well as the average speed and number of stops. This information was then used in conjunction with the synthetic cycle developing method illustrated by Dion et al. (2000) and the constraint of the maximum available straight test track (6,000 ft.) to construct the drive cycles. Figure 2 shows the developed cycles used in this study for the rural and urban conditions, respectively.

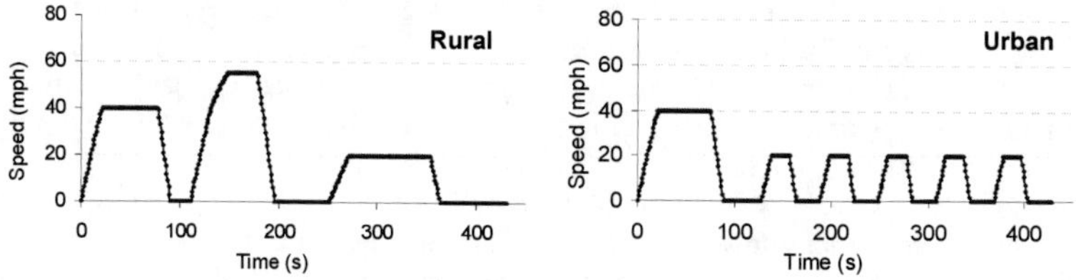

Figure 2. Synthetic driving cycles representing rural and urban driving conditions.

Both cycles consisted of three components; 1) cruising, 2) acceleration and deceleration, and 3) idling. The cruising portion of the rural cycle included a 40 mph, a 55 mph segment, and a 20 mph cruise segment. The 55 mph component reflected driving on a highway. The 55 mph maximum speed is set by law and some of the buses were equipped with governors that prohibited the buses to go faster than 55 mph. The 40 mph element represented driving on rural arterials and the 20 mph component represented driving conditions on rural access roads (possibly dirt roads). The cruising part of the urban cycle consisted of a 40 mph component for urban arterials and five short 20-mph elements reflecting driving short distances in neighborhoods and picking up schoolchildren.

To maintain consistency between each run, an experienced school bus driver was used for all the test runs. Because only one driver was used in the test, the driver's natural acceleration and deceleration behavior was consistent between the tests. The driver was instructed to accelerate and decelerate as she would usually do. The idling portions of the cycles reflected the amount of time that buses remained idle due to traffic control, congestion, and picking up and off loading children. A researcher on board the bus provided instructions to the driver for following the target driving profile. These instructions included: 1) the starting of acceleration and the target speed, 2) the time interval of each cruising speed and when to begin deceleration, and 3) the duration of idle period. The researcher was checking the speed at each step periodically to ensure that the driver was following the instructions.

Texas Transportation Institute's study of *School Bus Emission Reduction Programs* (Zietsman 2004) indicated that 28% of total miles driven by school buses in Texas occurred in rural areas and the rest (72%) were in urban areas. These proportions were used to build up the Texas' representative emission rates.

Fuel Changing

One of the main steps of the tests was the fuel line purging and fuel changing. The fuel line of a diesel school bus consists of a feed line and a return line. The feed line carries the fuel from the tank to the engine and the return line carries the excessive fuel that is not used in the engine back to the tank. To optimize control over the fueling process the feed and return line were disconnected from the vehicle's tank and were instead connected to an auxiliary 5-galon tank that was put inside the bus. A separate tank was used for each fuel to minimize possible contamination. To purge the previous fuel from the fuel system the engine was kept running while the feed line was connected to the new fuel tank and the return line was inserted into a separate purge tank. Experience showed that a purge was completed when approximately two gallons of diesel were collected from the return line after which the return line was re-connected to the test fuel tank.

Test Protocol

The study team developed a test protocol that would provide the best opportunity to test emissions differences resulting from using different fuel mixes. The effect of fuel mix was captured by driving the test vehicles equipped with each fuel mix using the developed synthetic rural and urban cycles. Each test scenario included three runs and for each of the runs the emissions, engine, and speed data were collected on a second-by-second basis. Each test scenario was repeated three times resulting in 90 total runs (3 runs × 5 buses × 3 fuels × 2 cycles = 90). 10 minutes of warm-up time was used before each test in order to reaching the hot stabilized state.

Impact of Biodiesel on On-road Emission Rates

The data recorded by the SEMTECH-DS were in second-by-second format. From the entire array of information that was recorded (emissions, ambient conditions, and vehicle parameters) the following information was extracted and used in this study:

- Engine parameters (if a Vehicle Interface was available) such as engine speed, throttle position, and engine load for data quality checking;
- Second-by-second vehicle speed from GPS in mph; and
- Emission mass rates in grams per second (g/s).

The data were then cleaned and observations that were not part of the desired drive cycles were removed. The length of the idle segments for each section of the cycles were examined and the extra records (i.e., the extra data recorded beyond the number of seconds in target driving cycles) were eliminated to provide cycles which were comparable to each other. The results are discussed for two aggregation levels: aggregated Texas Fleet mix, and individual buses.

Texas Fleet Mix

The emissions mass rates (g/s) were converted to accumulated emissions rates (grams per mile [g/mi]) for each run and cycle. The weighted averages of these accumulated rates were then calculated for each fuel according to Texas school bus model year mix provided in TTI's school bus report (Zietsman 2004). These results were subsequently weighted based on rural versus urban miles covered (rural 28%, urban 72%) (Zietsman 2004) to provide the estimated average mass emissions rates for the Texas school bus fleet. Table 2 summarizes these results.

Table 2: Average mass emission rates for Texas fleet.

Pollutant		TxLED	B20 market	B20 soy
NOx	Average (g/mi)	14.43	14.19	14.53
	Change from TxLED (%)	0.0%	-1.6%	0.7%
HC	Average (g/mi)	1.186	0.843	0.884
	Change from TxLED (%)	0.0%	-28.8%	- 25.4%
CO	Average (g/mi)	3.15	3.89	4.17
	Change from TxLED (%)	0.0%	23.6%	32.6%
CO_2	Average (g/mi)	1324.1	1343.8	1420.3
	Change from TxLED (%)	0.0%	1.5%	7.3%

Table 2 clearly demonstrates that on the average B20 does not considerably affect NOx emission for the Texas fleet mix and the Texas rural/urban composition (in both cases they are less than 2 percent). To support this conclusion, a statistical analysis was performed on the individual bus data. For each of the rural and urban cycles, a paired t-test was utilized for this purpose on the weighted (according to Texas fleet mix) differences between Texas Low-Emissions Diesel fuel emissions and the two biodiesel fuels. In all cases, it was found that the differences were statistically insignificant at a 99 percent significance level.

The total hydrocarbon emission of the school buses dropped notably (28.8% and 25.4% reduction) for both B20 mixes. These figures are consistent with the EPA suggested hydrocarbon benefits (21%reduction) of using B20 fuel. It appeared that B20 caused a small increase (1.5% and 7.3% increase) for CO_2 emissions. These results are close to EPA's observation that no considerable change in tailpipe CO_2 reading as the result of using B20. On the other hand, CO emissions showed significant increase (23.6% and 32.6% increase) which was in contrast with EPA's conclusions (11% decrease).

Individual Buses

Figure 3 illustrates the averaged emission rates (g/mi) for each individual bus. Each row of the figure explains one measured tailpipe emission. The first column of graphs demonstrates the observed emission rates for the rural cycle while the second column of graphs shows the emissions for urban cycle.

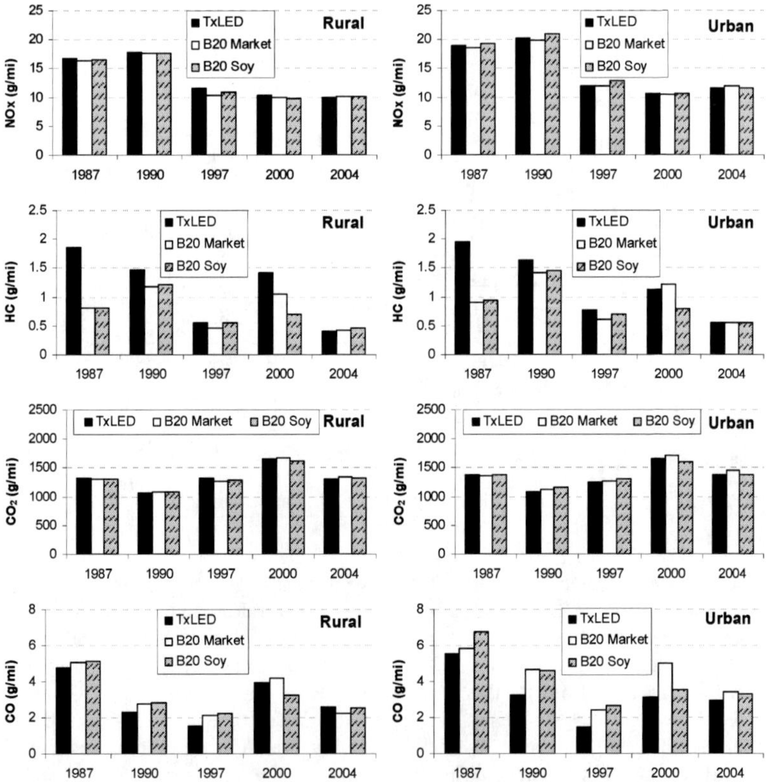

Figure 3. Average individual bus mass emissions rates.

The first row is showing NOx emission rates for each individual bus. With two exceptions, the NOx emission did not show considerable changes between the tested fuels. The exceptions to this trend were model year 1987 and 1990 for the urban drive cycle. It should be noted that these buses were built before 1994. This observation suggests that biodiesel might increase NOx emission for buses older than 1994 that are driven in stop-and-go situations.

As demonstrated by the second row of Figure 3, the benefit of using B20 for reducing HC is considerably higher for older buses (older than 1994). For buses newer than the model year 2000, the change in HC from using B20 is either small or slightly increasing. An exception to this trend was bus 30 (model year 2000) which showed significant reduction in HC level for B20 blends. It was also observed that this bus' level of HC for the base fuel was suspiciously higher than bus 6 (model year 2004) and bus 20 (model year 1997). Normally one expects to see lower HC emissions for newer models for the same fuel, and the fact that that HC emission for this bus is more than three times the HC emission for an older bus suggests that this can be attributed to the possible fuel injector problem of the bus. The possible problem of fuel injector can cause incomplete combustion and produces noticeably higher HC emissions than normal. Therefore, using a biodiesel might provide better combustion and greater reduction of HC level which might not have been achieved if the fuel injector was not defective. Since the researchers were not sure the HC reading of bus 30 (model year 2000) is abnormal, it was decided not to exclude this bus from the results, especially considering that the testing conditions were the same for all the tested fuels.

The third row of Figure 3 illustrates that B20 did not have significant effect on observed tailpipe CO_2 level, for all the vehicles and driving conditions. This again is in agreement with EPA conclusion on CO_2 emissions. It is also observed that CO_2 has the lowest changes due to model year (the only bus that is considerably different from others is bus 3, the same bus that possibly has an injector problem). In fact the CO_2 mass rates for the newest bust (bus 6, model year 2004) is almost the same as the oldest bus (bus 3, model year 1987). This might be explained by the fact that CO_2 is not a regulated pollutant by EPA and, therefore, there is no mandate to lower CO_2 from engines and engine manufacturers have not developed and implemented measures to lower CO_2.

The fourth row is dedicated to the effect of B20 blends on CO emissions. The figure illustrates that B20 had an increasing effect on CO mass emission rate from almost all the buses. This is in contrast to previous studies including EPA (2002) study. The only bus that showed a slight decrease of CO level as a result of using B20 was the newest bus (Bus 6, model year 2004). These results challenges the widely supported and accepted assumption that biodiesel lowers tailpipe CO emissions for all the heavy duty engines. This suggests more comprehensive study to investigate the combinational impact of multiple factors such as temperature, humidity, and driving condition on the CO emissions to clarify the cause of observed pattern.

Conclusions

The paper presents a unique application of portable emissions measurement system (PEMS) for the characterization of the impact of using biodiesel in the form of B20 on the emissions from school buses. The study uses the emissions and GPS data obtained from the state-of-the-art SEMTECH-DS PEMS unit to quantify the impact of mentioned factors on the emissions of a sample of school buses selected according to the Texas fleet mix.

The analyses presented in this paper demonstrate that using biodiesel has no considerable effect on NOx and CO_2 emissions from the school buses, both in Texas fleet mix aggregated and in individual bus level. The biodiesel lowers the HC emissions by 25 to 29 percent while it increases the CO emissions by 23 to 33 percent.

Any study has some limitations that could be used to expand the scope of the analysis or to apply improved methods for future work. A few limitations of this study imply recommendations for future work: - The focus of the study was only gaseous emissions, while diesel engines particulate matter (PM) is also of great importance. - While the developed cycle served the purpose of this study, testing school buses in real-world conditions (on-road testing) for an extended time period provides more confidence in the results in terms of emissions rates. - The overall vehicle sample size of 5 is relatively small compared to the in-use fleet. Having a bigger sample size would increase the statistical confidence in the results.

Acknowledgements

This study is an extension to TTI's report for Capital Area Council of Governments (CAPCOG); *School Bus Biodiesel (B20) Emissions Testing* (Farzaneh et al. 2006). The authors would like to thank the three sponsoring organizations; Capital Area Council of Governments (CAPCOG), Capital Area Metropolitan Planning Organization (CAMPO), and Texas Commission on Environmental Quality (TCEQ). A number of other organizations and individuals contributed to the success of this effort. The support of the Caldwell ISD was critical. Technical support was provided by our colleagues John Storey and Doh-Won Lee, of Oakridge National Laboratory. Finally, the authors thank the following TTI staff for their hard work in colleting the data used in this analysis: Ed Brackin, Mark Ojah, Aaron Rand, and Meghan Wieters.

References

Dion, F., Van Aerde, M., and Rakha, H. (2000). "Mesoscopic Fuel Consumption and Vehicle Emission Rate Estimation as a Function of Average Speed and Number of Stops." *79th Annual Meeting of Transportation Research Board* Washington, DC.

EPA. (2002). "A Comprehensive Analysis of Biodiesel Impacts on Exhaust Emissions." *EPA420-P-02-001, United States Environmental Protection Agency.*

EPA. (2006). "Clean School Bus USA." United States Environmental Protection Agency, http://www.epa.gov/otaq/schoolbus/basicinfo.htm.

Farzaneh, M., Zietsman, J., Perkinson, D., and Spillane, D. (2006). "School Bus Biodiesel (B20) NOx Emissions Testing." *R08-04/05-TTI-2, CAPCOG,* Austin, TX.

Frey, H. C., and Kim, K. (2005). "Operational Evaluation of Emissions and Fuel Use of B20 Versus Diesel Fueled Dump Trucks." *FHWA/NC/2005-07, U. S. Department of Transportation.*

Hearne, J. S. (2003). "School Bus Idling and Mobile Diesel Emissions Testing: Effect of Fuel Type and Development of a Mobile Test Cycle," *Rowan University,* Glassboro, NJ.

Hinch, J., McCray, L., Prasad, A., Sullivan, L., Willke, D., Hott, C., and Elias, J. (2002). "School Bus Safety: Crash Worthiness Research." *National Highway Traffic Safety Admininstration (NHTSA).*

Hosatte, P., and Lagacé, C. (2003). "Biodiesel Demonstration and Assessment with the Société de transport de Montréal (STM)." Montreal, QC, Canada.

Kilcarr, S. (2003). "EPA's take on '07." Fleet Owner, http://fleetowner.com/ar/fleet_epas/.

Proc, K., Barnitt, R., and McCormick, R. L. (2005). "RTD Biodiesel (B20) Transit Bus Evaluation: Interim Review Summary." *NREL/TP-540-38364,* N*ational Renewable Energy Laboratory.*

Schumacher, L. G., Weber, J. A., Russell, M. D., and Krahl, J. G. "An Alternative Fuel for Urban Buses – Biodiesel Blends." *Second Biomass Conference of the Americas*, Portland, OR.

Zietsman, J., Wieters, K.M., and Bochner, B.S. (2004). "School Bus Emissions Reduction Program." *Texas Department of Transportation.*

Impact of Alternative Vehicle Technologies on Measured Vehicle Emissions

H. Zhai,[1] H. C. Frey,[1] N. M. Rouphail,[2] G. A. Gonçalves,[3] and T. L. Farias [3]

[1] Department of Civil, Construction and Environmental Engineering, North Carolina State University, Campus Box 7908, Raleigh, NC 27695-7908, USA
[2] Institute for Transportation Research and Education, NCSU, Centennial Campus, Campus Box 8601, Raleigh, NC 27695-8601, USA
[3] Department of Mechanical Engineering, Instituto Superior Técnico, Av. Rovisco Pais, 1, 1049-001Lisbon, Portugal

Introduction

Emerging vehicle technologies such as hybrids, flex-fuel, electric, and fuel cells are projected to exceed 25% of total light-duty vehicle sales by 2030 (EIA, 2006). Because of their higher energy efficiency, hybrid electric vehicles (HEVs) produce lower emissions of greenhouse gases such as CO_2 than comparable gasoline vehicles. Measurements of selected advanced gasoline and diesel direct injection vehicles and gasoline-electric hybrid vehicles indicate that the HEV had the lowest emissions and highest fuel economy (Graham, 2005). One study concluded that replacing all conventional vehicles with HEV's could cut total CO and NO_x emissions in half (Yun *et al.*, 2007).

Flex-fuel vehicles that can use either gasoline or ethanol 85 (E85), a blend of 85% ethanol and 15% gasoline by volume have sensors that can analyze the fuel-air mixture and adjust the fuel injection and timing (CEPA, 2004). Since ethanol is an oxygenated fuel, the use of ethanol may reduce emissions of products of incomplete combustion including CO and HC. However, the oxygenate may tend to increase NO_x emissions because of lean combustion (Reuter *et al.*, 1992). The use of blends with a high percentage of ethanol in the mix was found to produce increased emissions of NO_x and aldehydes as the ethanol content increased (Guerrieri *et al.*, 1995). However, measurements of flex-fuel Chevrolet Lumina vehicles indicated that the use of E85 may decrease vehicle CO, HC and NO_x emissions, but increase aldehydes emissions when compared to gasoline vehicles (Kelly *et al.*, 1999). Durbin *et al.* (2006) showed that NO_x emissions increased with increasing ethanol content for some fuels, but were unaffected by ethanol content for other fuels, depending on fuel volatility. Hochhauser (2006) found that use of ethanol fuel may increase vehicle permeation emissions of VOC. Therefore, flex-fuel vehicles do not assure lower emissions for at least some pollutants compared to dedicated gasoline vehicles (CEPA, 2004).

The objective of this study is to develop an advanced modeling system to quantify influences of land use and vehicle technologies on on-road vehicle emissions such as CO_2, CO, and NO_x. The main focus here is to demonstrate a methodology for assessing how differences in driving cycles affect link-based average emission rates for selected vehicle technologies. The methodology requires second-by-second data, with a preference for real-world in-use data where possible. The methodology is illustrated here based on data from one flex-fuel vehicle that was measured during actual driving using a portable emission measurement system (PEMS) on both gasoline and E85 and for one HEV that was tested on a chassis dynamometer (Nam *et al.*, 2005). Testing for the flex-fueled vehicle was conducted by Instituto Superior Técnico (IST) in Lisbon, Portugal. The flex-fuel vehicle is a European 2006 flex-fuel Ford Focus wagon with a 1.8 liter engine (92 kW/6000rpm, 165 Nm/4000rpm). By comparison, the U.S. version of the Focus has a 2.0 liter engine (97 kW/6000rpm, 175 Nm/4000rpm). The HEV is a 2001 Toyota Prius with a 1.5 liter gasoline engine.

Methodology

A modal approach based on vehicle specific power (VSP) is used to relate the measured speed profiles with average link-based speed estimated from transportation models, in order to quantify the impact of advanced vehicle technologies on on-road vehicle emissions.

VSP has been identified as a useful explanatory variable for emission estimation (Jimenez-Palacios, 1999). It takes into account aerodynamic drag, tire rolling resistance and road grade. Based on coefficient values for a generic light duty vehicle, VSP is calculated as (EPA, 2002):

$$VSP = \mathrm{v}\times[1.1\times a + 9.81\times(\sin(\mathrm{atan}(grade))) + 0.132] + 0.000302\times \mathrm{v}^3 \quad (1)$$

where: VSP = vehicle specific power (m^2/s^3); v = vehicle speed (m/s); a = acceleration (m/s^2); *grade* = road grade (dimensionless fraction); 0.132 = rolling resistance term coefficient; and 0.000302 = drag term coefficient. Using second-by-second speed, acceleration and road grade values, instantaneous VSP values are computed from Equation (1). The continuous estimates of VSP values are then grouped into discrete VSP modes. Fourteen discrete VSP modes were defined for light duty vehicles in previous work (Frey *et al.*, 2002). For the HEV, a prediction is first made to determine whether the internal combustion engine (ICE) will be "off" for some combination of speed and acceleration. A set of "rules" for conditions under which the engine is off were determined based on analysis of the second-by-second dynamometer data using categorical and regression trees (CART). When the engine is on, VSP modal definitions are used for estimating average emission rates. Otherwise, the tailpipe emissions are zero. VSP modal average emission rates are estimated from PEMS and dynamometer tests databases for all vehicle technologies.

Speed profiles collected previously by NCSU in a real-world field study are reported in terms of a link mean speed and roadway class. For each point in the second-by-second link speed profile (associated with a narrow range of speeds) a VSP value is estimated, and aggregated into a time distribution of VSP modes across the link. Link emission rates are estimated as the product of VSP modal average emission rates multiplied by the fraction of time spent in the corresponding VSP mode (Frey *et al.*, 2006). Link-based average emission rates are estimated based on multiple speed profiles for a given speed level and facility type. For the HEV, the time spent in a VSP mode is accounted for only when the ICE is on. Link average emissions are estimated using the same set of selected speed profiles for all vehicle technologies. Two groups of ten arterial speed profiles were used: low (10-20 km/h) and medium (30-40 km/h). Because no dynamometer test data were available for HEV at high speeds, only low speed profiles were used. Emissions estimates are compared for three types of vehicles for each of two speed profile levels to quantify the differences between conventional and advanced vehicle technologies.

Results and Discussions

For the HEV, engine RPM less than 500 was selected as the criteria for the ICE being "off." The combinations of vehicle speed and acceleration associated with engine "off" status were identified using CART analysis. The resulting rules are summarized in Table 1. The engine is off only under threshold conditions of engine power demand that are speed and acceleration dependent. These rules are applied to each data point on the speed profile to estimate whether the ICE is off at a given time during a driving cycle. For example, if the measured speed is 32 kph and the measured power is 10 kph^2/s, the engine is "off" since power is below the applicable threshold of 11.6 kph^2/s. However, there were no data available regarding the operation of the HEV at speeds higher than 96 kph (60 mph).

Table 1 Hybrid vehicle engine-off rules

Speed (v, kph)	Acceleration (a, kph/s)	Threshold Power (v×a, kph^2/s)
0.0- 14.6	a<4.0	58.4
14.6-29.6	a<1.1	32.6
29.6-57.9	a<0.2	11.6
57.9-65.2	a<-0.2	-13.0

Modal average emission rates based upon PEMS and dynamometer test data are shown in Figure 1. Average emission rates increase with increasing VSP mode, except for E85 and HEV at the higher VSP modes. Because of the limitations of the test cycle, there are no HEV data for modes 12 to 14. For the HEV, the modal average emission rates of NO_x for modes 8 to 11 are not significantly different from one another. There is a possibility that the electric motor may have been activated at high VSP modes, which could lead to lower emissions for these modes. However, no data were available regarding second-by-second use of the battery and electric motor for the HEV. The wider confidence intervals for higher VSP modes associated with E85 and HEV measurements reflect the effect of smaller sample sizes and larger variability in the data compared to other modes.

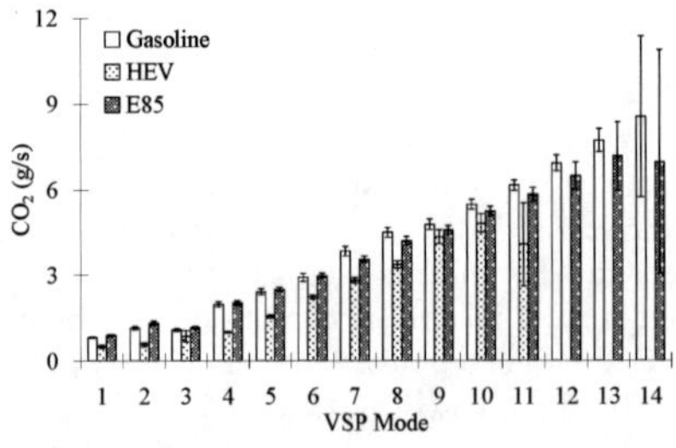

(1a)

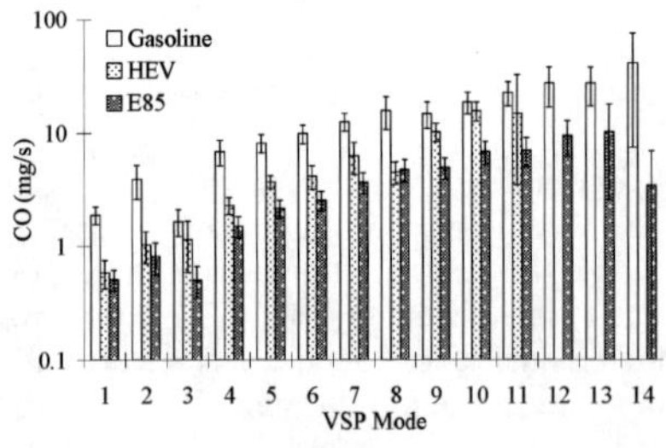

(1b)

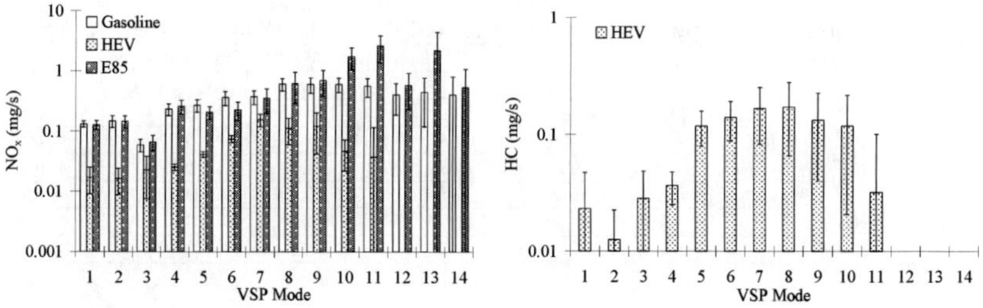

(1c) (1d)

Figure 1 Modal average emission rates for conventional and advanced light-duty vehicles

As shown in Figure 2a, the various arterial link speed profiles for a 30-40 kph average speed range have similar patterns. The average time distributions of VSP modes and 95% confidence intervals on the mean are shown for all vehicle technologies in Figure 2b. The time distribution of VSP mode is similar for all technologies for a given speed range. However, for the HEV, the percentage of time spent in each VSP mode shown in Figure 2b is only indicated when the ICE is on.

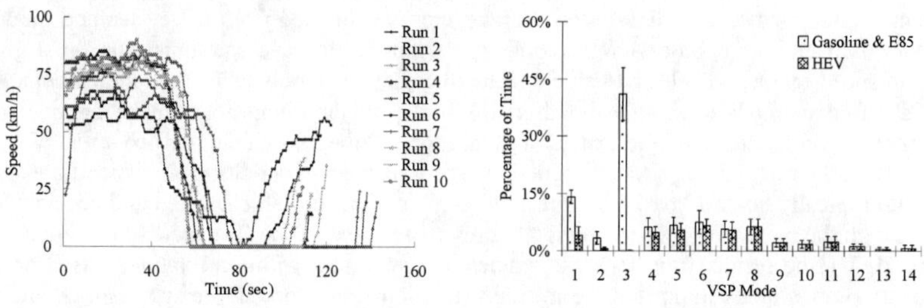

(2a) Speed Profiles (2b) Time Distributions of VSP Modes

Figure 2 Example of link speed profiles (30-40 kph) and time spent in each VSP mode

The resulting link average emission rates for all pollutants are summarized in Table 2. As an illustrative comparison, differences in total emissions for E85 or HEV are estimated for these selected link-based speed profiles with respect to the flex-fueled Ford Taurus with gasoline, as shown in Table 3. The HEV has substantially lower emissions of CO_2, CO, and NO_x emissions compared to the similarly sized gasoline-fueled vehicle for the low speed cycles, but the differences are less pronounced for the higher speed cycles. When operated on E85, the flex-fuel vehicle has lower CO emissions, and approximately the same CO_2 emissions for both the lower and higher average speed cycles. However, the NO_x emissions for E85 are approximately the same for the lower speed cycles but are higher for the higher speed cycles.

Table 2 Link average emission rates for selected flex-fuel and hybrid-electric vehicles

Link Mean Speed (kph)	Vehicle	Link Average Emission Rates			
		CO_2 (g/s)	CO (mg/s)	NO_x (mg/s)	HC (mg/s)
10-20	Flex-Fuel: Gasoline	1.62	4.75	0.17	n/a
	HEV	0.71	1.59	0.02	0.03
	Flex Fuel: E85	1.69	1.18	0.17	n/a
30-40	Flex-Fuel: Gasoline	2.15	5.52	0.22	n/a
	HEV	1.77	4.12	0.05	0.08
	Flex-Fuel: E85	2.14	1.85	0.29	n/a

Table 3 Emissions changes for substituting a Flex-Fuel Vehicle fueled with gasoline with either a flex-fueled vehicle fueled with E85 or a gasoline hybrid electric vehicle

Link Mean Speed (kph)	Vehicle	Changes in Total Emissions [a] (%)		
		CO_2	CO	NO_x
10-20	HEV	-56	-67	-88
	Flex-Fuel: E85	4	-75	0
30-40	HEV	-18	-25	-77
	Flex-Fuel: E85	-0.5	-66	32

[a] Percent change in total link-total emissions when comparing the indicated vehicle to a gasoline-fueled flex fuel Ford Focus. Link total emissions are the products of link average emission rates and travel time on the link.

Conclusions

In order to demonstrate a methodology for estimation of emissions from a variety of vehicle technologies for use with link-based average speeds estimated from travel demand models, VSP-based emissions models were applied to estimate link average emissions for a given mean speed range and selected vehicle technologies. The case studies here indicate that the VSP-based methodology for estimating link-based average emission can be applied to a variety of vehicle technologies, including data for a flex-fuel vehicle and for an HEV. For the HEV, the methodology was modified to take into account conditions in which the engine would typically be "off" based on threshold combinations of vehicle speed and acceleration. Although the methodology was applied to only two vehicles (one flex-fueled Ford Focus and one HEV), the results from the case studies are consistent with expectations based on the literature regarding comparisons of average emissions rates for E85 versus gasoline and for HEV versus a light duty gasoline vehicle. Based upon this preliminary demonstration of the methodology, the methodology to estimate link-based network-wide emissions based on vehicle activity derived from transportation demand models is recommended.

Acknowledgements

The work at NCSU was supported by U.S EPA STAR Grant R831835 via the UNC- Chapel Hill. Any opinions, findings, and conclusions or recommendations expressed are those of the author(s) and do not necessarily reflect the views of EPA. The work at IST was supported by MMSAFU project (POCI/ENR/57450/04) funded by the Fundação para a Ciência e a Tecnologia as well as FLAD (Luso–American Foundation). The authors also would like to acknowledge the support of Ford Lusitana and TOTAL Portugal.

References

California Environmental Protection Agency, (2004). *Fact Sheet: 2004-4-08 California Vehicle Emissions,* http://www.arb.ca.gov/msprog/zevprog/factsheets/calemissions.pdf>(February 28, 2007).

Durbin, T.D., Miller, J.W., Younglove, T., Huai, T., and Cocker, K. (2006). "Effects of Ethanol and Volatility Parameters on Exhaust Emissions." *CRC Project No.E-67,* Prepared by University of California-Riverside for the Coordinating Research Council (CRC), Alpharetta, GA.

Energy Information Administration, (2006). "Annual Energy Outlook 2006 with Projections to 2030." *Report DOE/EIA-0383*, Washington D.C.

Environmental Protection Agency, (2002). "EPA's Onboard Analysis Shootout: Overview and Results." Ann Arbor, MI., EPA420-R-02-026.

Frey, H.C., Unal, A., Chen, J., Li, S. and Xuan, C. (2002). "Methodology for Developing Modal Emission Rates for EPA's Multi-Scale Motor Vehicle & Equipment Emission System." Prepared by North Carolina State University for U.S. EPA, Ann Arbor, MI, August, EPA420-R-02-027.

Frey, H.C., Rouphail, N.M. and Zhai, H. (2006). "Speed- and Facility-Specific Emission Estimates for On-Road Light-Duty Vehicles based on Real-World Speed Profiles." *Trans. Res. Rec.*, 1987:128-137.

Graham, L. (2005). "Chemical Characterization of emissions from advanced technology light-duty vehicles." *Atmos. Environ.*, 39 (13): 2385-2398.

Guerrieri, D.A., Caffrey, P.J., and Rao, V. (1995). "Investigation into the vehicle exhaust emissions of high percentage ethanol blends." *SAE Special Publications, Gasoline: Additives, Emissions, and Performance*, paper no. 1095: 85-95.

Hochhauser, A. M. (2006). "Summary of Ethanol Impacts on Permeation." *California Air Resources Board fuels Workshop*, Sacramento, California, August.

Jimenez-Palacios, J. (1999). "Understanding and Quantifying Motor Vehicle Emissions with Vehicle Specific Power and TILDAS Remote Sensing." *PhD Thesis*, MIT, Cambridge, MA.

Kelly, K., Eudy, L. and Cobum, T. (1999). "Light-Duty Alternative Fuel Vehicles: Federal Test Procedure Emissions Results." *National Renewable Energy Laboratory,* Battelle, NREL/TP-540-25818.

Nam, E. K. and Mitcham, A. (2005). "Dynamometer and On-board Emissions Testing of the Honda Insight and Toyota Prius." *SAE 2005 World Congress & Exhibition*, Detroit, MI., paper no. 2005-01-0681.

Reuter, R. M., Benson, J. D., Burns, V. R., Gorse, R. A. Jr., Hochhauser, A. M., Koehl, W. J., Painter, L. J., Rippon, B. H. and Rutherford, J. A. (1992). "The effect of adding oxygenated compounds to gasoline on automotive exhaust emissions." *SAE Special Publications,* Auto/Oil Air Quality Improvement Research Program, 920:391-412.

Yun, J. and Miller, T. (2007). "Hybrid Electric Vehicles: Evaluating Emission Reductions and Cost Benefits for University Motor Vehicle Fleet." *2007 Annual Meeting of the Transportation Research Board,* Washington D.C., paper no. 07-2598.

Transportation Planning and Climate Change: New York State

M. P. Gaber[1]

[1]Federal Highway Administration, Office of Natural and Human Environment, 1200 New Jersey Avenue SE, Washington, DC 20590; PH (202) 366-2078; email: mark.gaber@gmail.com

Abstract

The transportation sector is responsible for roughly one third of national carbon dioxide emissions, making it a prime candidate for emission reductions. While no national policy to control greenhouse gases has yet been developed, many state and local governments have taken actions to reduce emissions. Twenty-eight states have adopted climate action plans, and 128 city and county governments are participating in an initiative to reduce emissions. While California has regulated tailpipe emissions of greenhouse gases, New York has enacted requirements on the transportation planning side.

The 2002 New York State Energy Plan required metropolitan planning organizations (MPOs) in the state to conduct a greenhouse gas energy analysis as part of the transportation planning process for regionally significant projects. This paper provides an overview of the transportation/climate change connection and its importance to New York, summarizes the requirements of the Energy Plan and examines the analyses conducted by three of the thirteen MPOs in New York. It synthesizes their findings, and assesses what effects the greenhouse gas analyses have had on the transportation planning decision-making process and the potential for achieving greenhouse gas reductions from such exercises.

Further, this paper will compare the methods used by the MPOs to quantify greenhouse gas emissions and to conduct the analysis, and recommend areas where further methodological development is needed or available for future planning documents.

With political and scientific attention increasing, the transportation sector may face some form of emission reduction targets in the future—whether mandatory or voluntary. Ultimately, this paper provides a case study on the potential for New York

State's Energy Plan requirements to serve as model for gaining emission reductions through the transportation planning process.

Introduction: Climate Change and Transportation

In February 2007, the United Nations Environment Programme's (UNEP) Intergovernmental Panel on Climate Change (IPCC) released the summary findings of its fourth assessment report on the physical science basis of global climate change. The report strongly supports the scientific consensus that global climate change is a real and consequential phenomenon. Further, the IPCC authors have determined that "[m]ost of the observed increase in globally averaged temperatures since the mid-20th century is *very likely* due to the observed increase in anthropogenic greenhouse gas concentrations" (IPCC, 2007). The authors define "very likely" to indicate a likelihood of greater than 90 percent—a change from the former designation of "likely", or greater than a 66 percent likelihood. The new report also ties the rising global average sea level and decreasing northern hemisphere snow cover to global climate change.

The anthropogenic contributions to global climate change result from emissions of greenhouse gases into the Earth's atmosphere. Simply stated, the greenhouse gases trap excess heat in the Earth's atmosphere, thus warming the planet. Carbon dioxide is the gas emitted in the largest quantity, but others, such as methane, nitrous oxide, and fluorinated gases, contribute to the warming (Environmental Protection Agency (EPA), Climate Change, 2007). These emissions are primarily produced by the burning of fossil fuels. The US Energy Information Agency (EIA) reported that, in 2005, the transportation sector accounted for thirty-three percent of US carbon dioxide emissions—the largest contributor of the end-use sectors. Additionally, since 1990, there has been an average annual growth of 1.5 percent in transportation sector carbon dioxide emissions (EIA, 2006).

The United States is responsible for twenty-five percent of global carbon dioxide emissions (EIA, 2004), which means that the US transportation sector is responsible for eight percent of global emissions. The US Department of Transportation's (DOT) Center for Climate Change and Environmental Forecasting acknowledged the transportation sector's contribution to greenhouse gas emissions in its 2006-2010 Strategic Plan. Importantly, the DOT also addressed the potential for impacts on the transportation sector as a result of climate change. "Transportation will also be affected by climate change, which has the potential to create significant weather irregularities, including sea level rise and more intense storms that could severely affect the safety and security of national transportation infrastructure" (DOT, 2006).

New York is one of several state governments that have been pioneering policies to reduce energy consumption and to integrate climate change considerations into decision-making processes. New York City is the largest US city, and its coastal location is threatened by the potential for sea level rise, making this an important issue for the state and local governments to address.

Climate Change Impacts in New York

New York City is one of the most significant centers of business in the world. As with most cities, New York City's location near waters makes it a natural site for trade and economic activity. The city has over 500 miles of coastline, of which transportation infrastructure covers a significant portion, including transit, tunnels, roadways, bridges, and other systems (Zimmerman, 2002). A significant portion of the city's transportation infrastructure is below or just slightly above sea level, which increases the threat of damage from flooding. According to Zimmerman (2002), the city has twenty-seven transit facilities; twenty-one surface transportation facilities, including roads, bridges, and tunnels; six marine facilities; and two airports; all that are ten feet or fewer above sea level. According to the IPCC (2007), global average sea level rose at an average rate of 1.8 millimeters per year from 1961 to 2003, but with a much higher rate of 3.1 millimeters per year from 1993 to 2003. The IPCC (2007) predicted a .18 to .59 meters (.6 to 1.9 feet) rise in sea level from 1980 to 2099.

The threat of rising sea level to the city's transportation infrastructure is broad in its economic, health, and safety effects. Submerged infrastructure is obviously not useable, but it can also not withstand submerged conditions for long, as the material corrodes and the structure are impaired (Zimmerman, 1996). Transportation infrastructure is critical to public health and safety, serving as the conduits for emergency response vehicles. The flooding and storm damage the occurred after Hurricane Katrina in New Orleans in 2005 demonstrate the difficulties that emergency management officials face in reaching threatened populations when transportation infrastructure is compromised.

The IPCC (2007) reports that, of the twelve warmest years on record since 1850, eleven occurred between 1995 and 2006. Additionally, it reports that the best estimates for expected temperature increases over the course of the twenty-first century range from three to seven degrees F. One of the consequences of such an increase in temperature is increased stress on transportation infrastructure. According to Zimmerman (1996), bridges and other road surfaces withstand temperature ranges of 120 degrees F, from -20 degrees to 100 degrees F. While temperatures above 90 degrees F. are not uncommon in the city during the summer months, temperatures above 100 degrees F. are relatively rare (NOAA, 2006). As temperatures increase, however, impairment of the road facilities from heat stress could be a concern (Zimmerman, 2006).

Clearly New York City, and by extension, New York State, have a significant stake in taking steps to mitigate the threats posed by global climate change. Having been the center of the 9/11 attacks, the city is familiar with disasters. According to Dolfman and Wasser (2004), in their assessment for the Bureau of Labor Statistics, the 9/11 attacks resulted in over $2 billion in lost wages during the four month period following the disaster in Manhattan alone. Severe flooding and the loss of important transportation infrastructure could be expected to cause similarly significant economic, environmental, and human damage. Acknowledging the threat posed by climate change, the state of New York developed its 2002 State Energy Plan with the goal of reducing greenhouse gas emissions.

New York State Energy Plan

In 2002, the New York State Energy Research and Development Authority (NYSERDA) released its State Energy Plan, a wide-ranging set of policy goals aimed at providing the state with efficient, clean, affordable, and reliable energy resources. The plan details fifteen policy recommendations, including seven with environmental focuses. It places a large emphasis on renewable fuels research and use and increased energy efficiency. The Plan calls for increased efficiency in transportation through support for public transit, transportation management, intelligent transportation systems, and capital construction.

The recommendation that this paper focuses on is to reduce the State's greenhouse gas emissions to five percent below 1990 levels by 2010 and to ten percent below 1990 levels by 2020. Specifically, the Plan calls for analyzing energy consumption of the transportation system as a part of the transportation planning process:

Examining and analyzing the transportation system's energy consumption and air emissions when long-range plans and Transportation Improvement Programs are adopted would enhance this commitment. This examination could be on a build/no build basis and include public review. If a plan or a program increases air emissions or uses more energy than doing nothing at all, additional measures or modifications to the plan or program could be considered to minimize the increases as much as practicable. This review would be in addition to existing federal and State requirements to address transportation conformity regulations in air quality non-attainment and maintenance areas (NYSERDA, p. 2-90, 2002).

While the Plan calls for the "build" scenarios to result in lower emission than the "no build" scenarios, it does not link this transportation planning requirement to the more ambitious goal of actually reducing greenhouse gas emissions below 1990 levels. Indeed, NYSERDA's 2005 State Energy Plan Annual report indicates that, through 2005, New York's greenhouse gas emissions actually increased by seven percent over 1990 levels. The state notes, however, that the actions it took to curtail emissions resulted in one-half of one percent fewer emissions than would have occurred without acting (NYSERDA, 2005).

The Plan identifies a large group of transportation planning strategies that can be included in transportation plans and TIPs to reduce emission. These strategies include enhanced bicycle and pedestrian programs, improved intelligent transportation systems (optimized traffic signals, incident response, corridor management), speed limit reduction, congestion pricing, transportation management planning for employers (providing telecommuting options, vanpooling, and flex time), and improved public transit.

The New York Department of Transportation carried out the directive in the Energy Plan by requiring MPOs to conduct a greenhouse gas energy analysis on their transportation plans.

Greenhouse Gas Energy Analysis

Any metropolitan region with over 50,000 people is required by federal transportation planning regulations to have an MPO. There are thirteen MPOs in New York of varying sizes. In a report commissioned by the U.S. Department of Transportation in 2005, ICF Consulting interviewed staff from the state MPOs about the Energy Plan requirements. According to ICF (2005), many MPOs do not view the Energy Plan requirements as mandatory, instead seeing them as voluntary actions. Indeed, some MPOs, particularly the smaller regions, did not complete analyses. In the interviews, many MPOs expressed confusion as to why they were given such a visible role in the state assessment of energy consumption, given their perceived inability to control increases in energy use. Some MPOs are resisting the requirement, as they do not believe the analyses will be used in the decision-making process—indeed, the analyses were conducted after the plans were complete. Additionally, many MPOs were concerned with the increased workload required by conducting the analyses, stating that they required two to four person weeks on average to complete. For small MPOs, this can be a significant burden.

This paper will assess the greenhouse gas energy analyses conducted by three of the MPOs of different sizes: the New York Metropolitan Transportation Council (NYMTC) based in New York City, the Capital District Transportation Council (CDTC) based in Albany, and Ithaca-Tompkins County Transportation Council (ITCTC) based in Ithaca.

New York City

NYMTC released its greenhouse gas energy analysis for its 2005-2030 transportation plan in November 2006. Table 1 presents data on NYMTC's past and future carbon dioxide emissions, gross regional product, carbon dioxide intensity, and energy intensity, as provided by ICF Consulting in a report for the US Department of Transportation.

Table 1-NYMTC Carbon Dioxide Emissions and Energy Intensity

	1990	2001	2010	2020
Gross Metropolitan Region Product (Million 2001 Dollars)	\$350,993	\$597,638	\$779,782	\$1,047,962
CO_2 Emissions (MMTCO$_2$)	44.0	47.0	53.8	59.6
CO_2 Intensity (MTCO2 per million 2001 Dollars)	125.3	78.7	68.9	56.9
Energy Use (Trillion Btu)	609.4	639.0	740.2	830.1
Energy Intensity (Thousand Btu per 2001 Dollar	1.7	1.1	0.9	0.8

Source: ICF, 2005

While Table 1 demonstrates that NYMTC is reducing its greenhouse gas and energy use intensity, as measured against gross metropolitan regional product, it also shows that actual greenhouse gas emissions and energy use are increasing and are projected to increase through 2020. This indicates that NYMTC is not on track to meet the Energy Plan greenhouse gas reduction goals. To meet the Energy Policy goals, NYMTC's greenhouse gas emissions would need to be at or below 41.8 million metric tons in 2010, and at or below 39.6 million metric tons in 2020. This means that NYMTC is projected to emit 51 percent more emissions in 2020 than it would if the Energy Policy greenhouse gas goals were met.

Table 2 summarizes the greenhouse gas emissions from direct energy consumption of the "build" versus the "no-build" scenarios.

Table 2-NYMTC Daily Greenhouse Gas Emissions from Direct Energy (2005-2030 Regional Transportation Plan)

	Build		**No-Build**		**Build Compared to No-Build**	
Year	**VMT**	**Direct Greenhouse Gas Energy (tons)**	**VMT**	**Direct Greenhou-se Gas Energy (tons)**	**Difference in VMT**	**Differ-ence in Direct Energy**
2002 Base Year			182,193,403	97,408		
2007	185,989,846	101,091	187,766,434	102,051	1,776,588	960
2010	188,061,642	103,204	189,790,440	103,984	1,728,798	780
2020	196,918,361	106,794	200,053,115	108,643	3,134,754	1,849
2025	201,785,908	108,928	205,046,401	110,959	3,260,493	2,031
2028	204,499,647	110,723	208,335,055	113,104	3,835,408	2,381
2030	205,672,794	111,374	209,480,633	113,840	3,807,839	2,466

Source: NYMTC, 2006

While the Energy Plan calls for actual reductions in greenhouse gas emissions, its section on transportation only indicates that the plans and TIPs should show less emissions under the build than under the no-build scenarios. NYMTC's 2005-2030 Regional Transportation Plan meets this goal in each analysis year, with 2,466 fewer tons of daily greenhouse gas emissions projected for 2030 under the build scenario than under the no-build scenario. It is important to note, however, that the analysis indicates that there will be an increase of 13,966 tons of greenhouse gas emissions per day, or roughly 5,100,000 tons per year, from the 2002 base year in 2030 if the Plan is enacted. This represents a 14 percent increase in emissions over 2002 levels. Furthermore, NYMTC did not compare emission level to those caused by transportation sources in 1990. Without that information, it is impossible to determine the increase that will happen relative to the 1990 levels, the year that the Energy Plan uses for its greenhouse gas reduction goals.

NYMTC concludes its greenhouse gas analysis by stating that "NYMTC's TIP and Plan are consistent with the 2002 State Energy Plan, and the forecasted reduction in future energy consumption for years 2010, 2020, and 2030 illustrate the regional focus and commitment to reducing greenhouse gas emissions" (2006, p. 9). This statement is true if viewed as a comparison of the implementation of the plan versus a no-build scenario, but it is not true for the Energy Plan's overall goal of actually reducing greenhouse gas emissions.

Albany

CDTC released its *New Visions 2025 Plan* for public comment in June 2004. It included the greenhouse gas analysis required under the Energy Plan. Table 3 presents data on CDTC's past and future carbon dioxide emissions, gross regional product, carbon dioxide intensity, and energy intensity, as provided by ICF Consulting in a report for the US Department of Transportation.

Table 3-CDTC Carbon Dioxide Emissions and Energy Intensity

	1990	2001	2010	2020
Gross Metropolitan Region Product (Million 2001 Dollars)	$19,375	$31,330	$40,878	$54,937
CO_2 Emissions (MMTCO_2)	4.2	4.7	5.5	6.0
CO_2 Intensity (MTCO2 per million 2001 Dollars)	215.0	150.0	133.4	109.7
Energy Use (Trillion Btu)	58.9	66.6	77.3	85.4
Energy Intensity (Thousand Btu per 2001 Dollar	3.0	2.1	1.9	1.6

Source: ICF, 2005

Like NYMTC, Table 3 indicates that CDTC has reduced and is projected to continue reducing its greenhouse gas intensity. However, actual greenhouse gas emissions are projected to continue rising, counter to the goals of the Energy Plan. To meet the Energy Plan goals, CDTC's greenhouse gas emissions would need to be at or below 4 million metric tons in 2010, and at or below 3.8 million metric tons in 2020. This means that CDTC is projected to emit 58 percent more emissions in 2020 than it would if the Energy Plan greenhouse gas goals were met.

Table 4 summarizes the greenhouse gas emissions from direct energy consumption of the "build" versus the "no-build" scenarios.

Table 4-CDTC's Yearly Greenhouse Gas Emissions (New Visions 2025 Transportation Plan)

	Build		No-Build		Build Compared to No-Build	
Year	Daily VMT	Carbon Emitted per Year (tons)	Daily VMT	Carbon Emitted per Year (tons)	Difference in Daily VMT	Difference Carbon Emitted (tons)
1990			17,740,000			
1996			20,470,000			
2003			23,498,000	800,912		
2008	23,167,000	788,290	24,774,000	846,046	1,607,000	57,756
2015	23,780,000	786,122	26,526,000	888,323	2,746,000	102,201
2021	24,942,000	815,998	27,756,000	927,828	2,814,000	111,830
2025	25,651,000	825,983	28,539,000	938,944	2,888,000	112,961

Source: CDTC, 2004

Like NYMTC, CDTC meets the Energy Plan goal of reducing greenhouse gas emissions with the implemented build scenario relative to the no-build scenario. But CDTC also indicates that actual greenhouse gas emissions will rise 37,693 tons per year from 2008 to 2025—a 5 percent increase. Like NYMTC's analysis, the CDTC analysis does not present greenhouse gas emissions for 1990, so a comparison to the Energy Plan targets is not possible from the information provided in the CDTC transportation plan.

Ithaca-Tompkins County

ITCTC released its 2025 Long Range Transportation Plan in December 2004, and it included a greenhouse gas analysis. ITCTC is one of the smallest MPOs in New York. Table 5 presents data on ITCTC's past and future carbon dioxide emissions, gross regional product, carbon dioxide intensity, and energy intensity, as provided by ICF Consulting in a report for the US Department of Transportation.

Table 5-ITCTC Carbon Dioxide Emissions and Energy Intensity

	1990	2001	2010	2020
Gross Metropolitan Region Product (Million 2001 Dollars)	$1,934	$3,045	$3,973	$5,339
CO_2 Emissions ($MMTCO_2$)	0.4	0.4	0.5	0.5
CO_2 Intensity (MTCO2 per million 2001 Dollars)	213.4	139.6	123.7	101.0
Energy Use (Trillion Btu)	5.8	6.0	7.0	7.6
Energy Intensity (Thousand Btu per 2001 Dollar	3.0	2.0	1.8	1.4

Source: ICF, 2005

ITCTC, like the other MPOs, shows a reduction in energy intensity through 2020, but it also fails to meet the Energy Plan goals of a reduction in greenhouse gas emissions relative to 1990 levels. While it shows no increase from 2010 to 2020, this can partially be explained by the small size of the MPO. To meet the Energy Plan goals, ITCTC would need to be emitting at or below .38 million metric tons of greenhouse gases in 2010 and at or below .36 million metric tons in 2020. Thus, in 2020, ITCTC is projected to emit 39 percent more greenhouse gas emissions than it would if it met the Energy Plan goals.

Table 6 summarizes the greenhouse gas emissions from direct energy consumption of the "build" versus the "no-build" scenarios. Unlike NYMTC and CDTC, ITCTC only conducted its analysis for 2004 and 2025.

Table 6-ITCTC Yearly Greenhouse Gas Emissions

Year	Build	No-Build	Build Compared to No-Build
	CO_2 Emitted (Metric Tons)	CO_2 Emitted (Metric Tons)	CO_2 Emitted (Metric Tons)
2004		115	
2025	131	133	2

Source: ITCTC, 2004

ITCTC, like the other two MPOs assessed, shows a reduction in greenhouse gas emissions for its build compared to its no-build scenario. Likewise, it shows a total increase in greenhouse gas emissions from current levels to 2025 levels, in this case 16 metric tons, or a 14 percent increase. It too does not provide information on 1990 levels.

Comparison of Three MPOs

The trends of all three MPOs assessed in this paper are the same. All of their analyses indicate that implementing their long range transportation plans will result in fewer greenhouse gas emissions than would occur without the projects in the plans. However, all three envision greenhouse gas emissions from the transportation sectors increasing over current, and therefore, 1990, levels. It is interesting to note that, although it is responsible for the largest share of the state's emissions, the New York City metropolitan region is significantly less energy intensive than both the Albany region and the Ithaca-Tompkins County region.

In interviews with New York MPOs, ICF Consulting found that "most MPOs see few circumstances in which these energy/CO_2 assessments could influence decisions in a significant way" (2005, p. 16). Two of the MPOs assessed in this paper, CDTC and ITCTC, included their analyses in their long range plan, while NYMTC conducted its analysis after the plan was released. Thus NYMTC did not use its analysis as part of the decision-making process of which projects to include in the plan. For future planning cycles, MPOs could better incorporate climate change considerations by using the information gained in these first-round analyses as a part of the decision-making process.

Quantitative Needs

Carbon dioxide, the primary greenhouse gas, is generally emitted in a manner directly proportional to fuel consumption. As such, the calculation of greenhouse gas emissions from the implementation of transportation plans requires determining the amount of fuel consumed by the vehicles that will use the roadways contained in the plan. While conceptually simple, it is more complex in practice and requires a model that can make adjustments based on fuel type and vehicle mix, among other factors. If this data is not available for an MPO for its facilities, then the calculation becomes more difficult (ICF, 2006). This difficulty is demonstrated by the calculations conducted by the three MPOs assessed in this paper—Ithaca-Tompkins County, the smallest MPO studied, provided the least detailed quantitative assessment of the greenhouse gas emissions expected from its transportation plan. Further, assessing the greenhouse gas emissions on a regional scale will not assist transportation decision-makers in project selection if they wish to choose projects that will result in fewer greenhouse gas emissions. Project-level emissions data, while less reliable than regional data, are nonetheless needed if MPOs are to select projects based on their expected greenhouse gas contributions.

While there are several off-the-shelf models currently available for use in quantifying greenhouse gas emission from transportation projects, the EPA is in the process of developing its Motor Vehicle Emission Simulator (MOVES) model. The MOVES model will be much more robust than current models and will be more sensitive to the factors, such as vehicle mix and fuel type, that affect greenhouse gas emissions. The official version of the model is tentatively scheduled for release within the next year (EPA, MOVES, 2007).

Conclusion

The New York Energy Plan is the only in the country that requires MPOs to assess the greenhouse gas emissions that will result for the implementation of their long range transportation plans. As such, it has served as a pilot for how climate change considerations can be included in the transportation planning process. While the greenhouse gas analyses that were conducted show that emissions will continue to rise, and the reduction goals will not be met (at least within the transportation sector), they also show that the build scenarios result in fewer emissions than the no-build scenarios. The analyses could be better utilized in future rounds of plan updates if the information gained from this first round is used to inform the decision-making process to select a mix of projects that most minimizes greenhouse gas emissions. Overall, the analysis requirements represent an important step in bringing climate change considerations into the transportation planning process.

References

Capital District Transportation Committee (CDTC). (2004). "New visions for capital district transportation: new visions 2025 amendment." Retrieved from, http://www.cdtcmpo.org/rtp2025/2025.pdf (>February,2007)

Department of Transportation (DOT). (2006). "Strategic plan: 2006-2010." Retrieved from, http://climate.volpe.dot.gov/plan/splan_2006.pdf (>February, 2007)

Dolfman, M. and Wasser, S. (2004). "9/11 and the new york city economy: a borough-by-borough analysis." Retrieved from, http://www.bls.gov/opub/mlr/2004/06/art1full.pdf (>February, 2007)

Energy Information Agency (EIA). (2004). "Greenhouse gases, climate change, and energy." Retrieved from, http://www.eia.doe.gov/oiaf/1605/ggccebro/chapter1.html (>February, 2007)

Energy Information Agency (EIA). (2006). "Emissions of greenhouse gases in the united states 2005." Retrieved from, http://www.eia.doe.gov/oiaf/1605/ggrpt/index.html (>February, 2007)

Environmental Protection Agency (EPA). (2007). "Climate change: greenhouse gas emissions." Retrieved from, http://www.epa.gov/climatechange/emissions/index.html#ggo (>February, 2007)

Environmental Protection Agency (EPA). (2007). "Moves: motor vehicle emissions simulator." Retrieved from, http://www.epa.gov/otaq/ngm.htm (>February, 2007)

ICF Consulting. (2005). "Estimating transportation-related greenhouse gas emissions and energy use in new york state." Retrieved from, http://climate.volpe.dot.gov/docs/nys.pdf (>January, 2007)

ICF Consulting. (2006). "Assessment of greenhouse gas analysis techniques for transportation projects." Retrieved from, http://www.trb.org/NotesDocs/25-25%2817%29_FR.pdf (>January, 2007)

Intergovernmental Panel on Climate Change (IPCC). (2007). "Summary for policymakers." Retrieved from, http://www.ipcc.ch/ (>February, 2007)

Ithaca-Tompkins County Transportation Council (ITCTC). (2004). "Long range transportation plan update." Retrieved from, http://www.co.tompkins.ny.us/itctc/lrp/index_2025.html (>February, 2007)

National Oceanic and Atmospheric Administration (NOAA). (2006). "Climatological data." Retrieved from, http://www.erh.noaa.gov/okx/climate_cms.html#Historical (>February, 2007)

New York Metropolitan Transportation Council (NYMTC). (2006). "Consistency assessment of NYMTC's 2006-2010 transportation improvement program (TIP) and 2005-2030 regional transportation plan with the new york state energy plan." Retrieved from, http://www.nymtc.org/PM2.5_docs/consistency_assessment.pdf (>January, 2007)

New York State Energy Research and Development Authority (NYSERDA). (2002). "New york state energy plan." Retrieved from, http://www.nyserda.org/Energy_Information/energy_state_plan.asp (>January, 2007)

New York State Energy Research and Development Authority (NYSERDA). (2006). "State energy plan: 2005 annual report and activities update." Retrieved from, http://www.nyserda.org/Energy_Information/sep_annual_report.pdf (>January, 2007)

Zimmerman, R. (1996). "Global warming, infrastructure, and land use in the metropolitan new york area: prevention and response." *In Eno Transportation Foundation (Ed.) Global Climate Change and Transportation: Coming to Terms* (pp. 55-64). Washington, DC: Eno Transportation Foundation.

Zimmerman, R. (2002). "Global climate change and transportation infrastructure: lessons from the New York area." Retrieved from, http://climate.volpe.dot.gov/workshop1002/zimmermanrch.pdf (>January, 2007)

Potential Impacts of Climate Change and Variability for Transportation Long-Range Planning and Investment

J. H. Suhrbier[1,2]

[1]Principal, Cambridge Systematics, Inc., 100 CambridgePark Drive, Suite 400, Cambridge, Massachusetts 02140; PH (617) 354-0167; FAX (617) 354-1542; email: jsuhrbier@camsys.com.

[2]This paper builds upon and in some cases incorporates work performed by Joanne Potter, Ken Leonard, Rob Hyman, Dan Beagan, David Hunt, and Alan Meyers of Cambridge Systematics, Inc.; Michael Savonis and Rob Kafalenos of the U.S. Federal Highway Administration; and Virginia Burkett of the U.S. Geological Survey.

Abstract

The U.S. Department of Transportation and the U.S. Geological Survey are performing a collaborative investigation of the potential impacts of climate change and variability on the transportation systems and infrastructure of that portion of the Gulf Coast extending from Mobile, Alabama to Galveston, Texas. The purpose is to understand the risks to transportation and to identify the range and timing of potential adaptation strategies that may be desirable. Changes in both average and extreme temperatures and precipitation are examined along with changes in sea level, land subsidence, and the frequency and intensity of hurricanes. Implications for long-range transportation planning and investment build on an assessment of potential impacts on individual modes of transportation. Adapting to climate change and variability can be accomplished within the framework of the existing transportation planning process, although changes in both the range of impacts considered and current institutional arrangements and partnerships may be desirable.

Introduction

Addressing climate change most commonly is thought of in terms of actions that can be taken to reduce or at least limit emissions of greenhouse gases. There is another aspect of climate change, though, that may be even more important and equally immediate for transportation agencies to address. This is to assess the impacts that various manifestations of climate change may have on both existing and planned transportation systems. What actions should transportation agencies be taking today and in the immediate future to *adapt* to conditions of climate change that the Intergovernmental Panel on Climate Change (IPCC) has called "unequivocal"? These include the manner in which existing transportation infrastructure is maintained and operated, and how improvements to this infrastructure are constructed. Equally importantly, strategies to adapt to climate change also include where new transportation facilities are located, the factors considered in conducting long-range transportation planning and in making transportation investment decisions, and even the manner in which the transportation planning process is carried out. As demonstrated by the results of Hurricane Katrina, considerations of climate change also are beginning to influence land use patterns. In the immediate future for the Gulf Coast states and in the longer-term for other areas, population, employment, and land use patterns very likely will be different than those assumed in current transportation plans or simple trends extended projections.

This paper describes results of a collaborative research effort of the U.S. Department of Transportation (DOT) Center for Climate Change and the U.S. Geological Survey (USGS) entitled, *Potential Impacts of Climate Variability and Change on Transportation Systems and Infrastructure – Gulf Coast Study*. The project is investigating the potential effects of climate variability and change on transportation infrastructure and systems in the central Gulf Coast extending from Mobile, Alabama to Galveston, Texas (Figure 1). The underlying purpose of the project is to increase the knowledge regarding the risks and sensitivities of transportation infrastructure to climate change and variability, the significance of these risks, and the range of adaptation strategies that may be considered to ensure the continuation of a robust and reliable transportation network. In addition to examining implications for long-range transportation planning and investment, an assessment also is being conducted of potential impacts on highways, transit, rail, ports and waterways, airports, pipelines, and emergency management.

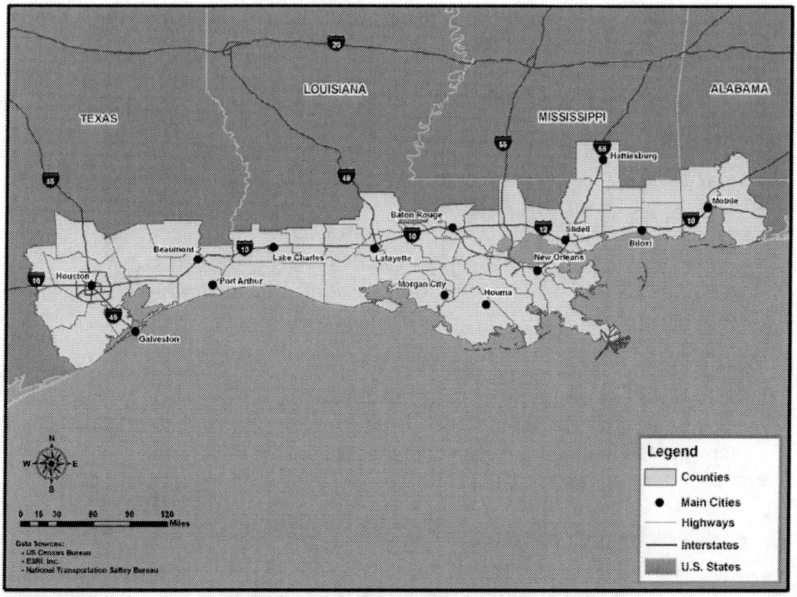

Figure 1. Central Gulf Coast Region

Approach

The work began with a review of existing science related to climate change and variability, including both historical data and available modeling. Specific analyses and modeling then were undertaken for the four-state Gulf Coast study area illustrated in Figure 1. These results next were translated into terms that would be easily understood by transportation professionals.

Building on these climate change findings, three approaches then were utilized to determine how state departments of transportation (DOTs) and metropolitan planning organizations (MPOs) currently are addressing issues of climate change and also how climate change might be addressed in the future. The approaches involved:

- Interviewing state DOT and representative MPO officials responsible for transportation planning within the study area;
- Obtaining and reviewing current vision and mission statements, long-range transportation plans, and transportation improvement programs for the states and selected MPOs within the study area; and
- Reviewing other recent documents from within the study area that address issues that are potentially related to the effects of climate change on transportation infrastructure development, operation, and management;

especially those related to recovery planning and reconstruction following Hurricanes Katrina and Rita.

Climate Change and Variability

While a few areas in the United States, such as Houston and Seattle, already are beginning to take climate change into consideration in their transportation decision-making, it is not uncommon to hear responses such as the following: "We already have plenty of hot days." "There just is too much current uncertainty." "Given extreme shortages in available transportation funding, climate change just isn't going to be a priority concern." "This is something that my successor's successor will have to address."

In contrast to the above quotes, both the USGS Gulf Coast assessment of climate change impacts and the recently completed fourth IPCC assessment indicate a need for more immediate concern and more direct action by transportation agencies, especially in the way new transportation investments are made. The following statements are taken from the IPCC February and April 2007 assessment reports:[1,2]

- Global atmospheric concentrations of carbon dioxide, methane, and nitrous oxide have increased markedly as a result of human activities since 1750 and now far exceed preindustrial values determined from ice cores spanning many thousands of years. The global increases in carbon dioxide concentration are due primarily to fossil fuel use and land use change.

- Warming of the climate system is unequivocal, as is now evident from observations of increases in global average air and ocean temperatures, widespread melting of snow and ice, and rising global mean sea level.

- The frequency of heavy precipitation events has increased over most land areas, with the likelihood of this trend continuing in the future estimated to be greater than 90 percent.

- Widespread changes in extreme temperatures have been observed over the last 50 years.

- There is observational evidence for an increase of intense tropical cyclone activity in the North Atlantic since about 1970, correlated with increases of tropical sea surface temperatures. The probability is greater than 66 percent that this increase in intense tropical cyclone activity will continue to occur.

- For the next two decades a warming of about 0.2 degrees Centigrade per decade is projected for a range of emission scenarios. Continued greenhouse gas emissions at or above current rates would cause further warming and induce changes in the global climate system during the 21st century that would very likely be larger than those observed during the 20th century.

- Anthropogenic warming and sea level rise would continue for centuries due to the timescales associated with climate processes and feedbacks, even if greenhouse gas concentrations were to be stabilized.

Looking specifically at the Gulf Coast region, the USGS analysis determined that:

- By 2100 compared to the present day, the average annual temperature of the study area is estimated to increase between two and eight degrees Fahrenheit.

- Twenty-five years from now, it is very likely that there will be an increase in the number of very hot days and record maximum temperatures. Climate change models project that the most extreme hot days could change more than the average temperature does over the course of the next century.

- A long-term trend towards increasing run-off is evident, with the trend line indicating a 36 percent increase in run-off over the 1905-2003 time period.

- It is expected that extreme precipitation events are likely to become more intense as the climate warms. Severe thunderstorms, in particular, may become more likely.

- The Gulf Coast study area is very sensitive to the effects of sea level rise. The rate of land subsidence along the existing coast is such that relative sea level rise is much larger than climate change induced sea level rise. For example, coastal Louisiana has lost 900,000 acres of land to erosion and inundation since 1920, and this rate is projected to increase in the future.

- Modeling and recent empirical evidence suggest that hurricane intensity will increase 10 to 20 percent over the next 50 to 100 years. For hurricanes of categories 3, 4 and 5, the hurricane return interval will decline from one strike every 22 years to one strike every 18 years.

The USGS assessment of climate change and variability concludes with the finding that, "Model results, climatic trends during the past century, and climate theory all suggest that extrapolation of the 20th century temperature record would likely underestimate the range of change that could occur in the next decades. Regional "surprises" are increasingly possible in the complex, nonlinear earth climate system, which is characterized by thresholds in physical processes that are not completely understood or incorporated into climatic model simulations."

Examples of Transportation Impacts

The described changes in climate affect all aspects transportation within the Gulf Coast study region, including maintenance, operations, and construction. All modes of transportation also are affected: highways, transit, rail, ports, airports, and pipelines. Design standards and maintenance practices may have to change for each mode. Improved drainage facilities may be needed. Bridges may have to be built with both a higher clearance and more strength. The results of Hurricanes Katrina and Rita, the most recent examples of intense tropical storms, indicated the need for improved emergency evacuation procedures, with special attention on the design and construction of new highway evacuation routes and the role of transit, rail, and airports. Buckling of rail track is a more likely occurrence under extremely high temperatures. Seventy-two percent of port facilities are vulnerable to a relative sea level rise of four feet. Similarly, three airports along the Gulf Coast would be inundated under this level of relative sea level rise, as would approximately 50 percent of the pipeline network.

The CSX was by far the most damaged of the railroads along the Gulf Coast as a result of Katrina. It took over five months and $250 million for the company to restore six major bridges, more than 40 miles of track, and its major rail yard in New Orleans. This four-state Gulf Coast area is home to the largest concentration of public and private freight-handling ports in the United States, measured on a tonnage basis. Four of the top five tonnage ports in the United States are located here, and eight of the top twelve. Study area facilities handle around 40 percent of the nation's waterborne tonnage. Much of the region's economy is directly linked to waterborne commerce; and in turn, this waterborne commerce supports a substantial portion of the United States economy.

A full description of the results of the transportation analysis is beyond the scope of this paper. An examination of storm surge, though, illustrates both the potential devastating impacts on existing roadways and the need to consider this factor in planning either new or improved roadway facilities.

To consider the impacts of storm surge inundation, highway facilities that pass through land areas with an existing elevation above sea level of less than or equal to 18 feet are generally considered to be vulnerable to storm surge inundation by 2050, while those facilities that are located in or pass through land areas with an existing elevation above sea level of less than or equal to 23 feet are generally considered to be vulnerable by 2100. The areas that are potentially vulnerable to surge inundation of less than 18 feet are shown in Figure 2. This figure also shows in red those Interstate and arterial highways that pass through these vulnerable areas. While the actual highways that would be flooded depends on the particulars of a given storm, its landfall location, direction, tidal conditions, etc., a substantial portion of the highway system is vulnerable to surge inundation, including roads in all four states in the study area. Fifty-one percent of all highways and 56 percent of the Interstates are in surge vulnerable areas of less than 18 feet.

While inundation from storm surges is a temporary event, during the period of inundation the highway will not be passable, and therefore cannot serve as an emergency evacuation or recovery route. After the surge dissipates, highways must be cleared of debris before they can function properly, and the expense of these cleanups is not typically included in state DOT budgets. Of particular concern is that all of the major east west highways in the study area, particularly I-10/I-12, are vulnerable to storm surge inundation, and during storm events and the recovery from these events, all long-distance highway travel through the study area may be disrupted.

The risk from surge inundation for National Highway System (NHS) Intermodal Connectors (IM) is even greater than that for all highways. For a storm surge up to 18 feet, 73 percent of IM Connector miles are located in areas subject to surge inundation.

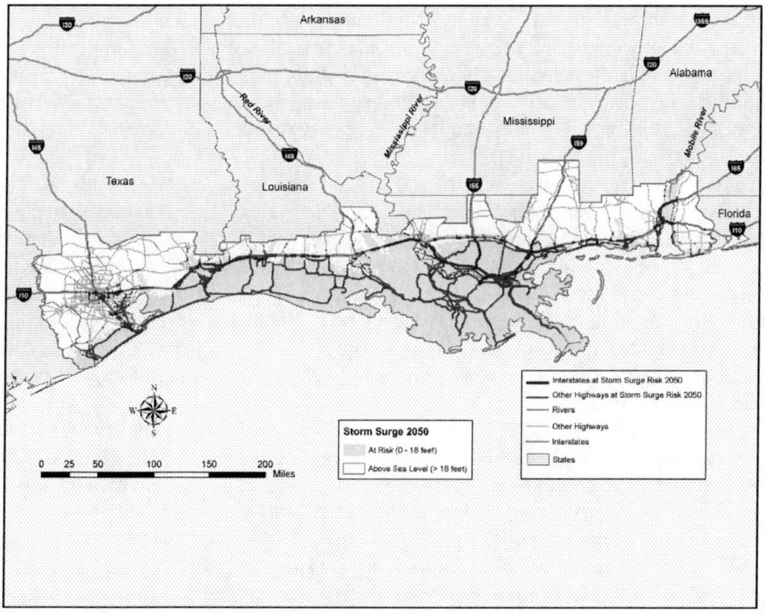

Source: Cambridge Systematics analysis of U.S. DOT Data.

Figure 2. Highways Vulnerable to Storm Surge at Elevation Currently below 18 Feet

Implications for Transportation Planning and Investment

One planning official that was interviewed stated that key leaders recognize the high likelihood that Hurricanes Katrina and Rita are related to climate change and that as a result, "there is now a whole new way of thinking." Another has stated that, "Y2K and 9/11 served as 'fire drills' for what transportation officials need to do regarding global climate change."

The 20- to 30-year time horizon of long-range transportation plans currently is far shorter than the 50 to 100 years used to project the impacts of climate change. A few of the interviewees said that to make reasonable choices within a 20-year timeframe, one had to consider the full lifetime of the facilities being proposed. Still, this does not necessarily extend as far as the 50- to 100-year period of climate change projections. The Florida DOT, though, has adopted a 50-year planning horizon for their Future Corridors planning initiative.

Changes in institutional arrangements either have been proposed or already made in Mississippi, Louisiana, and Texas. In some cases, this involves new leadership responsibilities for MPOs and state DOTs in planning and managing evacuation efforts. Recovery authorities were created in Louisiana and Mississippi that are addressing transportation issues, raising questions of how the work of these authorities relates to the ongoing MPO and state transportation planning and investment process. Interviewees also mentioned the increased importance of partnerships in addressing climate change, including, improved working relationships with the railroads, increased multistate cooperation, and changes at the national level as well as at the state, regional, and local levels of government.

Changed land use patterns, and the location of population and economic activity, resulting from severe hurricanes was mentioned as an important planning consideration in Mississippi, Louisiana, and Texas. One person stated that the previous set of assumptions on which the existing transportation long-range planning process are based are now obsolete; new estimates of the distribution of population, employment, and land use need to be developed. Population levels in New Orleans are not estimated to return to their prestorm levels for at least 20 years and the locational patterns of where people choose to live are likely to be different than they were before Katrina.

A disconnect exists between the long-range transportation planning and the programming of transportation funding conducted by state DOTs and MPOs from the long-range recovery plans facilitated in Louisiana and Mississippi by nongovernmental organizations (NGOs) such as the American Institute of Architects (AIA) and the Urban Land Institute (ULI). There has been a higher level of participation of transportation agencies in the recovery planning efforts conducted by the state recovery authorities that were established, but this participation has been greater for state DOTs than it has been for MPOs. FEMA's ESF Number 14 long-range planning process in Louisiana was characterized as poorly organized and conducted, and not adequately involving the affected urban areas. "This was worse than FEMA's recovery efforts." An important problem in a number of the long-range recovery plans is that cost and financial viability are not being given sufficient attention. For example, rail or guideway transit proposals are made without consideration of their ability to satisfy FTA New Starts funding criteria. From the perspective of transportation agencies, it may be more appropriate to consider these proposed transit systems as a candidate element for a possible long-range "vision" than as part of an adopted long-range transportation plan.

The Houston-Galveston Area Council (HGAC) is in the midst of developing a new vision and associated goals and objectives for that region. In the outreach activities conducted in support of this process, climate change has been raised as an issue that should be addressed. Important issues related to climate change also are being raised as part of the region's long-range planning process, including, coastal erosion, coastal protection, development in flood plains and the surge zone, wetlands, and use of retention areas as open space and public parks. In addition, consideration is being given to reporting greenhouse gas emissions as part of their annual scorecard reporting.

A central observation of an independent Working group for Post-Hurricane Planning for the Louisiana Coast[3] is the complex relationship among coastal wetlands and both the storm protection measures and navigation systems that are put in place, and the need to analyze and plan in a manner that simultaneously considers all three. Thus, changes in navigation channels now under consideration could have potentially important impacts on coastal ecology. While navigation is under the control of the Corps of Engineers, the report points out that changes in navigation channels also could result in changes in the location and types of activity occurring at individual ports. Further, changes in the ports will result in changes in settlement patterns (commercial and population) and the rail and highway infrastructure serving these ports. *"The many ports and channels that form the marine transportation network have already had adverse effects on wetlands and storm risks. Planning and operation of these facilities clearly must now be integrated with storm protection and restoration activities."*

Preservation of ecologically sensitive coastal wetlands areas is one way of minimizing damage from hurricanes. A variety of human activities are contributing to the current and projected rate of land subsidence, including but not limited to the location and management of navigation channels. The development of the full range of port, pipeline, shipping, and their supporting land transportation infrastructure can be screened for their potential to either directly or indirectly affect coastal areas. In essence, this is extending the concept of "secondary and cumulative effects" to include coastal ecology and storm protection. Similarly, strategies proposed to protect coastal areas should be screened for potential implications on the transportation system.

A major issue in recovery planning and reconstruction for the hurricane damaged areas in both Louisiana and Mississippi is the degree to which these areas are rebuilt as they were before, rebuilt a little differently (e.g., higher bridge heights, or rebuilt with fundamentally different land use and transportation arrangements that are explicitly designed to avoid or mitigate future storm damage. In both states, the consideration of fundamentally new land use patterns along the coastal area has received major attention, along with the changes in transportation that should be made to support these changed activity patterns. There is significant pressure in both states, though, to rebuild only a little differently. In Mississippi, land use issues were under debate prior to Hurricane Katrina, but were elevated in importance following the extensive flood-related damage incurred in the vicinity of the coast and the establishment by the Governor of a Recovery Commission. Three objectives drove this planning initiative: economic growth and revitalization, storm protection, and the environmental character of the coastal communities. As one person stated, "There always will be building and economic activity on the coast; the question is the type of development and the design of this development so that it can survive Hurricane Katrina-type storms. The new multistory condominiums with open parking on the lower floors were much less damaged than the older structures."

Planning Opportunities

In response to climate change, land use patterns may change within a region on both a subarea and regional basis. There also may be larger macro and even national changes in markets affecting the production and distribution of goods and commodities. It is important that these different scales of geographic changes be monitored and that the long-range transportation planning and investment process respond to these changes.

Environmental considerations have long played a role in the planning and development of transportation projects. Changes over time, though, have occurred in the manner in which environmental analyses have been conducted and the underlying legal framework in which these analyses are conducted. SAFETEA-LU, in Section 6001, defines eight planning factors that should guide a transportation planning process and the development of projects, strategies, and services (Figure 3). An interesting question is the manner in which climate change can be addressed in this list of eight planning factors and the associated consultative process. While climate change is not now directly named as part of any of the eight factors, six of the eight factors reflect considerations that are directly related to climate change. In addition to system preservation, these include protecting, enhancing, and mitigating impacts on the environment; system management and operation; access and mobility; safety; and economic vitality.

Compared to earlier legislation, SAFETEA-LU contains a new requirement that a transportation plan, "shall include a discussion of potential environmental mitigation activities and potential areas to carry out these activities, including activities that may have the greatest potential to restore and maintain the environmental functions affected by the plan."

Transportation plans, programs, and projects historically have been developed to meet the needs of future projected or planned land use, including population and employment patterns. In recent years, though, transportation and land use are being addressed in a much more interactive or coordinated manner. Rather than land use being viewed as driving transportation decisions, transportation investment and management decisions are being made collaboratively and in concert with growth management and economic development decisions. In this view, the manner in which transportation infrastructure is developed and managed is seen as one "tool" for helping to achieve desirable growth objectives.

In terms of introducing climate-related changes into the long-range transportation planning and investment process, the potential exists at each step. Long-range environmental quality, economic development, mobility, and other desired conditions such as safety commonly are defined as part of a vision and accompanying mission statement and then translated into goals, objectives, and performance indicators. Protection from climate change impacts potentially could be introduced at these stages as well. Given these defined goals and objectives, strategies then are developed that are specifically designed to meet the agreed upon goals and objectives, and evaluated using the appropriate performance measures. Again, strategies could be developed that address climate change and variability. Similarly, climate change protection and mitigation strategies could be evaluated with respect to their potential impact on the transportation system.

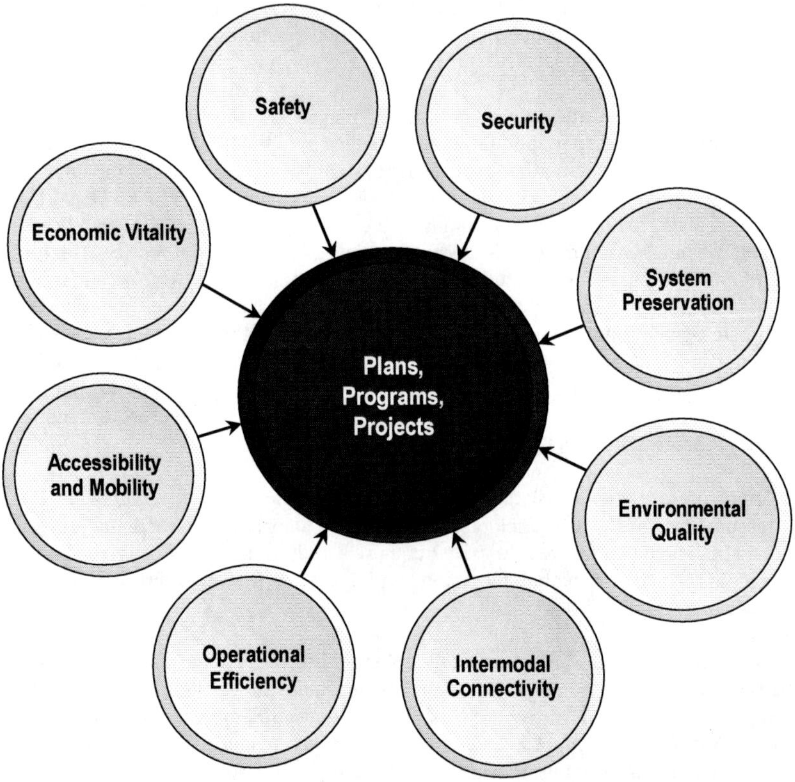

Figure 3. SAFETEA-LU Planning Factors

Note: The SAFETEA-LU legislation also contains provisions that long-range transportation plans be developed in consultation with agencies responsible for land use management, natural resources, environmental protection, conservation, and historic preservation. Further, these plans are to consider, where available, conservation plans or maps and inventories of natural or historic resources.

Conclusions

Based on currently available climate change information, there are potentially important implications of climate change for the manner in which transportation investments are planned, developed, implemented, and managed. When examined in terms of risk management, these implications are sufficiently significant that transportation planning organizations should develop an improved understanding of climate change issues and how their particular region is likely to be effected. Based on this understanding, climate change considerations can begin to be reflected in forecasts of population, land use, and economic activity and in agency decision-making. One person interviewed for this project stated that it will be far more costly if we choose not to respond than if we take actions now to adapt and protect our transportation investments.

Transportation planning and climate change professionals utilize languages and analysis approaches that are very different. Based on the interviews conducted in conjunction with the Gulf Coast study, transportation planners, for the most part, have not studied and have little or no understanding of climate change issues and the nature and magnitude of potential impacts of climate change and variability. Significant effort, therefore, may be required simply for each culture to understand the other and to establish effective working relationships.

Transportation planning has undergone, and continues to undergo, fundamental changes during a period when both human and funding sources are becoming increasingly constrained. Transportation agencies are being asked to be increasingly multimodal in their orientation and to simultaneously consider a broader range of transportation, environmental, economic, and other objectives. The mission no longer is simply to build, operate, and maintain a highway system. It is within this context, that transportation agencies are being asked to add yet another consideration and especially one that is as difficult, complex, and fundamental as climate change.

Existing institutional arrangements may not be sufficient for transportation agencies to fully address and respond to issues of climate change. An increased level of collaboration may be necessary for transportation planning and investment decisions to effectively respond to climate change issues, including, partnering with climate change specialists, the development of new institutional arrangements, a higher level of cooperative interstate planning than now exists, expanded public/private planning initiatives, and new state/Federal planning partnerships.

Responding to the potential effects of climate change, as demonstrated by the results of the Gulf Coast analysis, involves changes in the location of transportation facilities, housing, and business. Future growth can be managed so as to minimize if not avoid effects due to climate change. Consistent with emerging transportation and land use practice, these can be managed in concert with one another, with climate change added to the mix of criteria.

References

"Climate Change 2007: The Physical Science Basis – Summary for Policymakers." *Fourth Assessment Report of the Intergovernmental Panel on Climate* Change, Geneva, Switzerland. February.

"Climate Change 2007: Climate Change Impacts, Adaptation, and Vulnerability." *Working Group II Contribution to the Intergovernmental Panel on Climate Change Fourth Assessment Report*. April.

University of Maryland Center for Environmental Science. (2006). "A New Framework for Planning the Future of Coastal Louisiana after the Hurricanes of 2005." *Prepared by an Independent Working Group for Post-Hurricane Planning for the Louisiana Coast with logistical support from the National Research Council and the Institute of Water Resources of the U.S. Army Corps of Engineers*, Cambridge, Maryland. January.

Subject Index

Page number refers to the first page of paper

Author Index

Page number refers to the first page of paper